International Perspectives in Geography
AJG Library 6

Aims and Scope:

The AJG Library is published by Springer under the auspices of the Association of Japanese Geographers. This is a scholarly series of international standing. Given the multidisciplinary nature of geography, the objective of the series is to provide an invaluable source of information not only for geographers, but also for students, researchers, teachers, administrators, and professionals outside the discipline. Strong emphasis is placed on the theoretical and empirical understanding of the changing relationships between nature and human activities. The overall aim of the series is to provide readers throughout the world with stimulating and up-to-date scientific outcomes mainly by Japanese and other Asian geographers. Thus, an "Asian" flavor different from the Western way of thinking may be reflected in this series. The AJG Library will be available both in print and online via SpringerLink.

About the AJG

The Association of Japanese Geographers (AJG), founded in 1925, is one of the largest and leading organizations on geographical research in Asia and the Pacific Rim today, with around 3000 members. AJG is devoted to promoting research on various aspects of human and physical geography and contributing to academic development through exchanges of information and knowledge with relevant internal and external academic communities. Members are tackling contemporary issues such as global warming, air/water pollution, natural disasters, rapid urbanization, irregular land-use changes, and regional disparities through comprehensive investigation into the earth and its people. In addition, to make the next generation aware of these academic achievements, the members are engaged in teaching and outreach activities of spreading geographical awareness. With the recent developments and much improved international linkages, AJG launches the publication of the AJG Library series in 2012.

Editorial Board:

More information about this series at http://www.springer.com/series/10223

Kazuharu Mizuno • Lobsang Tenpa

Himalayan Nature and Tibetan Buddhist Culture in Arunachal Pradesh, India

A Study of Monpa

Springer

Kazuharu Mizuno
Department of Geography
Graduate School of Letters
Kyoto University
Kyoto, Japan

Lobsang Tenpa
Institute for South and Central
Asian Studies
University of Leipzig
Leipzig, Germany

ISSN 2197-7798 ISSN 2197-7801 (electronic)
International Perspectives in Geography
ISBN 978-4-431-55491-2 ISBN 978-4-431-55492-9 (eBook)
DOI 10.1007/978-4-431-55492-9

Library of Congress Control Number: 2015938781

Springer Tokyo Heidelberg New York Dordrecht London

Printed on acid-free paper

Springer Japan KK is part of Springer Science+Business Media (www.springer.com)

Preface

I learned for the first time that there was a mysterious area called the state of Arunachal Pradesh when I was still a high school student. We were being taught about the McMahon Line in a high school geography class. National borders normally appear as solid lines on maps, but extensive sections of the borders of Arunachal Pradesh were dotted on my atlas, which indicated the area was under dispute between India and China. I did not have any further information about the environment, people, or society in that region. It was a closed-off zone that foreigners were not allowed to enter, and I did not imagine that I would ever visit that part of the world.

During the British rule of India, the area was known as the North-East Frontier Tract (NEFT). India became independent from Britain in 1947. The NEFT was renamed the North-East Frontier Agency in 1951. It was renamed Arunachal Pradesh in 1972 and was considered a peripheral region with no particular value for development and was thus left untouched. The region was long regarded as nothing more than the unimportant southern outer edge of the Tibetan administrative territory. However, that peripheral area suddenly became the focus of political attention after the Chinese military invasion in 1962. The Chinese Army cut through Arunachal and almost reached Tezpur in Assam. The Indian government undertook regional economic development and reinforced its military defenses in Arunachal Pradesh. This change made the area even less accessible to foreign visitors.

I participated in a research project conducted by Dr. Kiyohito Okumiya called "Human Life, Aging and Disease in High-Altitude Environments: Physio-Medical, Ecological and Cultural Adaptation in Highland Civilizations," operated by the Research Institute for Humans and Nature of Japan. One of its study areas happened to be Arunachal Pradesh, which provided me with the opportunity to visit that mysterious region.

My first visit was in early July 2007. My main study area was an area occupied by the Monpa people, and I was captivated by the features of the region. In the early stages of my fieldwork, I investigated agriculture, forests, and rangelands. I then moved the hub of my research to the village of Thembang [them bang] in the Thembang Circle Area of the West Kameng district. This led to a shift in my research interests.

One of the stories that the villagers told me was astonishing. I heard how Thembang Dzong [them bang *rdzong*] was built in the village of Thembang as the second fort for a prince who had moved there from Tibet; however, the previous fort, Dirkhi Dzong [bde skyid *rdzong*], remained somewhere deep in the mountains. The older fort was feared by the villagers since they believed that their ancestral souls occupied the surrounding area. I then trekked through the mountains and located Dirkhi Dzong. The villagers also informed me that the present senior and junior clans were, respectively, descendants of the royal family and their subordinates.

That encouraged me to investigate the background and details of the clan histories. In addition, the information I received that some forts—*dzong* [*rdzong*]—were originally built as tax offices by the Tibetan administration encouraged me to conduct studies of local taxation systems. That necessitated my travel not only across the Thembang region but also into the wider area of the Tawang [rta wang/ dbang] and Kalaktang [kha legs steng] regions. The journey revealed the ways in which the Tibetan administration governed the area. The accumulated knowledge suggested further areas of investigation, and the state of Arunachal Pradesh now constitutes an exciting zone of exploration for me as a field researcher.

I summarized the findings of my fieldwork, which were published by Showado, Kyoto, in a book titled *Arunachal Pradesh, India: Nature of Assam Himalaya and Tibetan Society* (in Japanese) in March 2012. After the book's publication, I felt it was highly important to inform people about the features of the area. That book was reviewed by several academic journals. Lobsang Tenpa, a researcher from the Tawang region who is currently based at Leipzig University for his doctoral thesis, encountered one of the reviews. His research was focused on the taxation system used by the Tibetan administration for the Monpa people, and he took interest in me as a Japanese field researcher working on a similar theme. I happened to be working at Regensburg University when he sent me his first e-mail. I visited him in Vienna, and we had some interesting discussions. His knowledge of the topic and the area was clearly far-reaching. I was immediately convinced of the potential significance of collaborating with him. We agreed to revise and enhance my Japanese book and publish an English edition. To enhance our collaborative effort, we undertook further fieldwork separately in 2012 and 2013. Additionally, Lobsang Tenpa has been conducting fieldwork in the region since 2009, and we compared and shared our findings.

Lobsang Tenpa, a researcher in Tibetan and Himalayan Studies, was indispensable to the publication of this revised edition in English. Today, Monpa people learn Hindi at school, and most of them do not understand Tibetan. Tenpa studied at the Central University of Tibetan Studies, Sarnath, and two other universities in Delhi and is fluent in Tibetan language and literature. Moreover, his research interests include various topics related to Tibetan culture. He kindly agreed to my idea of collaboration on the book, and we started working on it in Vienna in 2012. His input filled gaps in my knowledge, and he corrected my misunderstandings. I invited him to my research office in Kyoto in July 2013. We held daily discussions over a week to compare and examine findings we had each collected from the field, and those efforts led to the completion of this book.

Not many books have ever been published about Arunachal Pradesh, and even fewer have been published on the Monpa people and their area. This is largely because international research collaboration has been scarce for the following reasons: First, the area was inaccessible to foreigners until the 1990s, and even today foreigners need to obtain special permission to enter the state. Second, there are a limited number of reports written by Indian researchers on the Monpa people, phonetic notations of Monpa terminology are often erroneous, and romanized Tibetan spellings are very rare.

In this English edition, we felt it necessary to present the terminology used by the Monpa people in Tibetan transliteration. The Monpa people belong to the Tibetan ethnic group, and the original Monpa words are presented in Tibetan orthography. The original Monpa (i.e., Tibetan) spellings include many unpronounced prefixes, and only written Tibetan can clearly and accurately convey their meanings. For example, "Dirang Dzong"—"Dirang Fort" in the Monpa language—is "rdi rang *rdzong*" when the Tibetan words are written using the Latin alphabet. With the exception of place names and proper nouns, the terminology used by the Monpa people is italicized in this English edition. The phonetic transcriptions of Monpa terminology used in the Tawang and West Kameng districts are followed by standard Tibetan, according to Wylie's (1959) transliteration, in square brackets. Some important terms are transliterated in every chapter for emphasis. See the glossary for further details about the transcription and transliteration of these terms.

All the contents of this book that are presented without references were based on interviews or participant observations undertaken in the field. The findings of many previous studies are at variance with what we observed. We present the findings from our fieldwork and references to those studies in this book. All uncredited photographs were taken by K. Mizuno.

We hope that this book will assist those with an interest in the area of the Monpa people in Arunachal Pradesh and will introduce its wonders to the outside world.

Kyoto, Japan Kazuharu Mizuno

References

Mizuno K (2012) Arunachal: nature of Assam Himalaya and Tibetan society. Showado, Kyoto (in Japanese) (The Prize of the Association of Japanese Geographers (2014))

Acknowledgments

Dr. Kazuo Ando of the Center for Southeast Asian Studies, Kyoto University, and Dr. Kiyohito Okumiya of the Research Institute for Humans and Nature of Japan and the Principal Investigator of the Research Project for High-Altitude Environments offered me the opportunity to visit Arunachal Pradesh. Mr. Passang Tsering [pa sangs tshe ring] supported me as an interpreter and local guide for a large part of the fieldwork. Mr. Bhupen Kalita, the guide for my second visit in February 2008, kindly responded to my request to observe agricultural fields in the Dirang Dzong [rdi rang *rdzong*] village. He took me to the household of one of his relatives, Mr. Rinchin Tsering [rin chen tshe ring], who was chairman of the Dung Kharpa Welfare Society [gdung mkhar pa*'i skyid sdug*]. After this encounter, Mr. Rinchin became a great source of information for my fieldwork.

Several people helped me with my fieldwork for this book. Regrettably, it is not possible to mention all of them here. However, I would like to express my gratitude to the following people: Mr. Dorjee Sonam [rdo rje bsod nams], the *tsorgen* [*gtso rgan*], village mayor, of Gyangkhar [rgyang mkhar] village; Mr. Lama Gombu [bla ma mgon po], the commissioner of the Tawang Monastery Museum; Mr. Yeshi Tsering [ye shis tshe ring] and Mr. Lekig Gombu [legs skyes mgon po], the first and third *tsorgen* of the village of Senge Dzong [seng ge *rdzong*], respectively; Mr. Jam Tsering [byams pa tshe ring], the second *tsorgen* of the village of Yewang [yid dbang]; Mr. Ahuja [a hu ja], the *tsespa lama* [rtsis pa bla ma] of the village of Gunthung [gung thung]; Mr. Pema Gombu [padma mgon po], the first *tsorgen* of the village of Lhou [lha'u]; Mr. Urgen Tsering [u rgyan tshe ring], the chairman of Lhou Junior High School; Mr. Tsering [tshe ring], the *tsorgen* of the village of Aodung [a'o gdung]; Mr. Rinchin Tsering, the *koenyer lama* [dkon gnyer bla ma] of Taklung Gompa [stag lung *dgon pa*]; Mr. Tashi Phuntso [bkra shis phun tshogs], the first *tsorgen* of the village of Lubrang [lu'brang/ grangs]; Mr. Netan [gnas brtan], the *minakpa lama* [mi nags pa bla ma] of the village of Aodung; Ms. Yeshi Wangmu [ye shis dbang mo], a teacher at Singsur Buddhist Nunnery [sing zur a ne *dgon pa*]; Khenpo Nyima Tashi [*mkhan po* Nyi ma bkra shis], head of Jangdokpalri Buddhist

Monastery [zangs mdog dpal ri *dgon pa*]; Mr. Phurpa Tsering [phur pa tshe ring], the *tsorgen* of the village of Manpat [sman pag], a village in the Lumla [klung la] Circle Area; and Mr. Pijush Kumar Dutta, a landscape coordinator for the World Wildlife Fund for Nature India (the Western Arunachal Landscape Program). These people as well as all those people whose names could not be mentioned here always welcomed my visits and questions. I also owe a very important debt to the people of the village of Thembang.

I would like to sincerely thank Mr. Takaaki Tanaka, the manager of the Study Group of Physical Geography, Kyoto University. He traveled with me for fieldwork in February, April, and May 2010. We explored mountains and valleys together and finally found Dirkhi Dzong covered with brambles in the final phase of the fieldwork. Thanks to Mr. Takaaki Tanaka, a few pieces of carbide were found among the ruins, providing useful clues about when the *dzong* was constructed. This date was unknown, even to the local people, and was identified here for the first time. His support in producing the figures in this book was also indispensable.

Dr. Yasuyuki Kosaka, Dr. Shinji Miyamoto, Dr. Kyoko Ishimoto, Dr. Akinobu Kawai, Dr. Koichi Usami, and other members of the Research Project for High-Altitude Environments provided insightful comments and suggestions. Dr. Naoji Okuyama, an expert in the Cultural History of Tibetan Buddhism, and Dr. Yumiko Ishihama, an expert in Tibetan Studies at Waseda University, offered constructive comments after the publication of the Japanese edition of this book. Dr. Satoshi Yokoyama, an expert in Southeast Asian Studies, and Ms. Kei Nagaoka, who is studying traditional medical practices in Tawang as a student at the Graduate School of Asian and African Area Studies, Kyoto University, provided invaluable suggestions. The English translation of the Japanese edition was supported by Dr. Hiroyuki Yoshida. Dr. Tomohiro Fujita assisted with the modification of the figures and checked the proofs.

We would also like to acknowledge Prof. Per Kjeld Sørensen of the University of Leipzig, a specialist in Tibetan and Himalayan Studies with a focus on cultural and historical subjects, for his contributions to this edition.

This book is published as volume 6 of AJG Library series "International Perspectives in Geography". We would like to express our heartfelt thanks to Prof. Yuji Murayama, a member of the editorial board of this series. Finally, our thanks to the anonymous reviewer of this monograph.

The funding for field research was provided by the Research Institute for Humans and Nature: Human Life, Aging and Disease in High-Altitude Environments: Physio-Medical, Ecological and Cultural Adaptation in Highland Civilizations, a Grant-in-Aid for Scientific Research (Project Nos. 17251001, 17255002, 23657020, 24251001) from the Ministry of Education, Science, Sports and Culture of Japan, and the Mitsui & Co., Ltd., Environment Fund 2007.

Kazuharu Mizuno

Contents

About the Authors

Kazuharu Mizuno: Professor of Department of Geography, Graduate School of Letters, Kyoto University

Born 1958 in Nagoya, Japan.

BA from Department of Geography, Faculty of Letters, Nagoya University.

Master's Degree from Graduate School of Environmental Science, Hokkaido University.

DSc from Department of Geography, Graduate School of Science, Tokyo Metropolitan University.

Specialized Fields

Physical geography, vegetation geography, environmental change, African studies

Research History

From 1981: Vegetation and environments of Japan Alps and Daisetsuzan Mountains, Japan

From 1992: Glacial fluctuation and vegetational changes in tropical high mountains (Mount Kenya, Kilimanjaro, and Andean Cordilleras)

From 2001: Natural environments and human activities in the Namib Desert, Namibia

From 2007: Nature, culture, and society of Arunachal Pradesh, India

Selected Publications

Mizuno K, Fujita T (2014) Vegetation succession on Mt. Kenya in relation to glacial fluctuation and global warming. J Veg Sci 25:559–570. doi:10.1111/jvs.12081

Mizuno K (2012) Arunachal Pradesh, India: nature of Assam Himalaya and Tibetan society. Showado, Kyoto (in Japanese) (The Prize of the Association of Japanese Geographers (2014))

Mizuno K (ed) (2010) Historical change and its problem on the relationship between natural environments and human activities in Southern Africa. African Study Monographs, Supplementary Issue, No. 40
Mizuno K (ed) (2005) Studies on the environmental change and human activities in semi-arid area of Africa. African Study Monographs, Supplementary Issue, No. 30

Lobsang Tenpa: Contractual Lecturer/PhD Candidate at the Institute for South and Central Asian Studies, University of Leipzig, Germany.

Born 1981 in Tawang, India.
BA from Central University of Tibetan Studies, Sarnath, India (2000–2003).
MA from University of Delhi (2005–2007).
MPhil from Jawaharlal Nehru University, Delhi (2007–2009).
Tibetan Studies Lecturer at University of Bonn, Germany (2009–2011).
Research Assistant at University of Vienna, Austria (2011–2013).

Specialized Field

Tibetan and Himalayan Studies, Buddhist Studies and Area Studies

Selected Publications

Tenpa L (2015) Notes on the (Tibeto-) Monpa and Sherdukpen treaties with British India in the 19th century, Bulletin of Tibetology. In: Paper presented at IATS UlaanBaatar, Mongolia 21–27 July 2013) (in press)
Tenpa L (2014) The centenary of the McMahon Line (1914–2014) and the status of Monyul until 1951–2. Tibet J XXXIX(2):41–86
Tenpa L (2013) The life and activities of Rgyal sras Blo bzang bstan pa'i sgron me, a 16th century Tibeto-Mon monk from Tawang. Tibet J XXXVIII(3 & 4):3–20
Tenpa L, Tempa T (2013) A brief history of the establishment of Buddhism in Monyul: Tawang and West Kameng Districts, Arunachal Pradesh, India. Department of Karmic and Adhyamik, Government of Arunachal Pradesh, Itanagar
Tenpa L (2012) The 1913 Mongol-Tibet treaty and the Dalai Lama's proclamation of independence. Tibet J Special Issue XXXVII(2):3–29

Chapter 1
Nature, Society, and People in Monyul (Monpa Area)

People of Dirang playing games and enjoying a sunny summer's day: locals commonly spend sunny days together at markets in towns and villages

K. Mizuno, L. Tenpa, *Himalayan Nature and Tibetan Buddhist Culture in Arunachal Pradesh, India*, International Perspectives in Geography 6,
DOI 10.1007/978-4-431-55492-9_1

Abstract The 24th Indian state, Arunachal Pradesh, has been part of the country since 1914, when the McMahon Line agreement between Tibet and the British Indian government at Simla was signed. Arunachal Pradesh means "Land of the Rising Sun." However, the McMahon Line is not accepted as a boundary by the Chinese government. Foreigners were prohibited from entering this area until the 1990s, and Indian military units are permanently stationed at key points in this mountainous state. Until recently, this area was veiled in mystery. Within Arunachal Pradesh, the region traditionally known as Monyul is currently divided into the Tawang and West Kameng districts. A range of mountains runs between the Dirang region (West Kameng district) and the Tawang region (Tawang district). This poses an obstacle for transport and is the major reason for differences in language, culture, and society between the two regions. The focus of this chapter is on the ethnicity of the Monpa [mon pa] people who have developed their own societies and cultures in this Monyul region, where the Arunachal Himalaya Mountains are the most prominent physical feature.

Keywords Ethnic group • Language • Monpa

1.1 Environment and Distribution of Ethnic Groups in Monyul (Monpa Area)

The state of Arunachal Pradesh is located in the northeastern part of India, to the north of Assam state, east of Bhutan, and south of Tibet (Fig. 1.1). The name Arunachal Pradesh translates as the "Land" (*achal*) of the "Rising Sun" (*arun*). There has been a long history of conflict over the ownership of this area between India and China. Foreigners were prohibited from entering the state until the 1990s, and therefore the area has been veiled in secrecy until recently. Even today, foreigners wishing to enter the area require special permission and an approved guide as an escort.

Arunachal Pradesh contains 25 major tribes and 100 sub-tribal groups distributed throughout the state who speak around 50 main dialects. The Monpa,[1] the people in the "Land of Mon" (mon *yul*) where the suffix *pa* denotes "people" and *yul* refer to "area", reside in the western region. They are among the major ethnic groups and historically have close ties with Tibet and the Tibetan Buddhist faith. Within Arunachal Pradesh, the Monpa area traditionally known as Monyul is currently divided into the Tawang and West Kameng districts. Religious temples are common in Monyul (Monpa area).

The climate of the West Kameng district is affected by the mountainous terrain along the southern slopes of the Eastern Himalayas, which are divided by long, deep, narrow valleys through which the Kameng River and most of its tributaries flow in a southward direction, eventually joining the Brahmaputra (Choudhury

[1] The phonetic spellings of Monpa terminologies in the Tawang and West Kameng districts are followed by standard Tibetan using Wylie's (1959) transliteration in brackets. The terminologies used by the Monpa people are italicized, except for place names and proper nouns.

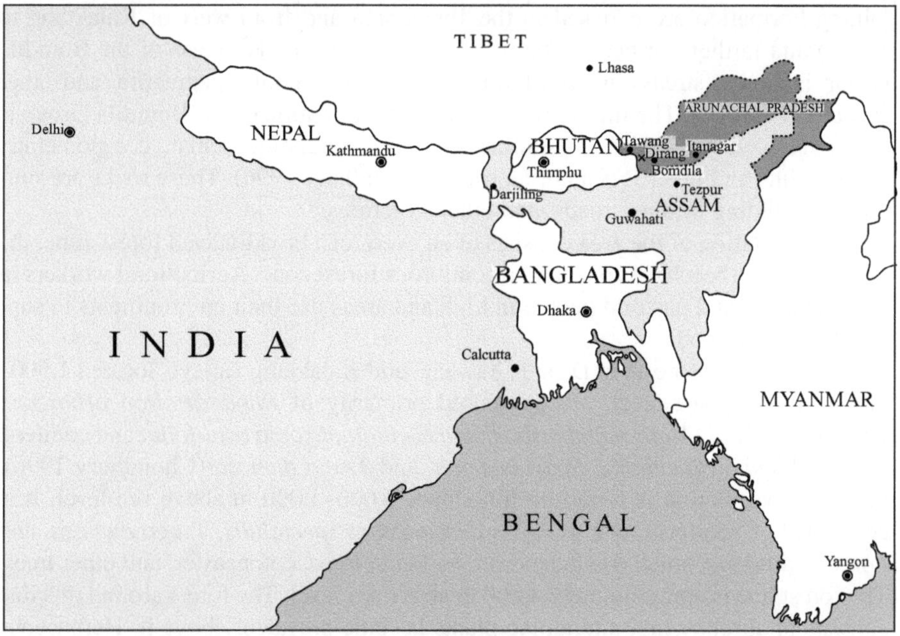

Fig. 1.1 Location of the state of Arunachal Pradesh

1996). The average annual precipitation between 2000 and 2005 was as follows: 1,610 mm at Tawang [rta wang/ dbang] (3,025 m above sea level); 1,788 mm at Bomdila ['bum'di la] (2,430–2,700 m above sea level); and approximately 1,000 mm at Dirang [rdi rang] (1,700 m above sea level) (Kri 2010). The southern half of West Kameng receives more than 70 % of its rain from June to September, whereas the northern half receives 50–60 % of its annual precipitation during the same period. In the pre-monsoon season, from March to May, slightly more than 20 % of the precipitation is received to the southern half and less than 20 % by the northern half. Thunderstorms occur primarily during the pre-monsoon season, from March to May. Hill fog is a common occurrence during the monsoon season. In the valleys, fog frequently appears on winter mornings (Choudhury 1996).

The annual mean maximum temperature at Bomdila is 20.8 °C, and the annual mean minimum temperature is 6.0 °C (Gopalakrishanan 1994). In 2002, the average daily maximum temperature at Tawang was 19 °C in January and 29 °C from July to August; the average daily minimum temperature was –4 °C in January and 11 °C in July (Norbu 2008). At Dirang, where the altitude is approximately 1,300 m lower than Tawang, the air temperature is presumably 6–8 °C higher than at Tawang.

In terms of geology, Precambrian rocks cover more than 60 % of the total area of the East Kameng, West Kameng, and Tawang districts. Rocks of the Sela [ze la] (*la* means "pass") Group are exposed from slightly west of Dirang to the Sela Pass and farther northwest, constituting the major part of the Tawang Valley. The lithology of the Sela Group consists of such rocks as high-grade schist and gneiss, migmatite granite, and pegmatite from the Archaean age. Bomdila gneiss and the overlying

Dirang Formation are exposed in the Tenga area and from west of Kalaktang to Dirang and farther northeast. The lithology of the Bomdila gneiss of the Bomdila Group is augen, streaky gneiss, biotite gneiss, amphibolite, pegmatite, and other Precambrian rocks. The lithology of the Dirang Formation of the Bomdila Group is quartzite, phyllite, schist, crystalline limestone, graphitic schist, conglomerate, amphibolite, and other Precambrian rocks (Choudhury 1996). These rocks are suitable for building bridges, roads, and other structures.

The vegetation of the area consists of an evergreen broad-leaved forest zone, the aestatilignosa (beechen) zone, and a coniferous forest zone. Agricultural workers in lowland areas and pastoral people in highland areas use their environments to support their livelihoods.

The temperate forests in Dirang, Tawang, and Kalaktang valleys, located 1,500–3,000 m above sea level, are composed primarily of *Rhododendron arboreum*, *Quercus griffithii*, *Alnus nepalensis*, *Populus ciliata*, *Lyonia ovalifolia*, and conifers, such as *Pinus wallichiana*, *Taxus baccata*, and *Tsuga dumosa* (Choudhury 1996). Subalpine vegetation is found on hill slopes 3,000–4,000 m above sea level. It is composed of conifers, such as *Abies densa*, *Abies spectabilis*, *Tsuga dumosa*, and *Larix griffithiana*, small rhododendron, berberis, *Rosa*, cotoneaster, and other trees. The forest line is approximately 4,000 m above sea level. The forest around this line consists of shrubby firs, and alpine plants become prevalent above it. Herbaceous species, including anemone, primula, potentilla, *Pedicularis*, *Cassiope*, and *Aconitum*, dominate the alpine zone, which is more than 4,000 m above sea level.

A total of 1,300 orchid species have been reported in India, and about 600 are found in the northeastern region (Choudhury 1996). Over 76 species belonging to 65 genera with known medical uses have been collected in this area. The starchy corms of *Fritillaria cirrhosa*, known as *yathu* [*ya thu* or *a bi kha*], are considered to be of great medicinal value by the local people, who apply the crushed paste of the corms to relieve acute muscular pain (Choudhury 1996). Such species as *Berberis aristata*, *Coriaria nepalensis*, *Symplocos theaefolia*, *Geranium nepalense*, *Rubia cordifolia*, *Engelhardtia spicata*, *Taxus baccata*, *Galium mollugo*, and *Galium triflorum* are used for dyeing and tanning. The roots of *Valeriana hardwickii* and *Valeriana wallichii* are employed to produce perfumes (Choudhury 1996).

A range of mountains runs between the Dirang region (West Kameng district) and the Tawang region (Tawang district) (Fig. 1.2). This poses an obstacle for transport and is the major reason for differences in language, culture, and society between the two regions. The Sela Pass is a natural place for fortification and is strategically important to India (Fig. 1.3). Indian military units are stationed on both sides of the Sela Pass, and a strained atmosphere prevails in the area. Landslides occur at several locations on the road between the Arunachal highlands and Assam lowlands in the rainy season, and access is blocked when this occurs (Fig. 1.4). The Sela Pass is also closed owing to heavy snowfalls in winter (Fig. 1.5).

The local people have long lived under these severe natural conditions. The focus of the present study is on the West Kameng district (total population in 2011: 83,947) and the Tawang district (total population in 2011: 49,977). These districts, in which the Monpa people reside, are in the westernmost part of Arunachal Pradesh. This study aims to provide information on their environment and society.

Fig. 1.2 Arunachal Himalaya, which divides the Dirang [rdi rang] area from the Tawang [rta wang/dbang] area (background), and migrating workers from Nepal undertaking road construction

Fig. 1.3 Sela [ze la] Pass (elevation 4,200 m) is the inter-district boundary line between the Tawang and West Kameng districts. Many Indian military units are stationed on either side of Sela Pass to defend their country against a possible Chinese military action

Fig. 1.4 Landslides occur at many locations on the road between the Arunachal highlands and Assam lowlands in the rainy season

Fig. 1.5 Sela [ze la] Pass in winter: most drivers do not put snow chains on their tires, which often results in gridlock

1.2 Types of Monpa People

The word "Monpa" has been used by Tibetans to denote "people living in lowlands." The Monpa people are speakers of Tibeto-Burman languages and can be divided into three subcategories. The Tawang Monpa live in the Tawang region (Tawang district), the Dirang Monpa in the Dirang region, and the Kalaktang [kha legs steng] Monpa in the Kalaktang region in West Kameng (Fig. 1.6, Table 1.1).[2]

Fig. 1.6 Ethnic distributions in the Tawang [rta wang/ dbang] and West Kameng [nub Kameng] districts in the state of Arunachal Pradesh. The boundary between the Dirang Monpa [rdi rang mon pa] and Tawang Monpa [rta wang/ dbang mon pa], including Mago-Thingbupa [dmag sgo thing bu pa], basically corresponds with the district border. Map based on local interviews carried out as part of this study

---

[2] This categorization is based on geography. Linguistically, the categorization would be Dakpa [*dags pa*] speakers in Tawang, Tshangla [*tshangs la*] speakers in the Dirang and Kalaktang regions, and Drokpa [*'brog pa*] speakers in the Mago-Thimbu and Senge Dzong-Nyukmadung regions. Sherdukpen, Aka [Hrusso], Miji [Sajolang], and Bugun are other tribal groups in West Kameng who speak their own dialects.

Table 1.1 Subgroups of the Monpa [mon pa] people

Ethnic group	
Tawang Monpa	Tawang Monpa: *Ungpa* (Agricultural)
	Dakpanangpa: *Ungpa* (*yulpa*) (Agricultural)
	Tshosumpa: *Ungpa* (Agricultural)
	Mago thingbupa: *Brokpa* (Pastoral)
	Pangchenpa: *Ungpa/Brokpa* (Agricultural and Pastoral)
Dirang Monpa	Dirang Monpa: *Ungpa* (Agricultural)
	Lish-Chug Monpa (Lishpa & Chugpa): *Ungpa* (Agricultural)
	Boot Monpa: *Ungpa* (Agricultural)
	Sengedzongpa & Myukmadungpa: Ungpa/Brokpa (Agricultural and Pastoral)
	Brokpa of Bhutan Border (Pastoral)
Kalaktang Monpa: Ungpa (Agricultural)	

Produced from interviews conducted locally

1. The language of the Tawang Monpa [rta wang/dbang mon pa] differs from that of the Dirang [rdi rang] and Kalaktang Monpa [kha legs steng mon pa]; the former language is known as Monkee or Dakpa [mon skad or dags pa] and the latter is known as Tshangla [tshangs la]
2. The language of the Dakpanangpa [dags pa nang pa] is slightly different from that of the narrowly defined Tawang Monpa, and the two have different pronunciations
3. The language and traditional attire of the Magthing Luksum [dmang thing lug gsum] are identical to those of the Jora Dzong [sbyor ra *rdzong*] Tibetans in Tibet
4. The language of the Pangchenpa [spang chen pa] is different from that of the narrowly defined Tawang Monpa and Dakpanangpa
5. The language of the narrowly defined Dirang and Kalaktang Monpas is almost identical to the language used in eastern Bhutan, which is known as Tshangla or, more commonly, Sharchokpa [shar phyogs pa]
6. The language of the Lishpa [rlis pa] and Chugpa [phyug pa] is different from that of the neighboring Dirang Monpa
7. The language of the Boot Monpa [bud mon pa] is identical to that of the Sherdukpens [sher gtug spen] in the Rupa and Shergaon valleys
8. The language and traditional attire of the Senge Dzongpa [seng ge rdzong pa], Nyukmadungpa [smyug ma gdung pa], and Lebrang [leb brang] close to the border with Bhutan are identical to those of the people of Merak [me rag] and Sakteng [sag steng] in Bhutan

The Tawang Monpa differ linguistically from the Dirang and Kalaktang Monpa (Table 1.2, Glossary). For example, "millet" is *kowp* or *khra* [*khre*] in the Tawang Monpa language, whereas its equivalent in the Dirang Monpa language is *khongpu* (*kong bu*). In the Tawang language, "maize" is *oshum* [*a shom*] or *boma* [*bo ma*], though the Dirang Monpa refer to it as *pinthang*. In the Tawang language, "rice," "paddy," and "cooked rice" are called *dep* or *depu* [*deb*], *nu* [*yu*] or *nun*, and *toh* [*lto*] or *topcha* [*lto chas*], respectively; in the Dirang language, they are *khu* [*khu*], *ra* [*ra*] or *raha* [*ra ha*], and *tochang* [*lto chang*] or *toshang* [*lto shang*], respectively; in Tibetan, they are *bras*, *bras sog*, and *bras btsos pa*, respectively. "Buckwheat" is *drawo* [*bra bo*] in the Tawang language and *gruntsun* [*bra'o zan*] in the Dirang language. "Noodles" are called *thukpa* [*'thug pa*] in both languages. While "cheese" is *chura* [*phyu ra*] in both languages, "butter" is *mar* [*mar*] in the Tawang language but *mo* [*moam*] in the Dirang language. "Butter" is also *mar* [*mar*] in Tibetan. The

Table 1.2 Terms in English and their phonetic transcriptions in the languages of the Tawang [rta wang/ dbang], Dirang [rdi rang], and Kalaktang [kha legs steng] Monpa as well as their Tibetan transliterations

Term	Tawang	Dirang & Kalaktang	Tibetan
alcohol	chang	yu	chang/yu
buckwheat	drawa	gruntsun	bra bo/bra'o zan
butter	mar	mo/ moam	mar
cheese	chura	chura	phyu ra
cow	ba	wa	ba/ba phyugs
cooked rice	toh/topcha	tochang/toshang	lto/lto chas/lto chang/'bras btsos pa
chicken	khat	goga/gogadama	bya/bya mo
family	mathtsan	pheisonga	ma tshang/nang mi/mi tshang
horse	ta	kuruta	rta/chibs rta
maize	oshum/boma	pinthang/pinang	a shom
millet	kowp/khra	khongpu	khre/kong bu/mon chag
monkee	modskat	modskat	mon skad
noodle	thukpa	thukpa	thug pa
paddy	nu/nun	ra/raha	yu/'bras sog
rice	dep/depu	khu	'bras/deb
village	yui	dung	yul/dung/gdung

Tawang Monpa language is closer than the Dirang Monpa language to Tibetan because of the geographic proximity of Tawang to Tibet.

The Tawang Monpa people—in the broad sense of the term—can be divided into five subgroups: the Tawang Monpa in the limited sense; the Pangchenpa [spang chen pa]; the Dakpanangpa [dags pa nang pa]; the Tshosumpa [tsho gsum pa]; and the Mago-Thingbupa [dmag sgo thing bu pa]. The Pangchenpa inhabit the area around Zemithang [bye ma'i thang]; the Dakpanangpa live in the region of Dakpanang; the Tshosumpa inhabit the area of Tshosum[3]; and the Mago-Thingbupa live in the regions of Mago [dmag sgo], Thingbu [thing bu], and Lukguthang [lug dgu thang]. The Dakpanangpa speak the Tawang language with an accent that is slightly different from that of the other Tawang Monpa groups. The Mago-Thingbupa are pastoral people and share their language and native costumes with Tibetans in Jora Dzong [sbyor ra rdzong] in Tibet. The language of the Pangchenpa is different from that of the Tawang Monpa and the Dakpanangpa, but it is identical to that of the Lebo Tshoshi [legs po tsho bzhi] people of Tibet.

The Dirang Monpa in the broad sense of the term can be divided into three subgroups: the Dirang Monpa in the more limited sense; the Lish-Chug [sli phyug] Monpa (the Lishpa and Chugpa); and the Boot [bud] Monpa. The language of the narrowly defined Dirang Monpa is almost identical to that used in Eastern Bhutan. This particular language is called Tshangla [*tshangs la*] and is widely spoken in the

[3] It is also known as La'og yulsum [la'og yul *gsum*], which refers to a current Tawang sub-divisional area. See Table 3.1 for the names of the Monyul region's 32 tsho(s).

Dirang and Kalakteng regions of West Kameng and five districts in Eastern Bhutan (Andvik 2010).

The Dungkharpa [gdung mkhar pa] and Sharchokpa [shar phyogs pa] groups of the Dirang Monpa people emigrated from Eastern Bhutan. The Lish-Chug Monpa are believed to have originated from the Assam area; however, this remains to be verified. According to one theory, the Lishpa emigrated from Balemu, Nekteng [nags steng], and Bhairabkund [Kariapara *duar*] to avoid a tax imposed during the Darrang kingdom in Udalguri, Assam. The Lishpa are also believed to be descendants of immigrants from Tibet. Their language is very different from that of the surrounding Dirang Monpa. "Cooked rice" in the Dirang Monpa language is *tochang* [*lto chang*] or *toshang*, as noted above, whereas its equivalent in the Lishpa language is *hoi* and in Chugpa *mama*. "Horse," "cow," and "chicken" are *kuruta*, *wa*, and *goga*, respectively, in Dirang Monpa, but they are *sita* [*si tek* or *chibs rta*], *shui* or *soi*, and *wakhi*, respectively, in Lishpa and *sita*, *shui* or *soi*, and *hokhi*, respectively, in Chugpa (Glossary). The language of the Lishpa is similar to that of the Chugpa, though there are minor differences between them. "Where are you going?" is "Awba?" in Lishpa and "Hakuawba?" in Chugpa. According to local tradition, the Chugpa also originated in the west, and they initially settled in Sangti [sangs rdi] before gradually moving to their present location (Dhar 1995).

The Boot Monpa are inhabitants of the villages of Jirigaon [byi ri sgang], Salari [sa la ri], Khoitum/Khoitam [khul gtam], Rahung [ra hung], and Khoina [khu'i sna]; they share their language with that of the Sherdukpen [gsher stug spen] people of the Rupa and Shergaon regions. The customs and ceremonial practices of the Boot Monpa also resemble those of the Sherdukpen. However, the Boot Monpa are considered part of the Monpa since they have largely been assimilated over a long period of interaction (Dondrup 2004).

Senge Dzongpa [seng ge rdzong pa], Nyukmadungpa [smyug ma gdung pa], and pastoral people in Lubrang [lug grangs], close to the border with Bhutan, share a language and native costumes with the people of Merak [me rag] and Sakteng [sag steng] in Bhutan. Some words used in Senge Dzong and Nyukmadung are similar to those employed by the Tawang Monpa; other words are similar to those used by the Dirang Monpa. For example, "village" is *yui* [*yul*] in the Tawang Monpa's language as well as in Senge Dzong and Nyukmadung, but it is *dung* [*dung*] in the Dirang Monpa language. Similarly, "cow" and "horse" are *ba* [*ba*] and *ta* [*rta*], respectively, in the Tawang language, whereas *wa* [*ba*] and *kuruta* are the corresponding words in the Dirang language. By contrast, "maize" is *phinthang* [*phinang*] in Senge Dzong and Nyukmadung as well as in the Dirang language; its equivalent in the Tawang language is *oshum* or *boma*. Similarly, "millet" is *khongpu* [*kongpu*] in the Dirang language but *kowp* or *khra* in the Tawang language. There are also situations where each of the three is different from the other two. "Chicken," "rice," and "family" are *khat*, *dep*, and *mathtsan* [*ma tshang*], respectively, in the Tawang language, *jahu* [*bya*], *bre* [*'bras*], and *zatshan* [*za tshang*], respectively, in Senge Dzong, and *goga* or *gogadama*, *khu*, and *pheisonga*, respectively, in the Dirang language (Wangchu 2002; Dondrup 1993; Dasgupta 2007; Dutta 1999).

The Dirang Monpa and Kalaktang Monpa refer to themselves as Tshangla (Andvik 2010). They are called Sharchokpa, meaning "People of the East," by the Tawang Monpa. The Monpa in the Dirang area and those in the Kalaktang area are also called Drangnangpa [sbrang nang pa] and Rongnangpa [rong nang pa], respectively (Norbu 2008). The Dirang Monpa and Kalaktang Monpa refer to the Tawang Monpa as Brami [*brag mi*]. *Brami* means "people of the mountains"; *brag* means "rock" or "mountain" and *mi* means "people." *Drakmi* is the phonetic spelling in Tibetan.

Although the Tawang and Dirang areas are geographically adjacent to each other, they are divided by wall-like mountain ranges. The only traffic route connecting them is the Sela Pass. The mountain ranges affect the languages and cultures of the Dirang Monpa and Tawang Monpa. The Dirang Monpa are distributed to the south of the mountain ranges, and the Tawang Monpa to the north. That division is also the borderline between the Tawang and West Kameng districts (Fig. 1.6). In addition to the Tshangla-speaking Monpa, other ethnic groups such as Aka (Hrusso), Bugun, Miji (Sajolang), Sherdukpen, and Drokpa speakers can be found in West Kameng.

With the exception of the Boot Monpa, who are animists and shamanists, the Monpa practice Tibetan Buddhism or Bon. The Sherdukpen also follow Buddhism, whereas the Bugun and Miji practice semi-Buddhism as well as their traditional faith. The Aka (Hrusso) adhere to their own traditional religion.

As detailed above, a number of ethnic groups form a cultural and linguistic mosaic in the Monpa area, and the lingua franca is currently Hindi. Assamese was formerly used as a common language. The Tibetan script is employed by monks to read Tibetan Buddhist sutras. However, most people are unable to read such characters. A comparatively large number of people in the Tawang area are Tibetan speakers: some have relatives in Tibet and can understand Tibetan well.

References

Andvik E (2010) A Tshangla grammar. Brill, Leiden

Choudhury SD (ed) (1996) Gazetteer of India, Arunachal Pradesh: East Kameng, West Kameng and Twang District. Government of Arunachal Pradesh, Shillong

Dasgupta K (2007) An introduction to central Monpa. Government of Arunachal Pradesh, Itanagar

Dhar B (1995) Monpá: Chug. In: Dutta P, Ahmad SI (eds) People of India, Arunachal Pradesh. Seagull Books, Calcutta, pp 223–227

Dondrup R (1993) Brokeh language guide. Government of Arunachal Pradesh, Guwahati

Dondrup R (2004) An introduction to Boot Monpa language. Government of Arunachal Pradesh, Itanagar

Dutta DK (1999) The Monpas of Kalaktang: Arunachal Pradesh. Government of Arunachal Pradesh, Guwahati

Gopalakrishanan (1994) Arunachal Pradesh (land and people). Omsons Publications, New Delhi

Kri S (ed) (2010) State gazetteer of Arunachal Pradesh. Government of Arunachal Pradesh, Itanagar

Norbu T (2008) The Monpa of Tawang. Government of Arunachal Pradesh, Itanagar

Wangchu L (2002) An introduction to Tawang Monpa. Government of Arunachal Pradesh, Itanagar

Wylie T (1959) A standard system of Tibetan transcription. Harv J Asiat Stud 22:261–267

Chapter 2
Arrival of a Tibetan Prince in Monyul and Establishment of a Clan

The fortress Dirkhi Dzong [bde bskyid *rdzong*] dates back to around the fourteenth century and it was located on a ridge. Thus it was exposed to all kinds of attack. Subsequently a new fortress Thembang Dzong [them bang *rdzong*] was built at a higher altitude during the seventeenth century

K. Mizuno, L. Tenpa, *Himalayan Nature and Tibetan Buddhist Culture in Arunachal Pradesh, India*, International Perspectives in Geography 6,
DOI 10.1007/978-4-431-55492-9_2

Abstract A Tibetan prince in exile arrived in the Monyul region, the land of the Monpa, in the ninth century. He had been ousted by his brothers, became the king of Monyul, and constructed several fortresses there. His subordinates and soldiers also settled in the Monyul region. The master–subordinate relationship established at that time is still active today. Descendants of the king and his family constitute the senior clans in Monyul and those of the subordinates and soldiers make up the junior clans. This chapter also examines the origin of the clans in the Monyul region. The Dirang Monpa and the Sherdukpen in this area have clan systems and the systems play an important role for local communities and land ownership. The focus of this chapter is on the historical processes by which the king's fortresses (Dzong [*rdzong*]) were built. It also focuses on the historical formation of the clans in the area of the Dirang Monpa.

Keywords Clan • Tibet • Royal family line • King's fortress • Tax collection

2.1 Arrival of a Tibetan Prince and Origin of the Clans

The Tibetan Empire flourished when Songtsan Gampo [srong btsan sgam po] (AD 629–49) of the Yarlung [yar lung] dynasty was on the throne. He is traditionally regarded as the 33rd king after the very first Tibetan king, Nyatri Tsanpo [gnya' khri btsan po] (third century BC). The Yarlung dynasty originally began as a royal family that dominated the central south area of Tibet. The actual existences of the first to the 32nd kings have been greatly debated because written records are available only from the seventh century. However, Tibetan historical sources record the successive descendants of the kings. It is reasonable to consider the first to 32nd kings to be fictitious characters created by Tibetan historians. Those historians regard Songtsan Gampo as the 33rd king; however, it is more realistic to assume that Songtsan Gampo was only about the 14th king (Yamaguchi 2004). Tibetan historians state that there were 26 kings from the eighth king, Drigum Tsanpo ['gri gum btsan po], to the 33rd king, Songtsan Gampo. However, according to the Chinese T'angshu, there were only six kings during that period (Yamaguchi 2004). Thereafter, the 38th king (according to the traditional list), Trisong Detsan [khri srong lde'u btsan] (755–97), established and vigorously promoted Buddhism as the state religion. Conventional Bon practitioners saw the rise of the foreign religion as a danger and banned it while the king was still young (Yamaguchi 2004).

Norbu (2008) states the follows. During the reign of the 41st king, Tri Ralpachen [khri ral pa can] (806–38) or (806–41 in Yamaguchi 2004), the king's brother Lang Darma [glang dar ma, 'u dum btsan po] (803–41) or (809–42 in Yamaguchi 2004) and several ministers with an anti-Buddhist agenda planned the king's assassination (Fig. 2.1). However, Lang Darma realized that he would be unable to take over the throne even after the king's death because of the king's other brother, Lhase Tsangma [lha sras gtsang ma]. Lang Darma abandoned the plan to assassinate the king and prepared instead a new plan to have Lhase Tsangma expelled. During the reign of

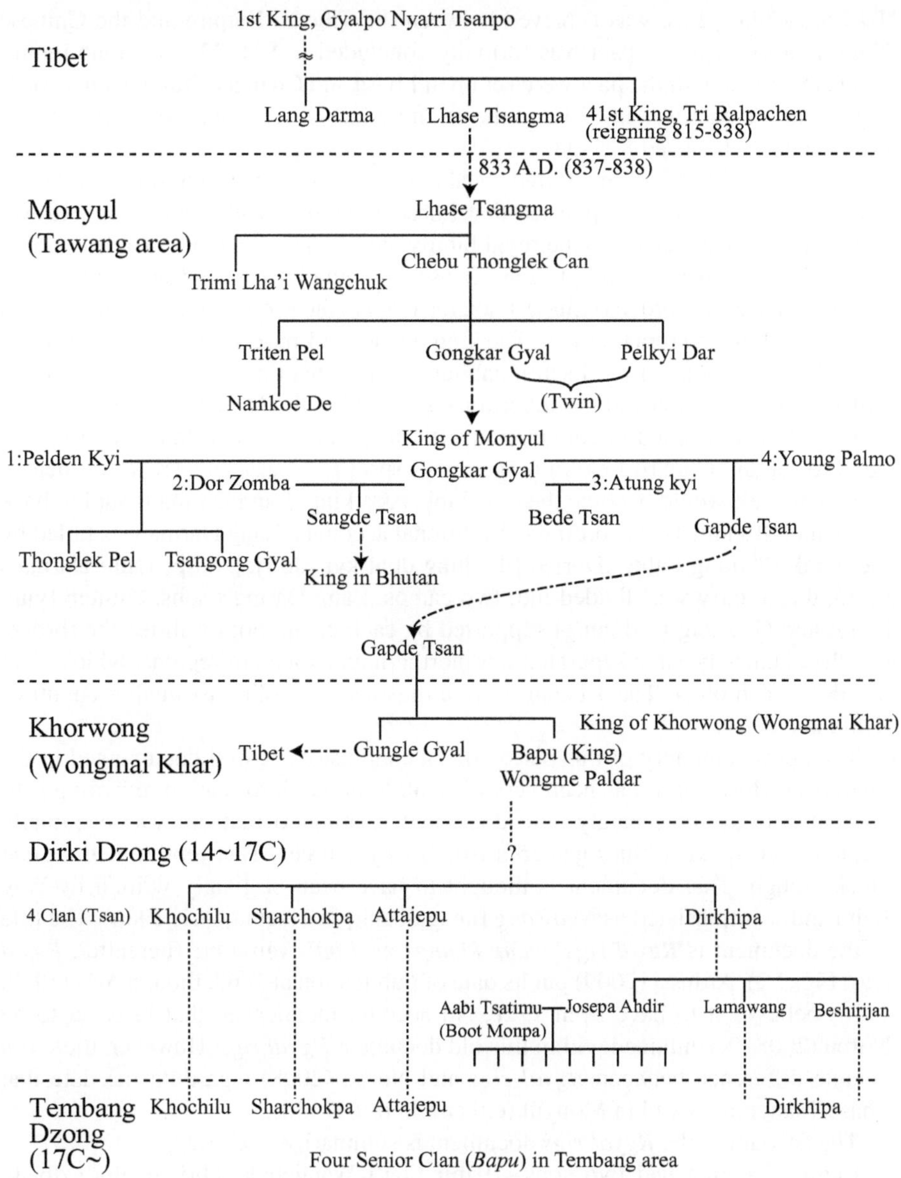

Fig. 2.1 Genealogy of clan formations in the Thembang (them bang) area of the Dirang (rdi rang) region. Monyul [mon yul] and Khorwong ['khor 'ong] sections are from both *Rgyal rigs 'byung khung gsal ba'i me long* (*Rgyal rigs* for short), which was compiled by Wag Indra [ngag dbang], and oral sources. Its date of publication is considered to be either 1668 according to John Ardussi (2009) or 1728 according to Micheal Aris (1986). The Dirkhi Dzong [bde skyid *rdzong*] and Thembang Dzong [them bang *rdzong*] sections were based on local interviews carried out as part of this study

Tri Ralpachen, peace was achieved between the Tibetan Empire and the Chinese Tang dynasty. A peace pact was formally concluded in 821–22, and stone monuments inscribed with the pact were set up in Lhasa and Changan. Yamaguchi (2004) states that the pact was in 833 and the reign period of Tri Ralpachen was 815–41, not the widely believed 815–38.

Thereupon, Lang Darma offered royal diviners and prophets enormous bribes to deceive the king. They prophesied that Lhase Tsangma would indulge in extravagance and bring adversity to the royal family and, eventually, to the whole country. The diviners advised the king to send Lhase Tsangma into exile. The king believed their prophecies and ordered Lhase Tsangma to go to the Monpa area, which was then known as Mon, and remain there. The king instructed Lhase Tsangma to investigate the living conditions of the local inhabitants during his time in Monyul. Thus, the exiled Lhase Tsangma and his subordinates arrived in the Monpa area in 837–38.

Lang Darma worked in conjunction with the great ministers Bah Gyetore [dbas rgyal to re] and Chokro Lekdra [cog ro legs sgra] to assassinate Tri Ralpachen so that he himself would become the next king. As a king, Lang Darma is said to have persecuted Buddhists. According to traditional accounts, Lang Darma was killed by the monk Lhalung Palkyi Dorjee [lha lung dpal kyi rdo rje]. After Lang Darma's death, the country was divided into two camps. Lang Darma's sons, Yumten [yum brtan] and Oedsung ['od zung] supported by each camp, both claimed the throne, and the country became separated into northern and southern regions, which were hostile to each other. The Tibetan Empire thus became split into smaller countries (Ishihama 2004).

The above summary of this period of Tibetan history is well documented in the literatures. However, it has been very difficult to obtain information regarding subsequent periods. Fortunately, one document relating to that era was eventually located after tracking clues gathered from many sources. That represented a great breakthrough. That document is thought to have been originally written by Wag Indra and was translated as *Ngawang* [ngag dbang] in Tibetan (Aris 1986). The title of the document is *Rgyal rigs'byung khungs gsal ba'i sgron me* (hereafter, *Rgyal rigs*; Fig. 2.2). Ardussi (2009) put its date of publication at 1668, though Aris (1979, 1986) believes it to have been 1728. An account identical to that referred to by Norbu (2008) is contained within this old document *Rgyal rigs*. However, there is a notable difference between *Rgyal rigs* and Norbu (2008) regarding the date that Lhase Tsangma arrived in Monyul (either 833 or 837–38).

The account in the *Rgyal rigs* document is summarized below.

Lhase Tsangma had two sons—Trimi Lha'i Wangchuk [khri mi lha'i dbang phyug] and Chebu Thonglek Tsan [gces bu mthong legs btsan]. The latter had three sons of his own: Triten Pel [khri brten dpal]; Gongkar Gyal [gong dkar rgyal]; and Pelkyi Dar [dpal kyi dar] (Fig. 2.1). Gongkar Gyal and Pelkyi Dar were twins. Triten Pel had a son named Namkoe De [gnam bskos sde]. One of the twins, Gongkar Gyal, became the king of Monyul (presently the Monpa area), and he had four wives: Pelden Kyi [dpal 'dren skyid]; Dor Zomba [rdor 'dzom pa]; Atung Kyi [a thung skyid]; and Yang Palmo [g.yang dpal mo]. Pelden Kyi had two sons: Thonglek Pel [mthong legs dpal] and Tsangong Gyal [Btsan gong rgyal]. Dor Zomba, Atung Kyi,

Fig. 2.2 Compiled by Wag Indra, the document *Rgyal rigs'byung khungs gsal ba'i sgron me* (*Rgyal rigs* for short) is translated as *Ngag dbang* in Tibetan. Ardussi (2009) believes that it was written in 1668, though Aris (1986) puts this date at 1728

and Yang Palmo all had a son each with the king, and their names were, respectively, Sangde Tsan [gsang sde btsan], Bede Tsan [sbas sde btsan], and Gapde Tsan ['gab sde btsan]. Sangde Tsan became king of a region in Bhutan. Gapde Tsan was taken to the place known as the Khorwong ['khor dbang] valley under Thembang [them bang] to be enthroned, even though poverty was prevalent in that area. Khorwong was also known as Wongmai Khar [dbang ma'i mkhar]. Gapde Tsan had two sons named Gungle Gyal [dgung las rgyal] and Wongme Palder [dbang ma'i dpal dar]. These two sons went back to Tibet owing to the poverty in Khorwong. However, subsequently, the second son, Wongme Palder, returned to Khorwong with two guards. Upon his return, the *tsorgen* [*gtso rgan*] (village mayor or *Goanbura* in Assamese) named Agyal [a rgyal] told him, "You are a descendant of the king, you should govern this area" (*Rgyal rigs*: 27a–28b).

There is an oral tradition in the village of Thembang that is related to the above account detailed in *Rgyal rigs*. The following is a summary of interviews (Mizuno, February 2010) with people in the village.

> A king was once expelled by his Boot Monpa subordinates because he married a woman from a junior clan. The Boot Monpa subordinates then needed a new king. They went around villages with the leg of a yak [g.yag], a goat, and a sheep to find a person who could choose the leg of the yak. It proved difficult to locate someone who was able to make the

Fig. 2.3 Lake Mugye-Tchang [mu ge'chag] is a lake close to Sela Pass, known locally as Lake Paradise or Lake Sela, is the source of many legends

right choice, but they eventually found such a person. The Boot Monpa subordinates wanted that person to become their king, but they did not tell him this immediately. Instead, they took him hunting and finally told him their intention at the site of Lake Mugye-chang [mu ge 'chag]. It is a legendary lake near Sela Pass and referred to locally as Lake Paradise or Lake Sela (Fig. 2.3). A teardrop then rolled from each of the selected person's eyes and formed two lakes, which merged to become one large lake. The party continued traveling. After passing Tawang, the subordinates started telling their new king, "This land is yours" and "This land also belongs to you." They decided on place names based on what had happened on the hunt. For example, a place where the new king felt cold was named Changla (cold), and another place where someone held an iron case containing the king's clothes was named Tungri (iron case).

Thus, this individual became their new king, and he is said to have been Gapde Tsan, as mentioned in *Rgyal rigs*. His son, Wongme Palder, was frequently referred to by the elders during the interviews in the village of Thembang. It is reasonable to assume that Wongme Palder is a key person in the history of the village. Several generations after Wongme Palder, the unknown king of Monyul had four sons; from the eldest to the youngest, their names were Khochilu [kho phyi lu], Sharchokpa [shar phyogs pa], Attajepu [a ta rje bu], and Dirkhipa [bde skyid pa]. They became the founders of the four clans in the Thembang area (Table 2.1). A clan is called *tsan* [btsan] in the Dirang Monpa language. Wongme Palder was more powerful than the king of the Darrang kingdom in Assam, and he collected taxes from residents of that

Table 2.1 Clans of the Thembang [them bang] and Dirang [rdi rang] areas

Old time	Present time	Clan			
Ruling family	Ruler clan (Senior clan) (Bapu)	Khochilu	Sharchokpa	Attajepu	Dirkhipa
Servant	Subject clan (Junior clan) (Gilla)	Lhofsonga Merakpa	Sharmu	None	×Chumsonga Nimsonga
Subject	Boot Monpa				

Old time	Present time	Clan				
Ruling family	Ruler clan (Senior clan) (Khangtangphu)[a]	Bagipa	Dungkharpa	Serthipa	Jamsempa	Phai chulupa
Servant	Subject clan (Junior clan) (Gilla)	Grangpa	Kompa	Merakpa	Barkhapa	Tsumklapa

Produced from locally conducted interviews

[a]Khangtangphu = Big bone

1. Bapu [sba bu] is the Hindi equivalent of "father"
2. The Merakpa [me rag pa] are people whose ancestors emigrated from Bhutan
3. The Chumsonga [chum song ga] disappeared, and the Nimsonga [ni ma song ga] were introduced by Ata Devo Dorjee [a tha bde bo rdo rje]
4. The term Khangtangphu [rkang thang phugs] means "thick bone"
5. The Dungkharpa [gdung mkhar pa] partially originate from the Sharchokpa [shar phyogs pa], who migrated from the Thembang area to the Dirang area

kingdom. The king of the Darrang region told Wongme Palder, "You are the true *Bapu* [sba bu]" ("father" in Hindi), and from that time the descendants of Wongme Palder were all known as Bapu. Members of the four royal clans are also considered Bapu and are ranked as senior clans.

The royal family had servants, and subordinates (soldiers) consisting of the Boot Monpa. The descendants of the servants formed junior clans called *gilla* [gyi la]. The gilla of Khochilu formed the clans called Lhosonga [lho song ga] and Merakpa [me rag pa]. The Merakpa clan consists of people who originally came from Bhutan, while their relatives still live in Bhutan. The gilla of Sharchokpa formed the Sharmu [shar mo] clan. A clan called Chumsonga [chum song ga] formerly existed and was known as the gilla of Dirkhipa, though it has not survived. Instead, another clan called the Nimsonga [nyi song ga] is now the gilla of Dirkhipa. Attajepu does not have a gilla.

2.2 Construction of Dirkhi Dzong Fortress

Several generations after Wongme Palder, the king of Monyul constructed the fortress Dirkhi Dzong [bde skyid *rdzong*] on a ridge at an elevation higher than Khorwong (Wongme Khar) (Figs. 2.4, 2.5, 2.6 and 2.7). The following is a summary of interviews (Mizuno, May 2010) carried out in the village.

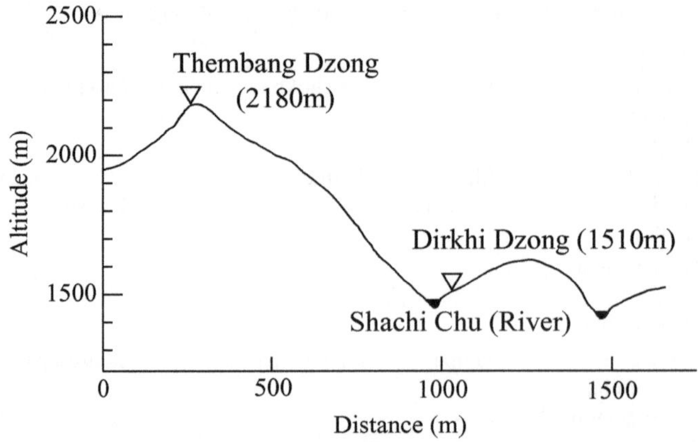

Fig. 2.4 Topographical cross-section showing the positions of Dirkhi Dzong [bde skyid *rdzong*] and Thembang Dzong [them bang *rdzong*] (Produced from a topographical map)

Three brothers of the Dirkhipa clan lived in Dirkhi Dzong. They were Josepa Ahdir [jo sras pa a 'dir], Lamawang [bla ma dbang], and Besirijan [be si ri byan] (Fig. 2.1). Josepa Ahdir had a wife, Aabi Tsatimu [abi tsa ti mu], from the junior clan called the Rahungpa [ra hung pa], a group of Boot Monpa and inhabitants of Rahung village. Josepa Ahdir had nine sons, who regularly harassed and robbed the Dirkhipa and their subordinates. Lamawang and Besirijan paid the Khoinapa [khos sna pa] people, a group of Boot Monpa inhabitants of the village of Khoina, and the brothers of the wife of Josepa Ahdir to kill all nine sons at the Khorwong River, located just below the fortress. They then stoned Josepa Ahdir to death near the fortress. Some people claim that the reason for this killing was that the two brothers, Lamawang and Besirijan, wanted to remove the bloodline of the Boot Monpa to secure their own royal bloodline. After the killing of Josepa Ahdir, the Dirkhipa clan consisted exclusively of the descendants of the two brothers. The body of Josepa Ahdir was buried where he was killed. Since then, passersby have thrown pebbles at that point, and so it is now marked by a small mound (Fig. 2.8).

When Dirkhi Dzong was attacked by the Miji tribe, the Khochilu clan fled to Yayung of the Nafra area, the Sharchokpa clan to Munda (the area under Bomdila), and the Attajepu clan to Redhipongman [re rdi spong ma] in Bhutan. Only the Dirkhipa clan remained at the fortress to fight back. The Dirkhipa moved into the forest between Thembang and Cherong to offset their numerical inferiority and managed to kill the chief of the Miji tribe. The Miji force was then forced to retreat. The Dirkhipa returned to the fortress and continued to occupy the site. The fortress was referred to as Dirkhi Dzong from then on. The Dirkhipa told their brother clans to return, and the Sharchokpa came back, followed by the Khochilu, and later the Attajepu. The royal clans decided to build a new fortress at a site with a higher elevation than Dirkhi Dzong, which became known as Thembang Dzong [them bang *rdzong*]. For construction of both Dirkhi Dzong and Thembang Dzong, Boot Monpa worked as laborers. Before the war attacked by the Miji, the people of Munda, where the Sharchokpa clan was living in exile, had been paying a tax of 2,000 bows and arrows and 20 bags of dried chicken to the Bapu, the royal clan. The Dirkhipa abolished that taxation, which had been used to maintain Dirkhi Dzong, in recognition of the Munda people having provided refuge for the Sharchokpa during the war.

Identifying the site of Dirkhi Dzong in the Thembang area has proved difficult. The state government of Arunachal Pradesh decided to undertake an investigation to

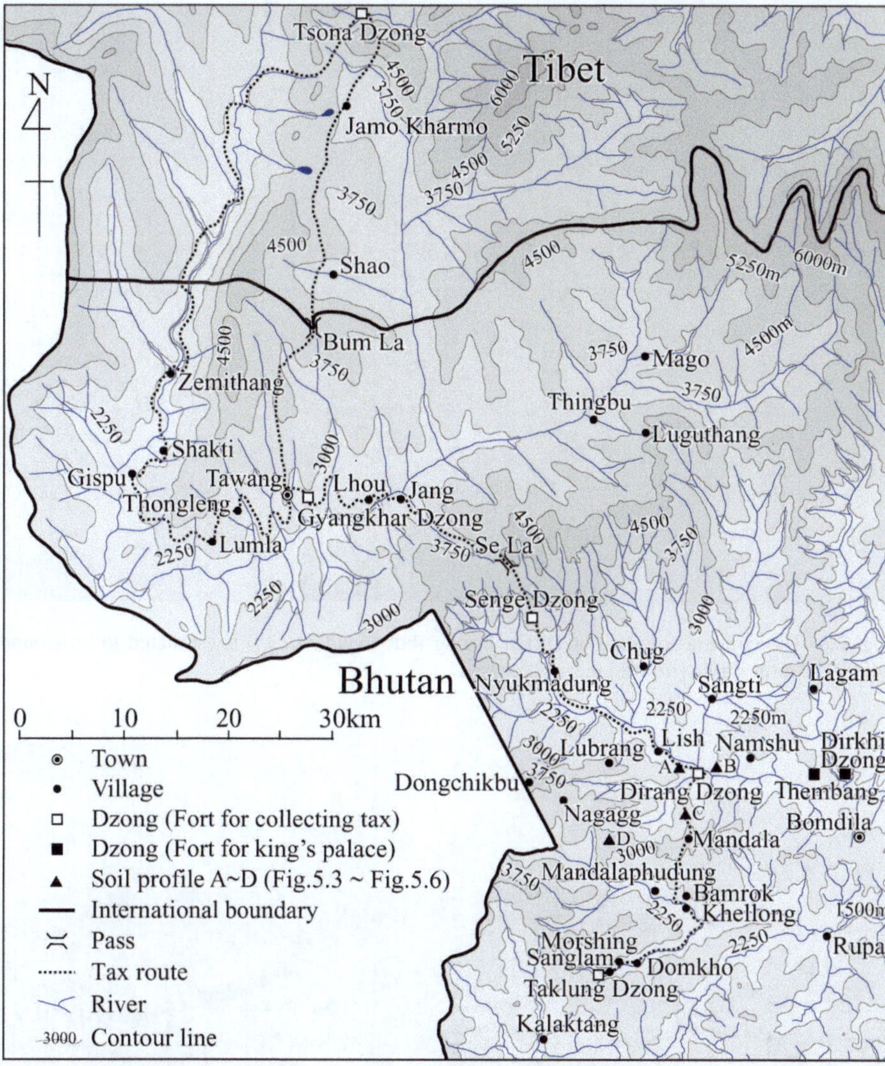

Fig. 2.5 Locations of *dzong* [*rdzong*] and tax transportation routes. The routes were drawn using information acquired locally through interviews and based on topographical maps. Tawang Monastery is in Tawang. The *dzong* at Senge Dzong [seng ge *rdzong*] was a rest house for *dzong-pon* [*rdzong* dpon]

locate Dirkhi Dzong in 2004. The government's investigation team wanted the Thembang villagers to guide them to the site. However, the villagers were reluctant to do so since they believed that Dirkhi Dzong was the home of their ancestral spirits, and they were afraid of being cursed through having disturbed the site. Despite this, the state government made it mandatory for every household to offer one person who would act as a guide. According to the participants, the only outcome of the

Fig. 2.6 The construction date of Dirkhi Dzong [bde skyid *rdzong*] is estimated to be around 1400. The term *dzong* [*rdzong*] means "fortress"

Fig. 2.7 Remains of Dirkhi Dzong [bde skyid *rdzong*]: the holes in the wall are thought to be ventilation openings for the kitchens or crenels from which to shoot arrows

Fig. 2.8 Site where Josepa Ahdir [jo sras pa a 'dir] is buried: a small mound has grown as a result of passers-by dropping pebbles upon it

first expedition to locate the site was taking photographs of Dirkhi Dzong. No serious investigation was subsequently undertaken.

We requested one of the participants of the 2004 investigation to act as a guide for us, and we managed to locate Dirkhi Dzong in 2010. It was the first time for a foreigner to do so. After departing from Thembang Dzong and trekking across hills and valleys toward Dirkhi Dzong, we came upon an open patch of land in the vicinity of a stream. That was the site of Khorwong (Wongme Khar), which Wongme Palder had occupied as the new king (Fig. 2.9). Beyond the open patch and above a steep slope covered with briers and brambles, Dirkhi Dzong was located on a small ridge. The fortress ruins were also covered with briers and brambles, and we could not make a proper investigation of the site.

According to the local tradition, Dirkhi Dzong consisted of 24 dwellings in its final years as a settlement. It is impossible to determine the chronology of Dirkhi Dzong based on records and traditional sources. All incidents are reported as being matters of "the past." The fortress was made of stone, however wooden frames which are common in other fortresses *dzong* were not present. Whether they were absent from the outset or removed at a later date is unclear. When the guide, our student, and we examined the site, our student found a piece of charcoal weighing approximately 1 g in a gap in a stone wall of the structure. Radiocarbon dating (Beta Analytic Radiocarbon Dating Laboratory, USA) indicated that the charcoal dated from approximately AD 1400, more concretely the result indicated a figure of

Fig. 2.9 The site of Khorwong ['khor 'ong] or Wongme Khar [wang ma'i mkhar] is where Gapde Tsan ['gab sde btsan] from the Tawang area ruled as the new king, probably in the early medieval period. The people crossing the bridge are Boot Monpa on a hunting expedition: they are basically agricultural people, though they are also engaged in hunting

550 ± 40 years prior to 1950 (Table 2.2). The date of the construction of Dirkhi Dzong has not been revealed before. This result confirms that the fortress was already being used in the period between 1400 and 1500. In interviews conducted in Thembang, some interviewees said that Wongme Palder had built Dirkhi Dzong, others said that his son was the builder. If the fortress dates from approximately 1400, the family tree shown in Fig. 2.1 suggests that it must have been built several generations after Wongme Palder.

2.3 Move to the New Thembang Dzong Fortress

Dirkhi Dzong was vulnerable to foreign enemies, such as the Miji and Sherdukpen (Fig. 1.6). It was also at a great distance from water resources and located in an area with limited agricultural potential. A new settlement at higher elevation was needed, and the new fortress of Thembang Dzong was subsequently built (Figs. 2.4, 2.5 and 2.10).

Thembang Dzong is enclosed by stone walls, and has a gate at its western and eastern sides. Dwellings were located within the stone walls (Fig. 2.11). According to

Table 2.2 Estimated dates of *dzong* [*rdzong*] obtained by means of radiocarbon dating of materials

Date of Dirkhi Dzong					
beta ID	Sample name	Material	^{14}C date		Measuring method
279546	Dirkhi1	Charcoal	550	±40	Radiocarbon dating (AMS)

Date of Thembang Dzong					
beta ID	Sample name	Material	^{14}C date		Measuring method
276486	Tembang1	Wood	260[a]	±50	Radiocarbon dating
279547	Tembang2	Wood	160[b]	±60	Radiocarbon dating
279548	Tembang3	Wood	280[a]	±40	Radiocarbon dating
279984	Tembang4	Wood	240[a]	±40	Radiocarbon dating
284119	Tembang5	Wood	190[b]	±40	Radiocarbon dating

All: Weighted average 232+/−20; Calendar era (2sigma)* 1645–1671AD, 1780–1796AD
[a]Weighted average 260+/−25; Calendar era (2sigma)* 1535–1545AD, 1634–1665AD
[b]Weighted average 181+/−33; Calendar era (2sigma)* 1654–1698AD, 1723–1817AD, 1843–1871AD

Date of Taklung Dzong					
beta ID	Sample name	Material	^{14}C date		Measuring method
279549	Taklung1	Wood	250[c]	±40	Radiocarbon dating
284120	Taklung2	Wood	130[d]	±40	Radiocarbon dating
284121	Taklung3	Wood	160[d]	±40	Radiocarbon dating
284122	Taklung4	Wood	190[d]	±50	Radiocarbon dating
284123	Taklung5	Wood	250[c]	±50	Radiocarbon dating
284124	Taklung6	Wood	250[c]	±50	Radiocarbon dating
284125	Taklung7	Wood	120[d]	±40	Radiocarbon dating

All: Weighted average 186+/−16; Calendar era (2sigma)* 1662–1682AD, 1738–1805AD
[c]Weighted average 250+/−26; Calendar era (2sigma)* 1637–1669AD, 1783–1792AD
[d]Weighted average 146+/−21; Calendar era (2sigma)* 1672–1704AD, 1717–1778AD, 1797–1883AD

*Dates after 1900 have been eliminated
^{14}C dates indicate the number of years prior to 1950

the information plate set up by the state government at the gate to the fortress, construction of Thembang Dzong dates from the eighteenth century. Radiocarbon dating of samples from the fortress has suggested several different possible dates for its construction. Five samples of wood were taken from part of the structure that is believed to be the oldest building in Thembang Dzong and is now just a ruin (Fig. 2.12). The results of the radiocarbon dating indicated two different periods: the older period was the sixteenth or seventeenth century, and the newer period was the seventeenth or eighteenth century (Table 2.2). The wooden parts of this oldest section of the structure were burnt and blackened. Records indicate that Thembang Dzong was engulfed by fire on three occasions in its history. The third fire occurred in 1959. Residents of Thembang Dzong village states that many buildings were

Fig. 2.10 The fortress of Thembang Dzong [them bang *rdzong*] is enclosed by stone walls and has a gate on its western and eastern sides

Fig. 2.11 The construction date of Thembang Dzong [them bang *rdzong*] is estimated as the seventeenth to eighteenth centuries. There are residential spaces within, and people still live there

Fig. 2.12 This dwelling is believed to be the oldest in Thembang Dzong [them bang *rdzong*], and only its surrounding walls remain

actually constructed in the period following the second fire, probably in the nineteenth century. It is, then, reasonable to assume that earlier date, the sixteenth or seventeenth century from the radiometric results was after the first fire, and the later date was after the second fire.

The *Rgyal rigs* text (1668 or 1728) includes records of taxation by Bapu after the time of Wongme Palder (Table 2.3). In the area that was formerly known as Magthing Luksum [dmag sgo thing bu lug dgu thang gsum] but is presently the Thingbu Circle area of the Tawang district, silver coins, known as *baitang* [*bey Tram*], were collected as a type of tax called *pungur* [spung 'gyur], cheese and butter were collected as *jhalane* [ja las gnas] tax, and yaks [g.yag] were collected as *tshapa* [tshwa pa] tax. Sheep and cows were collected as tax from an area called Drangnang Tshodruk [sbrang nang tsho drug]. Drangnang Tshodruk is the area between Sela Pass and Boot Monpa villages; it roughly corresponds to the present-day Dirang and Thembang Circle areas of West Kameng. Six cows and 12 sheep were collected as tax from the village of Chug, 8 cows and 30 sheep from the village of Lish, 6 cows and 10 sheep from Nyukmadung and Senge Dzong, 7 cows from Sangti, and 15 cows and 30 sheep from Dirang and Yewang [yid dbang]. The taxation on Dirang and Yewang was later abolished, probably in the nineteenth century as compensation for the Bapu's murder of villagers. Japanese pepper (*Zanthoxylum piperitum*), known locally as *khagei* [kha 'gas] (Fig. 2.13), and sacred scarves called *chongleywha* [chong las ba] were collected from an area that used to be called Phundung Thongtse

Table 2.3 Types of tax collected by the *bapu* [sba bu] from their territory

	Old area name	Collecting tax
1	Magthing Luksum (dmag sgo thing bu lug dgu thang) (Present Tawang area: Mago, Thingbu and Lugthuang)	Pungur: Baitang (Silver coin), Jhalane (Chese and Butter), Tshapa (Yak)
2	Drangnang Tshodruk (brang nang tsho drung) (Present Sela to Boot Monpa village Khoina)	Sheep, Cow Chug: cow 6, sheep 12, Lish: cow 8, sheep 30 Senge Dzong & Nyukmadung: cow 6, sheep 10, Sangti: cow 7 Dirang & Yaweng: cow 15, sheep 30*
3	Phudung Thongtse (phugdung mthong rtse) (Present Mandala Phudung)	Khagei, Chongleywha (Holy scarf)
4	Lego Geydar (las sgo rgyas dar) (Present Darrang District of Assam and Kalaktang area)	Rice, Cloth, Raw silk, Cotton threads etc.

Produced from information in *Rgyal rigs*
*Some villagers in the Dirang [rdi rang] area were killed by the bapu, and tax collection was discontinued in the Dirang and Yewang [yid dbang] areas by way of compensation

Fig. 2.13 Berries of Japanese pepper (*Zanthoxylum piperitum*), locally called *khagei* [kha'gas]: these are used to pay a form of tax called *khray* [khral]

[phu gdung mthong brtse]. Phundung Thongtse is the area around present-day Mandalaphudung [man da la phu gdung]. Japanese pepper has long been an important condiment in the Monpa area. Its pulverized berries are used as a spice, and even today it costs as much as 400 rupees (1 rupee = 0.02 US$) per kilogram in local markets. Rice, clothing, silk, cotton yarn, and other items were collected from an area that was formerly known as Lego Gyadar [las sgo rgya dar]; this corresponds to Kariapara *duar*, which is now called Bhairabkunda in the Udalguri district, Assam. The tax called *posa* is still collected from the Kariapara *duar* of the Udalguri area. The Assam state government sends about 280 rupees to the Bapu clans, the four senior clans, in late January every year (see Chap. 4, Sect. 4.4 for an additional discussion of *posa*).

In the Thembang area, the Bapu of Thembang used to collect land rent in the form of cheese and butter from pastoral areas, such as Chandar [phyan dar], once a year. This old system remains. Chandar pays rent to the village of Thembang once every 2 years and to the village of Namshu [nam shu] every year. This particular type of land rent is called *tsarin* [*rtswa rin*], where tsa [*rtswa*] signifies "grass" and *rin* "amount" and "due." This land rent is paid by pastoral people to farmers for the use of grazing land (see Chap. 7, Sect. 7.7).

The village of Thembang currently consists of 67 households located within the fortress and 35 households outside. According to interviews (Mizuno, February 2010) conducted during fieldwork, the villagers practice the following tradition.

> A king of the Sherdukpen once governed the Boot Monpa. The two tribes shared the same language. When the king and his forces attacked Thembang Dzong, a 7-year-old boy in the fortress fired an arrow, which hit the king and killed him. The Boot Monpa cut the king's head from his body. The body was given back to the Sherdukpen, and the head was buried under the eastern gate of Thembang Dzong (Fig. 2.10). People spat on the ground under the gate when passing through.

A note related to this story can be found in Nagaraju and Tripathy (2006).

> The fortification of Thembang Dzong was possible only through the manual labor provided by the villagers of Khoina, Boot (now known as Jirigaon), Khoitam, and Rahung as well as by the gilla. Their services were also used for seed sowing (*ngyangben* [myang bon]), tree plantation (*ungashing cheewn* ['ung ga shing mkhan]), carrying goods (*rongbanghen* [rong bang mkhan]), constructing houses (*phaijanken* [khyim rgyag mkhan]) and fortification walls (*jangri* [byang ri]).
>
> Those villagers also had to provide military help for the Bapu. For example, when the soldiers of the Rupa village were retreating after their attack to the Thembang Dzong, the villagers of Khoitam beheaded the military commander of Rupa village. They sent the commander's head to their king (Bapu) and the body to the Sherdukpen Rajas (king),Thongs of Rupa. As a result of this incident, the people of the Rupa area call the villagers of Khoitam the *Songa-sha-boianken* [song ba sha bor mkhan], "dividers of human flesh."

The fort of Thembang (recorded Thebengia Bhutias in British records) proved advantageous for the kings there. The Tibetan rifle, which the Bapu possessed but others lacked, was another advantage. These two factors, combined with high agricultural production and a fairly large tax base extending up to modern-day Assam, contributed significantly to the establishment of Bapu authority and also to their successful resistance to the expansion of the Tibetan monastic order in their area in the eighteenth to early twentieth centuries (Nagaraju and Tripathy 2006).

References

Ardussi JA (2009) Historical introduction to the sources for the history of Bhutan. Arbeitskreis für Tibetische und Buddhistische Studien: Wien (Vienna). Reprinted by Motilal Banarsidass Publication, Delhi

Aris M (1979) Bhutan, the early history of a Himalayan Kingdom. Aris and Phillips Ltd., London

Aris M (1986) Sources for the history of Bhutan (Wiener Studien zur Tibetologie und Buddhismuskunde). Arbeitskreis für Tibetische und Buddhistische Studien. Universität Wien, Vienna

Ishihama Y (ed) (2004) 50 Chapters to study Tibet. Akashishoten, Tokyo (in Japanese)

Nagaraju N, Tripathy B (2006) Cultural heritage of Arunachal Pradesh. Indus Publishing Company, New Delhi

Norbu T (2008) The Monpa of Tawang. Government of Arunachal Pradesh, Itanagar

Rgyal rigs (1668/1728) Wag Indra (ngag dbang) Rgyal rigs'byung khungs gsal ba'i sgron me. In Michael Aris translated and edited (2009 [1986]): Sources for the History of Bhutan (Wiener Studien zur Tibetologie und Buddhismuskunde).

Yamaguchi Z (2004) Tibet: 2 (revised edition). Toyo Sensyo, Tokyo (in Japanese)

Chapter 3
Taxation by the Tibetan Authorities and the Formation of Fortresses

Gorsam Chorten [gor zam mchod rten], Zemithang: this Buddhist stupa is thought to have been built by Lama Pradhar [bla ma spre'u thar] in the eighteenth century; however, the thirteenth, sixteenth, and nineteenth centuries have also been suggested

K. Mizuno, L. Tenpa, *Himalayan Nature and Tibetan Buddhist Culture in Arunachal Pradesh, India*, International Perspectives in Geography 6,
DOI 10.1007/978-4-431-55492-9_3

Abstract The Tibetan government set up fortresses, locally called *dzong* [*rdzong*], at several locations in Monyul as administrative offices for the purpose of collecting taxes. The *dzong* also functioned as administrative centers for local communities. In addition, the Tibetan government constructed Tawang Monastery, which exerted an influence on many aspects of the lives of the local inhabitants. Taxes collected from local residents were used by the monastery, and a portion of the taxes was transported to Lhasa. In the early eighteenth century, two administrators *dzongpon*, one at Dirang and another at Taklung, were appointed for collection of tax as contribution towards the monastery etc. Since 1944, especially in 1946 just before the independence, British-ruled India tried to abolish taxation by *dzongpon* in the Dirang and Kalaktang areas. However, since the taxation contributed to Tawang Monastery or to locally resettled former officials, the local people themselves continued this practice until 1951. This chapter examines the types of items that were collected as taxes and how the taxes were conveyed to Tawang Monastery and Lhasa. The main focus of this chapter is on the history of the interrelationships among the local people in the Monpa area, Tawang Monastery, and the Tibetan government.

Keywords Tax • Fortress • Tibetan government • Lhasa • Tawang Monastery

3.1 Taxation and Construction of *Dzong* in the Monpa Area

Backed by the military resources of both the Tibetans and Mongols, Ngawang Lobsang Gyatso [ngag dbang blo bzang rgya mtsho], the great fifth Dalai Lama (1617–1682), achieved political supremacy over Tibet in 1642. The Dalai Lamas had traditionally been religious leaders, but the fifth Dalai Lama established a theocracy in his country. At that time, Monyul was under the control of the Tibetan administration in Lhasa. This situation was reinforced by the imposition of a tax called *khrey* [*khral*]. Fortresses called *dzong* [*rdzong*] were constructed in various locations in Monyul, serving as centers for administration and tax collection (Figs. 2.5 and 3.1). They included Dirang Dzong, Taklung [stag lung] Dzong, and Gyangkhar [rgyang mkhar] Dzong. Dirang Dzong, Taklung Dzong, and Gyangkhar Dzong were built by the Tibetan administration to collect tax from the Dirang Monpa, Kalaktang Monpa, and Tawang Monpa, respectively. The tax was conveyed to Tawang Monastery and then to Lhasa via Tsona [mtsho sna] Dzong (Figs. 2.5 and 3.1).

A commissioner was sent from Tibet or Tawang Monastery to a *dzong* to collect the tax. These officials were called *dzongpon* [*rdzong dpon*] in Tibet, Bhutan, and the Monpa area. The name is pronounced *dzongpen* in local languages; the suffix *dpon* signifies "officer" or "leader" in Tibetan as well as in the Monpa languages. In the early eighteenth century, two *dzongpon*, one at Dirang and another at Taklung, were appointed for collection of tax as contribution towards the monastery etc. The *dzongpon* were selected among the senior monks of Tawang Monastery.[1] The

[1] Directorate of Research, Government of Arunachal Pradesh (without date). The booklet states that it was in 1831 that two *dzongpon*, one at Dirang and another at Taklung were appointed.

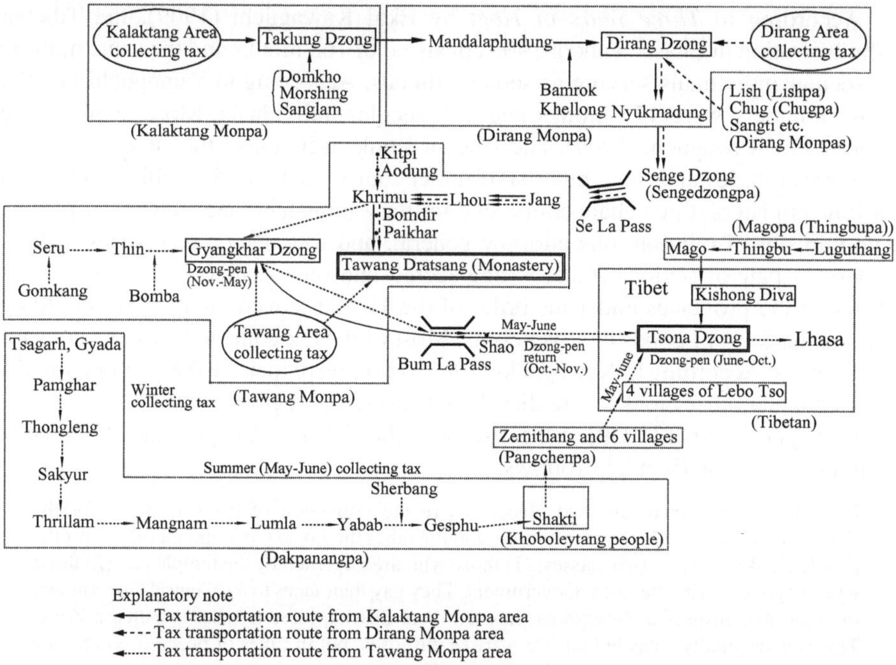

Fig. 3.1 Tax transportation routes before 1951 from Monpa [mon pa] regions to Lhasa, where the Tibetan administration was located (Produced based on information acquired through locally conducted interviews. Taxes from the Kalaktang [kha legs steng] and Dirang [rdi rang] areas were collected by *dzongpon* [*rdzong* dpon] of Taklung Dzong [stag lung *rdzong*] and Dirang Dzong [rdi rang *rdzong*], respectively, and transported to Tawang Monastery. Taxes were collected from the Tawang area by Tawang Monastery as well as by Gyangkhar Dzong [rgyang mkhar *rdzong*]. Taxes received at Gyangkhar Dzong were transported to Tsona Dzong [mtsho sna *rdzong*] over Bumla ['bum la] Pass in May or June by the *dzongpon*. Assistants of the *dzongpon* collected taxes from the Dakpanang [dags pa nang] and Pangchen [spang chen] areas and transported those taxes to Tsona Dzong in May or June. Tax collection in the Dirang and Kalaktang areas halted for a period in 1946, and it was completely terminated after Gyangkhar Dzong and Tawang Monastery were controlled by the government of India in 1951. The Tawang area still provides religious tribute to Tawang Monastery)

dzongpon of Gyangkhar Dzong was jointly appointed by the Tsona *dzongpon* and officials of Tawang Monastery. According to several British sources, both Tsona Dzong and Gyangkhar Dzong were used by the Tsona *dzongpon* as summer and winter residences (Chakravarty 1953), respectively, but this is not supported by Tibetan sources (Ye shes 'phrin las 1983). The *dzongpon* of Tsona Dzong commissioned Tawang Monastery to collect tax from the areas of the Dirang and Kalaktang. One or two *dzongpon*, who held 3-year terms, were sent to each area by the monastery. In each village, the *tsorgen* ([gtso rgyan]: village mayor) collected the tax and stored it in his dwelling. A *dzongpon* traveled around the villages and transported the tax from one *tsorgen*'s place to the next using villagers as helpers. In 1938 and 1946, before Indian independence, the British colonial administration objected to the tax-collection system operated by the *dzongpon* since it obliged villages to offer their services for free. However, the system was not abolished then.

According to *Three Years in Tibet* by Ekai Kawaguchi (1909), the Tibetan administration in Lhasa in about 1900 consisted of 165 monks and the same number of secular individuals serving as senior officials. According to Yamaguchi (1987), those numbers were 175 monks and 175 secular individuals. More details were reported by Kawaguchi (1909). The Tibetan monks were under the direction of four senior secretaries and the secular staff under four ministers. The cabinet consisted of four ministers, three chancellors, two military ministers, one steward of palace, one minister of religion, one attorney general, and four senior secretaries in holy orders. A pair of senior officials—one sacred and one secular—were sent to the *dzong* or the provinces under the order of the Tibetan government. Taxes could be paid in the form of goods or silver coins. Most of the tax collected was used to support monks. According to Kawaguchi (1909), there were then 24,000–25,000 monks in Lhasa, and other monks were distributed across the regions.

Kawaguchi (2012 [1909]: 145) discussed the *dzong* in Chapter 117, "Financial Administration of Tibet", as follows:

> In each province there are two places where the collection of taxes is made for the Government, one of which is the temple, and the other the Local Government office; for the people are divided into two classes: (1) those who are governed by the temple and (2) those who are governed by the Local Government. They pay their taxes to the Central Government through their respective Governors. In each local district, there is what is called a Zong. This was originally a castle built for warlike purposes, but in times of peace it serves as a Government office, where all the functions of Government are carried out, so taxes are also collected there. The Zong is almost always found standing on the top of a hillock of about three hundred feet and a Zongpon (chief of the castle), generally a layman lives in it. He is the chief Governor of the district and collects taxes and sends the things or money he has gathered to the Central Government. The Zongpon is not paid by the Central Government directly, but subtracts the equivalent of his pay from the taxes he has collected.

The account in the early twentieth century by Yamaguchi (1987) is as follows:

> There were a total of 52 fortress-like *dzong* in the provinces, with one or two resident officials called *dzongdoe* (*rdzong sdod*) at each of them. Depending on the area, the resident official was referred to as the chief *goba* (*sgo pa*) or tax officer, *tshukpa* (*tshugs pa*). The resident officials of the 24 most important areas, such as Tsona, enjoyed treatment almost equal to that of senior officials in Lhasa.

There was a *dzongpon* residence within a *dzong* in the Monpa area. The walls contained several crenels allowing guns to be fired from within. Apprentices called *gethruk* (*dge'phrug*), who worked under the *dzongpon*, were in charge of actual tax collection. A fully ordained monk called a *gelong* (*dge slong*) was the highest official working under the *dzongpon*. One of the *dzongpon* appointed in Dirang and Taklung Dzong had to be a *gelong* monk. Gyangkhar Dzong, Dirang Dzong, and Taklung Dzong had jails, and they were used to imprison people who refused to pay tax as well as other criminals.

There were four types of *dzong* in the Monpa area:

Daryab [mda' g.yab]: a *dzong* that functioned as a checkpoint; *mda' g.yab* means "border" or "balcony" in Tibetan.

Ta [lta]: a *dzong* that functioned as an observation post; *lta* means "to watch" in Tibetan.

Fig. 3.2 Chaksam [lcags zam] (*lcags* means "iron" and *zam* means "bridge") was built by Thangtong Gyalpo [thang stong rgyal po] (1361–1485) in the Tawang area

Chu [chu]: a *dzong* that functioned as a tower on the side of a bridge; *chu* means "river" in Tibetan.
Seng [g.yab]: a *dzong* that located on a cliff; *seng g.yab* means cliff in Tibetan.

Dirang Dzong is a *ta dzong*, whereas Gyangkhar Dzong and Taklung Dzong are categorized as *daryab dzong*. An example of a *chu dzong* is Chaksam [lcags zam], where *lcags* means "iron" and *zam* means "bridge." This was built by Thangtong Gyalpo [thang stong rgyal po] (1385–1464 or 1361–1485: a great Buddhist adept) in the Tawang area (Amundsen 2001) (Fig. 3.2). He is said to have built as many as 108 bridges in Bhutan, Tibet, and the Monpa area. Some of these remain as ruins in Bhutan and Monyul, and a reconstructed version of another of his bridges is still in use. The existence of iron bridges is mentioned in Kawaguchi's *Three Years in Tibet* (1909), which suggests that there used to be a number of iron bridges in that country.

3.2 Formation of Taklung Dzong Fortress

Taklung Gompa [*dgon pa*] is one of the oldest fortresses (*dzong*) *gompa* in the Monpa area (Figs. 3.3 and 3.4). It is located upon a small hill at an elevation of 2,186 m. Taklung Gompa is known to have been constructed by a monk named

Fig. 3.3 Taklung Dzong [stag lung *rdzong*] was built by a monk named Tenpe Dronme [bstan pa'i sgron me] in the Kalaktang [kha legs steng] region in the sixteenth century: it is one of the oldest fortresses cum *gompa* [dgon pa] in the Monpa area

Fig. 3.4 A wall of Taklung Dzong [stag lung *rdzong*]: this *dzong* consisted of three floors, but it was destroyed by an earthquake. It was later partially restored

Tenpe Dronme (bstan pa'i sgron me) in the Kalaktang region during the sixteenth century. Tenpe Dronme was a disciple of the second Dalai Lama, Gedun Gyasto [dge 'dun rgya mtsho] (1475–1543; Tenpa 2013). Taklung Dzong [stag lung *rdzong*] was originally constructed as a Buddhist monastery and was later used as a fortress for administrative and tax-collection operations. It was originally a three-story building, with each floor having three rooms (Fig. 3.5). One of the rooms on the ground floor (first floor) was used for storage, and the two remaining rooms were prison cells. The three rooms on the next floor were used as a prayer room, a kitchen, and a private room for the *dzongpon* (Fig. 3.5). The top floor was a *gompa*, which is the term used in the Monpa area to refer to a Tibetan temple.

The etymology of "Taklung" is as follows. Three monks including Tenpe Dronme were looking for a site to construct a *gompa*. They came across a tiger upon a large rock. Two of the monks wanted to kill the animal, but Tenpe Dronme stopped them. The site was thereafter named Taklung as a result of this incident since *stag* means "tiger" and *lung* signifies "rock" or "prophecy." A large rock may still be seen in front of Taklung Dzong. It is said that Tenpe Dronme built the *gompa* to defend the nearby village from a demon called *sinpo* [*srin po*] in the Monpa language.

Cereal grains and specialty goods were collected as tax. Examples of specialty goods include *Rubia manjith*, known locally as *laningru* [la nin 'bru] or *leinauri* [le'i na'u ri] (Fig. 3.6), *arba* of plant to use for color fixative, and a type of paper called *boi shuku* [*bod shog gu*] (Fig. 3.7) made from *Daphne papyracea* that is known locally as *shokshoku shing* [*shog shog bu shing*]. *Laningru* was used to produce a red dye for coloring the clothing worn by monks (Fig. 4.28). *Laningru* is decocted, and it is employed to dye clothes. *Arba* is decocted in the boiled water, and clothes are then soaked there for color fixing. The Monpa believe that *arba* prevents dyed clothes from fading. The native dresses of Monpa women are also dyed with a red stain (Figs. 3.8 and 4.24), though this is called lac and is produced from insects (*Laccifer lacca*). This colorant is not used for clothes worn by monks since they believe it to be produced from the blood of insects so it is inappropriate for them.

As noted above, taxes collected in the region were transferred to Tawang Monastery and the *dzongpon* were appointed from the monastery on 3-year terms. When a *dzongpon* traveled to transport taxes, the *koenyer lama* [*dkon gnyer bla ma*], the caretaker of the *gompa*, instructed *gethruk* (apprentice monks) to assist. It is reasonable to assume that the taxes collected at Taklung Dzong, notably cereal grains, were used primarily for expenditures related to the *Dungyur* [*gdung 'gyur*] and *Torgya* [*gtor rgyag*] festivals held at Tawang Monastery. The former is held every 3rd year and the latter every year. Any surplus money was taken by the *dzongpon*.

The taxation system of the *dzongpon* was completely eliminated in 1951. As Indian influences became more prevalent in the area after the Indian independence from Britain in 1947, the taxation system ended, and the local *dzongpon* returned to Tibet or to Tawang Monastery.

Taklung Dzong was destroyed and became a ruin as the result of a large earthquake in 1950. Around that time, the location of Taklung Dzong, on the border between the Assam region of British India and the Monpa area of Tibet, was of

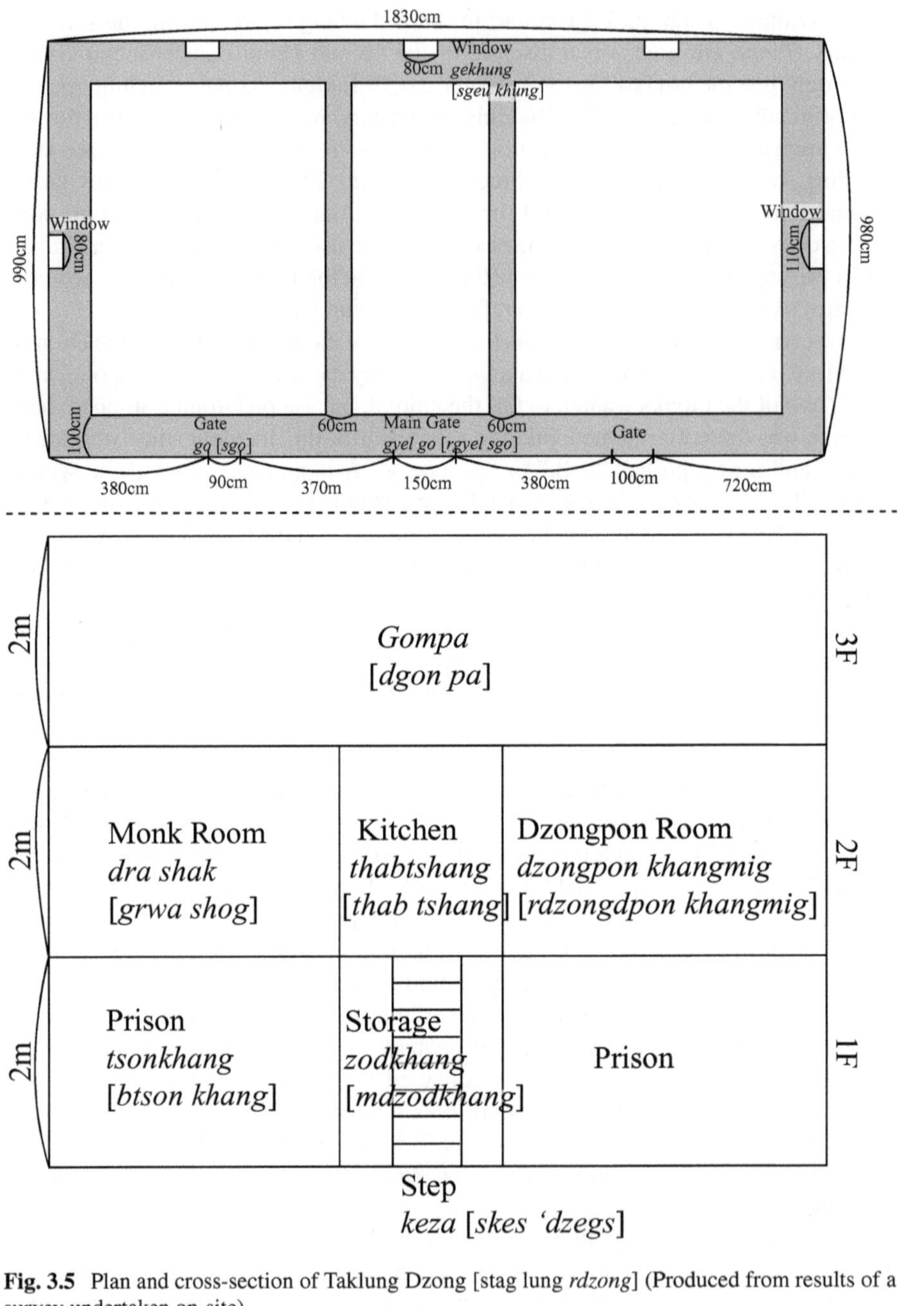

Fig. 3.5 Plan and cross-section of Taklung Dzong [stag lung *rdzong*] (Produced from results of a survey undertaken on-site)

Fig. 3.6 *Rubia manjith* is locally known as *laningru* or *leinauri* [la ning'bru or le'i nau ri]; it is *btsod* in Tibetan. The roots contain a colorant that is extracted and used to dye the red robes of monks

Fig. 3.7 The paper called *boi shuku* [bod shog] is produced from *Daphne papyracea* trees, which are locally called *shokshoku shing* [shog shog bu shing]

Fig. 3.8 The Monpa style jacket is locally called *toedung* [stod 'thung]. It is dyed red by means of a colorant called lac, produced from insects (*Laccifer lacca*)

significance. An old road mainly used by Bhutanese or Monpa traders was located close by, and assistants of the *dzongpon* collected tariffs at the entrance of the *dzong* (Fig. 3.9). Tawang, Dirang, and Sanglam [gsang lam], and Taklung *dzong* as well as the *gompa* in Rupa were all located on this trading route. That connected Udalguri in Assam and Tsona in Tibet. The route was regularly used by traders and others for commerce and pilgrimages (Narain Jha 2006). The amount of tariff levied for passing through Taklung Dzong was determined by the price and quantity of goods.

To provide the traditional Monpa dress, Bhutan traders visited the Tangla area (home of the Rava people) and Udalguri (home of the Bodo people in Assam) to acquire the silk known as *aliane*. They then returned to Bhutan and dyed the silk with a colorant made from *lac*, described above, before bringing the dyed silk back to Assam. Rava and Bodo people would then weave the final product, with the resultant fabric being sold in Monpa villages. When transporting the fabric from Assam to the Monpa area, the traders had to pay the road tariff at Taklung Dzong. From the silk fabric brought to their villages, the Monpa people produced the following materials: *shingka* or *shinka*, a one-piece dress for women; *toedung* [stod 'thung], a jacket worn over the *shingka*; and *ahlung*, white plain silk fabric. *Ahlung* was used to produce *dorna* [dor na], the trousers worn by Monpa men, and the shirts worn by both women and men. According to Wakita (2009), fabrics for the Monpa's *shingka* and *toedung* were produced mainly in eastern Bhutan and Assam. Import of the

Fig. 3.9 This old road close to Taklung Dzong [stag lung *rdzong*] was used by Bhutan or Monpa traders. Assistants of the *dzongpon* [*rdzong dpon*] collected tariffs at the entrance to the *dzong*. The entrance is visible in the photograph

lac-dyed fabric for such clothes began in the 1930s at the latest. Before these imports started, the locals wore woolen *shingka* made of a commonly-available yarn and other clothes made of white cotton and silk.

A *dzongpon* normally demanded his own private residence outside a village. It was said that he would often hit villagers with a whip, known locally as a *taicha* [*rta lcag*], if the dwelling was not sufficiently comfortable. It was also said that he ordered villagers to feed his horse, and if the bell around the horse's neck stopped ringing he would also hit them. It would appear that *dzongpon* were feared by the villagers. Similar information was also provided in other areas, such as Senge Dzong.

A monk appointed as a caretaker of a *gompa*, the *koenyer lama*, was sent from Tawang Monastery and resided in Taklung Dzong from 1947 to 1964 (Fig. 3.10). Around the same period, another *koenyer lama* was also sent from the monastery to Namshu Gompa in the Dirang area. The monks of Tawang Monastery were asked to work as caretakers at the 18 *gompa* (now only 12) under its supervision. A Tawang Monastery monk typically served as *koenyer lama* at three different *gompa* for 3 years each (i.e. 9 years altogether) before finally returning to his home institution. A monk sent to a village by the monastery was supposed to play a crucial religious and social role (Dorjee 2006). However, in the case of Taklung Gompa, the influence of Tawang Monastery disappeared following the death of Sang Tashi [sangs

Fig. 3.10 Taklung Gompa [stag lung rdzong *dgon pa*]

rgyas bkra shis], the very last *koenyer lama* appointed by the institution. Since then, the villagers have chosen a local person as their *koenyer lama*. The current *koenyer lama* is selected by the 21 villages of the area to be the caretaker of the *gompa*. The current *koenyer lama* is the fifth to be appointed in this way (at the time of writing in 2011). The most probable reason for terminating the residence system in the Dirang and Kalaktang areas was their physical remoteness. In some cases, appointed lamas married and settled down locally at their posts.

Currently, the locally appointed *koenyer lama* visits several of the 21 villages in one circuit while carrying the holy icon of Dorjee Chang [rdo rje 'chang]. This is said to be the figure that received Tenpe Dronme when he descended from the sky. The *koenyer lama* stays overnight at each village—except for the village of Boha, where he stays three nights for a ceremony that lasts 3 days. His usual accommodation in each village is the *tsorgen*'s dwelling, and the villagers gather there.

The *koenyer lama* places the icon on each villager's head. He provides holy necklaces called *sungma* [srung ma], which expel evil and enable the villagers to travel safely and enjoy good fortune and a long life, as well as incense sticks called *pos* [*spos*] and religious offerings called *torma* [gtor ma], made of barley and corn powder. Villagers bring gifts called *phok* [phogs] to the *tsorgen*'s dwelling. Each household contributes 7 *brey* [bre] of corn, 12 *brey* of rice, or other materials, including money as well as seasonal agricultural products such as oranges, sugar cane, and Japanese pepper (*Zanthoxyli fructus*). A *brey* is a cylindrical measuring cup that holds more than 600 g of cereal grain (see Sect. 7.4 for additional details regarding measurement). These items become a source of income for the *koenyer lama*.

When the *koenyer lama* travels to the next village, a few villagers escort him and transport the *phok*. The *tsorgen* informs the next village of the travel schedule of the *koenyer lama* beforehand to enable them to make preparations.

The *koenyer lama* is often a *tsespa lama* [*rtsis pa bla ma*] (see Sect. 4.1). Three days after the birth of a child, the *tsespa lama* is invited and asked to write down the child's name, birthday, and Chinese zodiac sign on a sheet of *boi shuku* (Fig. 3.7) paper. He uses an astrological book called a *kartsi* [*dkar rtsis*] to determine the child's name. During this event, a feast is held by the relatives and neighbors. Traditionally, the youngest son is supposed to take care of aged parents in the Kalaktang Monpa communities (Barua and Ahmad 1995).

3.3 Formation of Dirang Dzong Fortress

The present Dirang Dzong [rdi rang *rdzong*] fortress is considered to have been constructed in 1831 when *dzongpon* administrations were strengthened in Monyul. This *dzong* is a four-story building with crenels and contains a dwelling for the *dzongpon* at its center. There is a jail adjacent to the building (Fig. 3.11). The dwellings of villagers surround the *dzongpon*'s fortress and all are enclosed by walls (Fig. 3.12). When the plan to construct the *dzong* in Dirang village was conceived, the Buddhist villagers were so pious that they opposed the secular governance of Tawang Monastery (Sarkar 1978). A violent clash occurred when 12 leaders of the village families resisted the construction. The village leaders were killed, and a long *mani* [*MaNi*] wall was built as a monument to commemorate them (Fig. 3.13). *Mani* walls were built by the families of deceased individuals as a cenotaph and used as places of rest for travelers.

Tawang Monastery sent the *dzongpon* to Dirang Dzong with a 3-year term, and he collected tax from the large Dirang area. The tax collected was transported to the village of Nyukmadung by the villagers of Lish, Chug, and Sangti. Nyukmadung villagers then conveyed the tax to Senge Dzong. Finally, the collected tax arrived at Tawang Monastery (Figs. 2.5, 3.1 and 4.21). Tax collected in the Taklung Dzong was also sent to the monastery via Dirang Dzong.

The *khray* or *thre* [*khral*] tax used to be levied on 2–3 acres (8,000–12,000 m^2) of land in the Dirang area. A *khray* is a measure, and it is equivalent to 20 *brey* measuring cups. The cylindrical *brey* measuring cups are 8 cm in depth and 14 cm in diameter: 20 *brey* of cereal grain weighs approximately 12 kg (Fig. 7.12). Owner of small land could pay *khray* by fire woods or grass for horse etc. A local interview (Mizuno, August 2010) revealed that an owner of 12 acres (48,000 m^2) of cultivated land had to pay 80 *brey* of cereal grains as tax. The *dzongpon* was in charge of collecting taxes until 1951 (i.e. more than 3 years after Indian independence from Britain in 1947). The cereal grain collected as tax by Dirang Dzong was allocated primarily to the Monlam Chenmo [*smon lam chen mo*] festival held in Tawang

Fig. 3.11 Dirang Dzong [rdi rang *rdzong*], the present fortress is believed to have been constructed in 1831 and it was restored in the early 2000s. The photograph shows the central four-story building with crenels. It is used as the dwelling of the *dzongpon* [*rdzong* dpon]. The structure visible on the left-hand side of the photograph was used as a jail

Fig. 3.12 Dirang Dzong [rdi rang *rdzong*] composed of dwellings of villagers surround that of the *dzongpon* [*rdzong dpon*], which is enclosed by walls. The photograph shows the gate in the wall

Fig. 3.13 A violent clash occurred when 12 leaders of families from the village resisted the construction of Dirang Dzong [rdi rang *rdzong*]. They were all killed, and a long *mani* [MaNi] wall was built as a monument to commemorate them

Monastery during the 1st month of the Tibetan lunar calendar (around February and March), but these funds were also used in Tibet and the Monpa area (Sarkar 1975a, b, 1978, 1980, 1981). Anything remaining from the collected tax was presumably taken by the *dzongpon* of Dirang Dzong. *Three Years in Tibet* (Kawaguchi 1909) described the Festival of Monlam as the largest festival and prayer service in Tibet. It is held from the 3rd to 25th days of the 1st month of the Tibetan lunar calendar. Kawaguchi (1909) also noted that 25,000 to 26,000 monks traveled from local temples to Lhasa. It is reasonable to assume that the Festival of Monlam held in Tawang Monastery was also on a very large scale with large expenditure.

3.4 Formation of Gyangkhar Dzong Fortress

Villagers in the Tawang area paid tax not just to Gyangkhar Dzong [rgyang mkhar *rdzong*] but also to Tawang Monastery. Whereas cereal grains were paid as tax to the monastery, various kinds of local specialties were collected and paid to Gyangkhar Dzong (Fig. 3.14). These included paper called *boi shuku* from the village of Mukto [mog tog], inks of vegetable origin from the Dakpanang people, and dried vegetable and wooden containers from the people of Zemithang and Tsangporong

Fig. 3.14 Only the foundations of Gyangkhar Dzong [rgyang mkhar *rdzong*] remain today: as its main structure was dismantled, the stone was used to construct Nehru Gompa [*dgon pa*]

[gtsang po rong]. Tax collected at Gyangkhar Dzong was initially transported to Tsona Dzong via Bumla Pass by a party led by the *dzongpon* in May or June. The journey took 4 days and included Bumla Pass on the first night, Sha'og [sha'og] on the second night, Jamo Kharmo [bya mo dkar mo] on the third night, and finally Tsona Dzong on the 4th day (Fig. 2.5). At Tsona Dzong, the party was replaced by Tibetans, who took 10 days to reach Lhasa. Details of this Tibetan part of the travel route were apparently as follows: Tengshot [steng zhol] on the first night; Nyei Shatra [nyal sha khra] on the second night; Nyei Rithang [nyal ri thang] on the third night; Gyero [rgyal ro] on the fourth night; Karkhang Darding [kar khang dar lding] on the fifth night; Tuumbthru Drema Lhagang [ldum thru sgrol ma lha khang] on the sixth night; Tsethang [btse thang] on the seventh night; Samyas or Samyas Gyalgo [bsam yas rgyal sgo] on the eighth night; Dechin [bde chen] on the ninth night; and finally Lhasa on the 10th day. Although the journey between Samyas and Lhasa was within 1 day's traveling distance, an extra day was usually spent at Dechin because it was a scenic pilgrimage site. When Bumla Pass was blocked owing to snow, an alternative route was used via Dakpanang and Pangchen in the Zemithang area (Figs. 2.5 and 3.1).

Shar Nima Tshosum [shar nyi ma tsho gsum] or Laog yulsum [la 'og yul gsum] is the ancient name of the present Tawang Subdivision. Shar Nima Tshotsum was based on the administrative unit known as the *tsho* during the period when the *dzongpon* was responsible for collecting taxes. The *dzongpon* gave tax-related instructions to the *tsorgen*, the mayor of every village, such as Shar tsho [Shar tsho], Lhou tsho [Lha'u tsho], and Seru tsho [Bse ru tsho], in Monyul. Those *tsorgen* then passed on instructions to individuals known as *gomi* [sgo mi] and *thumi* [mthu mi] or *chopon* [bcu dpon] in the Dirang area. The *gomi* and *thumi* collected tax directly from the villagers during the winter cereal grain-harvesting season (between November and May) and stored the tax at the *tsorgen*'s dwelling. The appointments of *gomi* and *thumi* were chosen in turn from among the villagers each year. Although the task they had to complete was identical, the *gomi* apparently had greater responsibility than the *thumi*.

During the period of the *dzongpon*-operated taxation system, the villages constituted large administrative units called *tsho* or *ding* [tsho or lding]. Each *tsho* or *ding* was under one or two *tsorgen*. As Table 3.1 indicates, each large area that operated as one administrative unit contained a small number of villages: the Pangchenpa area, denoted as Pangchen Dingdruk [spang chen lding drug], consisted of six clusters of villages; the Dakpanang area, denoted as Dakpa Tshogey [dags pa tsho brgyad], consisted of eight clusters of villages; and the Tawang Monpa area, denoted as Shar Nima Tshosum, consisted of three clusters of villages. During that *dzongpon-operated* period, the other administrative units were as follows: the Mago Thingbu and Lukguthang areas, denoted as Magthing Luksum; the Senge Dzong and Dirang areas, denoted as Dangnang Tshoduk [sbrang nang tsho *drug*]; and the Kalaktang Monpa area, denoted as Rongnang Toemey [rong nang *stod smed*] or Rongnang Tshoshi [rong nang tsho *bzhi*]. The names of those administrative units were compiled during the interviews (Mizuno, February & May 2011) in the preparation of this book. They were later confirmed using Tibetan textual sources (Table 3.1). It should be noted that some of the villages in each of the large administrative units contained a number of settlements that today function independently as villages.

The contemporary Tawang and West Kameng districts, which cover all the historical administrative units mentioned above, apparently had 254 *tsorgen* in the *dzongpon*-operated taxation period. It is reasonable to assume that *tsorgen* at that time had greater authority in their administrative units than their present-day equivalents do. *Tsorgen* counted, recorded, and managed the tax collected and temporarily stored it in their homes. *Dzongpon* transported the tax collected from one settlement to another and finally stored it at Gyangkhar Dzong in May or June. To undertake this tax transport operation, the *tsorgen* selected a settlement in turn and authorized its occupants to convey the tax (Fig. 3.1). It was transported from Gyangkhar Dzong to the village of Shao via Bumla Pass, and then Shao villagers took charge of the onward route to Tsona Dzong. In this system, the settlement in charge of transporting the collected tax was called *ula* ['u lag]. Villagers used horses from their villages to carry the burden.

Taxes were collected under the aegis of the *tsorgen* of each of the eight villages (*tsho*) during winter (between November and May) in the Dakpanang area, which was then known as Dakpa Tshogey and spanned the area from the village of Gyada [rgya mda'] to that of Shakthi [shag sti]. The tax was collected and stored at the home of the *tsorgen* of Lumla [rlung la], which was considered the hub village,

Table 3.1 Administrative units used by *dzongpon* [*rdzong dpon*] to collect tax for Tsona Dzong [mtsho sna *rdzong*] and Tawang Monastery

Dzong (Fortress)	Tsho/ding	Sub-Tsho/ding	Villages in the Tsho/ding	Present status/ remarks
Tawang Gonpa Gyangkhar Dzong	3 Eastern Tsho of Nyima (Shar Nyima Tshosum)	(1) Shar Tsho	Kitpi, Sharnub, Audung, Khurdung, etc.	Tawang District
		(2) Seru Tsho	Seru, Trimu, Namet, etc.	
		(3) Lhou Tsho	Jang, Mukto, Gangkhar, etc.	
	8 Western Tsho of Dakpa (Dakpa Tshogey)	(1) Sakpret Tsho	Sakpret, etc.	
		(2) Mukho shaksum Tsho	Muktur, Shurbi, Khokem, Shakti, etc.	
		(3) Ungla Tsho	Lumla, Khozo, Pharmey, Ongla, Dugumba, etc.	
		(4) Sanglung Tsho	Narmaleng, Marmey, Namtsering, Sanghar, etc.	
		(5) Khabong Tsho	Kharung, Bomja, Bongleng, etc.	
		(6) Trilam Tsho	Thrilam, Mangnam, Sakyur, Burkung, etc.	
		(7) Thongleng Tsho	Thongleng, Kungba, etc.	
		(8) Pamakhar Tsho	Pamakhar, etc.	
	6 Ding of Pangchen (Pangchen Dingduk)	(1) Panchen Shoktsan Barding/ Nekording		
		(2) Panchen Shoktsan Toeding		
		(3) Pangchen Shoktsan Meeding		
		(4) Klubum/ Lhunpo Ding		
		(5) Muchoe Ding		
		(6) Latshe Kharmen Ding		
	1 Tsho of Shau Hro Jangdak		3–4 Villages	
	4 Tsho of Lekpo (Lekpo Tshozhi)	(1) Sinmo Tsho		Tibet Autonomous Region
		(2) Gomri Tsho		
		(3) Kyipa Tsho		
		(4) Shanglang Tsho		

Senge Dzong Dirang Dzong	6 Tsho of Drangnang (Drangnang Tshoduk)	(1) Senge Dzong/Nyukmadung Tsho	2–4 villages	West Kameng District
		(2) Chug Tsho	2–4 villages	
		(3) Lish Tsho	2–4 villages	
		(4) Sangti Tsho	6–8 villages	
		(5) Dirang Tsho	12–16 villages	
		(6) Namshu/Thembang Tsho	8–12 villages	
Taklung Dzong	4 Tsho of Rongnang (Rongnang Toemey)	(1) Toe Tsho	Murshing, Sanglam, Domkha, Phudung, etc.	
		(2) Mee Tsho	Bokhar, Shampong, Tsingki, etc.	
		(3) Shertuk Tsho	Sher (Shergoan), Tukpen (Rupa), etc.	
		(4) Rakhul Tsho	Rakhung, Khuldam, etc.	

Produced from locally conducted interviews, written sources of the 5th Dalai Lama's Edict of 1680, Thupten Choephel (1988: 24–43), and Yeshi Thinle (1983: 132–163) *ongla tsho* ['ong la tsho] (Fig. 3.1, Table 3.1). The tax was transported to Tsona Dzong via the Pangchenpa area, which was then known as Pangchen Dingdruk, and also via Lebo Tshoshi [Legs po tsho *bzhi*] in Tibet. It was transported by means of the *ula* system under the supervision of aides of the *dzongpon*, who, according to British sources, was simultaneously in charge of both Tsona Dzong and Gyangkhar Dzong. Since agriculture and stockbreeding were practiced in the Pangchenpa area, the tax collected consisted of cheese and butter. The *dzongpon* with the tax collected from the Tawang Monpa area (Shar Nima Tshosum) and his aides with the tax collected from both the Dakpanang and Pangchenpa areas, stayed together at Tsona Dzong from June to October. They made the short trip to transport the collected tax to Lhasa during that time. They then returned to Gyangkhar Dzong via Bumla Pass in October or November and stayed there from November to May (Fig. 3.1).

People belonging to the Dakpanangpa but with a slightly different language and living in the area between present-day Shakti and the border with the Pangchenpa area are called Khobleytang ['khob legs steng] (Fig. 3.1). Since this area is close to the border between the Dakpanangpa and Pangchenpa areas, the two local languages appear to have amalgamated.

The village mayor was referred as *gyepo* [rgan po], rather than *tsorgen* in Magthing Luksum, where pastoral people had established the Mago, Thingbu, and Lukguthang settlements. The *gyepo* instructed *gomi* or *thumi* to collect tax. The tax of cheese and butter was initially collected and stored at Tsona Dzong's subsidiary administrative office, Kishong Deba [skyid shongs sde pa], and then transported to Tsona Dzong (Fig. 3.1).

Tax collected in the area from Jang [byang] to Sela and Gomkang [sgom khang] was also transported via Zemithang if Bumla Pass was blocked by snow in May and June (Figs. 2.5 and 3.1). Villages in the Tawang area used to pay tax not just to Gyangkhar Dzong (i.e. to Lhasa via Tsona Dzong) but also to Tawang Monastery, although tax to Gyangkhar Dzong was terminated in 1951.

We conducted an interview (Mizuno, February 2011) with an 81-year-old part-time monk, a *minakpa lama* [mi nag pa bla ma] or simply a lama (in the Tawang area), living in the village of Aodung [a'u gdung] (2,192 m above sea level). He claimed to have been to Tsona at the age of 20 as an itinerant merchant. At that time, he transported dried vegetables, sweet potatoes, and chilies from his village and sold or exchanged them in Tsona, and he brought back silver coins, clothes, dried yak meat, cheese, butter, and rock salt and sold or exchanged them in his village. Many Monpa and Tibetan merchants visited Tsona around that time since large summer fairs were held once a month in May, June, and July. When the interviewee traveled (around 1950), he stayed in an army bunker the first night after departing from his village and in the tent of a pastoral worker who lived close to Shyo [zhol] the second night (Fig. 2.5). The final part of the trip was to Tsona Dzong by way of Kyakyenla [skyag gyen la] Pass.

Gyangkhar Dzong was a two-story fortress, and its *dzongpon* was resident there. In about 1960, approximately 50 years before the interview, the interviewee, several people from Aodung village, and others from the surrounding villages transported stone materials from Gyangkhar Dzong to Tawang and built Nehru Gompa. Wealthy

Fig. 3.15 Pema Gombu [Padma mgon po], the *tsorgen* [gtso rgan] (village mayor) of Lhou [lha'u]; He died five months after Mizuno's May 2011 interview

villagers used horses while other villagers carried the materials themselves. They managed to transport all the materials in 11 days by completing two round-trips from Gyangkhar Dzong to the destination. Today, only the foundation of Gyangkhar Dzong remains (Fig. 3.14).

Another interview with a monk of Tawang Monastery revealed that Major Ralengnao (Bob) Khathing, an administrative assistant officer of the independent Indian government, played a crucial role in the removal of *dzongpon* from the Tawang area to Tibet in 1951 and the subsequent abolishment of tax collection (Tenpa 2014). The same monk also stated that Pema Gombu [padma mgon po], the *tsorgen* of the village of Lhou, had been the leader of the *tsorgen* of the villages in the Tawang area in the meeting organized by Major Khathing.

We visited the Lhou (pronounced "hoh") village in February 2011 and requested an interview with the *tsorgen*, but this was not possible because of his poor health. Following a wish for his recovery, we made a second visit in May 2011. Initially the reply was negative, but he finally permitted us a short interview. Pema Gombu, 84 years old at the time of the interview, was born in 1927 and had been the *tsorgen* for more than 50 years (Fig. 3.15). We held a preparatory meeting with Pema Gombu's son, Urgan Tsering [u rgyan tshe ring], and had an interview with Pema Gombu himself in the presence of the son on the following day. The following is a summary of his account in the interview (Mizuno, May 2011).

Since 1944, especially at the beginning of Indian independence in 1946, British-ruled India tried to abolish taxation by *dzongpon* in the Dirang and Kalaktang areas, requesting that the *dzongpon* either remain in the *dzong* or return to Tawang. However, since the taxation contributed to Tawang Monastery or to locally resettled former officials, the local people themselves continued this practice until 1951. Because both the *dzongpon* of Dirang Dzong and Taklung Dzong were appointed from the office of Tawang Monastery, they were mostly natives of Monyul. Therefore, the *dzongpon* remained at the *dzong* as caretakers, moved to the *gompa* of the *dzong*, or returned to their respective villages in the Tawang district or to Tawang Monastery. However, the dominance of the *dzongpon* and the Tibetan administrator *kungo la-nyer* [*sku ngo bla gnyer*] at Tawang Monastery continued in the Tawang area until 1951. Well before Indian independence and the creation of the McMahon Line in 1914, Tibet had ceded the Monpa area to Britain in return for a "supply of British weapons" and their services as mediators in the China–Tibet border talks in the Kham/Amdo region. However, owing to concerns that Tibet might be occupied by China, it was British India's boundary policy to create a buffer state with natural boundaries between Tibet and British India.

Therefore, Major Khathing traveled from Dirang to the village of Jang in the Tawang region in 1951 and assembled the *tsorgen* (Shar tsho, Lhou tsho, and Seru tsho) of the Shar Nima Tshosum region, which was then the Tawang Monpa area. These villages covered wide areas and contained settlements that were the equivalent of contemporary villages. Those settlements sent *gomi* and *thumi* (assistants to the *tsorgen*) to the assembly. Major Khathing instructed Pema Gombu, *tsorgen* of the Lhou tsho, and others to see the *dzongpon* at Gyangkhar Dzong and encourage him to accept the Indian delegation's visit. Pema Gombu and others had a meeting with Ngyerpa Konchok La [*gnyer pa* dkon mchog lags], the assistant officer of the *dzongpon* of Gyangkhar Dzong, and informed him of the intention of the Indian delegation. After that meeting, Pema Gombu returned to the village of Jang and reported what had happened to Major Khathing. On February 6, 1951, 2 days after Pema Gombu's return to Jang, the delegation moved to Lebrang Tsangbu [bla brang gtsang bu].

On February 9, Major Khathing and the Tibetan officials began discussing territorial issues. The Tibetan official, Lanyer Thubten Choephel [bla gnyer Thub bstan chos'phel] (1988), and representatives from Tawang Monastery informed the Indian delegation that they had sent a messenger to Lhasa to inform the Tibetan government about the current state of affairs. On February 14, the *dzongpon* and his assistants visited Major Khathing in Tawang. They demanded that the delinquent taxes be paid and that the services of 20 people per day be provided free of charge; however, these proposals were rejected. Pema Gombu stated, "India is our nation, and Tawang is a part of India. We shall respect Indian laws." Shortly afterward, the Indian administrative assistant officer took over governance of the Tawang region; the *dzongpon* and his people accepted the change and gradually left the area. After that incident, the tax burden on the inhabitants of the Tawang area was reduced to that of Tawang Monastery. As well as Pema Gombu, *Tsorgen* of the village of Gyangkhar and of the village of Surbi [zhur bi] played important roles throughout this process.

Regarding the proportion of tax collected by the *dzongpon* that was delivered to the Tibetan government, Kawaguchi (1909) states that the *dzongpon* embezzled his income from the collected taxes. According to the interview with Pema Gombu, about a quarter of the total tax collected eventually reached the Tibetan government. His account was based on the following personal observations. Pema Gombu was angry with the *dzongpon* because his father had been killed by officials of Gyangkhar Dzong. All the *tsorgen* at that time were highly respected, and Pema Gombu's father was especially powerful and also popular among the villagers. His power and popularity frightened the *dzongpon*. Upon the return from his trip to Bhutan, the *dzongpon* treated him with alcohol, let him fall in asleep, and killed him. Pema Gombu then spent 16 days traveling on foot to Lhasa and reported the circumstances surrounding his father's death to the authorities there. It took 2½ years as well as frequent visits to Lhasa before the court made its decision. Pema Gombu won the case and received 600 silver coins from Lhasa and a pledge rendering his family immune to harassments by Tsona officials.

When shuttling between his village and Lhasa, Pema Gombu noticed a discrepancy. The tax collected in the Monpa area was transported to the Tibetan government in Lhasa; however, the amount arriving there was only a quarter of the original amount collected—even by the largest estimate. At that time, the taxes transported to Lhasa were used by the Sera [se ra], Drepung ['bras spung], and Ganden [dga' ldan] monasteries as well as by the Tibetan government. It is reasonable to assume that the other three-quarters of the original tax collected was appropriated primarily by the *dzongpon* and subsequently by senior officials and ministers of the government. Cereal grains, paper, and inks and dye compounds collected as tax were most probably sold and exchanged for silver coins.

Kawaguchi (1909) describes frequent cases of bribery. At that time, bribery was essential to becoming a high priest or senior official in Tibet. Naturally, those who managed to receive such high positions were limited to members of wealthy families. Families producing monarchic officials, such as *dzongpon*, were also predetermined and mostly aristocratic. However, some commoners were able to become high-ranking officials in both monastic and aristocratic positions. The latter were achieved mostly through marriages. The following passages are taken from Kawaguchi (2012 [1909]: 108–110):

> By this time the process of consultation had to undergo considerable modifications owing to the fact that the high Lamas who were to be reborn not infrequently omitted enlightening others about the places of their reappearance on the earth. These places had to be discovered, for the Tibetans firmly held, as they do even today, that high Lamas who die are sure to re-incarnated somewhere after the lapse of forty-nine days from the day of death. Hence the necessity to determine the place of such re-incarnation, and this task devolved onto the oracle-invokers of one of the four particular temples mentioned previously.
>
> The process as it is at present is essentially identical to that prevailing in former times, and is exceedingly strange. The mediums or invokers who perform this holy business behave themselves in such an extravagant way that the uninitiated would consider them to be stark mad…
>
> …At any rate the practice of invoking divine oracles extensively came into vogue from the time of the fifth Grand [Dalai] Lama, and is used for all matters great or small, from vexed international problems to trifling questions that admit of solution…
>
> …The Nechung who are under the direct patronage of the Hierarchy are generally very wealthy. This, taken in conjunction with another fact, that the re-incarnations of higher Lamas are generally sons of wealthy aristocrats, or merchants, and that it is only very rarely that they are discovered among the lowly, must be considered as suggesting the working of some collusion. I have even heard that some unscrupulous people corrupt the oracle-priests for the benefit of their unborn children, so as to have their boys accepted as Lamas incarnate when born. From a worldly point of view the expense incurred on this account not infrequently proves a good 'investment', if I may use the profane expression, for the boys who are the objects of the oracles have a good chance of being installed in the temples where their spiritual antecedents presided, which are sure to be large properties. This property goes, it need hardly be added, to the boys, after they have been duly installed. Whatever may have been the practical effect of incarnation in former times, it is, as matters stand at present, an incarnation of all vices and corruptions, instead of the souls of departed Lamas.

Fig. 3.16 Senge Dzong [seng ge *rdzong*] was used to accommodate the *dzongpon* [*rdzong dpon*]

> I once remarked to certain Tibetans that the present mode of incarnation was a glaring humbug and that it was nothing less than an embodiment of bribery. The Nechung [gnas chung] mentioned above refers to the state oracle serving the Dalai Lama and the Government. High ranking governmental officials of Tibet often consulted with the Nechung Oracle about political and religious matters, and received divine revelations.

Apart from the taxes, *khray* collected by Gyangkhar Dzong and Tawang Monastery, each household of the Tawang region also had to pay a tax of one scarf, called a *khada* [*kha btags*], as well as one silver coin called a *betang* [*dngul Trang*], to Drepung Monastery (Narain Jha 2006). In addition, a massive amount of paper for the printing of religious texts and production of prayer flags as well as incense sticks for rogation at a house or bethel, *choisam* [*mchod gshom*], of a monastery were necessary. Because of this demand, cottage industries were established to manufacture paper and incense sticks (Narain Jha 2006).

3.5 Formation of Senge Dzong Fortress

Senge Dzong [seng ge *rdzong*] is located between Dirang Dzong and Gyangkhar Dzong (Fig. 3.16). This *dzong* was constructed and used as an accommodation facility for the *dzongpon*. The settlement, which consisted of only 10 households around 1950, was located approximately 3,000 m above sea level on the way to Sela Pass. Today, the settlement includes approximately 140 households. However, it was too small for the *dzongpon* and his assistants, and so Senge Dzong was built. The *dzong* was not used for tax collection. Instead of paying taxes, people living around the *dzong* offered and took care of lodging facilities, fuel wood, food, and feedstuff for horses.

Nyukmadung villagers transported the collected taxes to Senge Dzong, where they stayed and ate at the houses of local villagers in accordance with the tradition of *neatsang* [gnas tshang] ("host") (see Sect. 7.4). The term *neatsang* is both a Monpa and Tibetan word, and it indicates mutually supportive and hereditary ties among families. For example, each agricultural family undertakes barter exchanges with a specific pastoral family through such ties.

The *dzongpon* and his party stayed in this *dzong* on journeys to and from Taklung Dzong and Dirang Dzong. Villagers of Senge Dzong transported the collected taxes to the next stop, Jang.

As noted earlier in Sect. 3.2, the *dzongpon* always ordered the villagers to feed his horse. According to our interview (Mizuno, February 2011) with the first *tsorgen* of the village of Senge Dzong (who was born in 1943, was 69 years old at the time of the interview, and about 4 or 5 years old at the time of the events in question), British officials and soldiers with Nepalese physiognomy arrived at Senge Dzong, where they pitched a camp. He claimed that their purpose in the area was to check the border between India and Tibet. Immediately after this visit, the tax collection and its transport were terminated in the Dirang and Kalaktang areas. Tax collection by the *dzongpon* from the Monpa area for Tawang Monastery and the Tibetan government ceased completely in 1951. It is reasonable to assume that the change was a result of Arunachal Pradesh becoming part of India when India became independent from Britain in 1947.

The villagers of the Tawang region still pay tribute to Tawang Monastery, but villagers in the Dirang and Kalaktang regions do not. The Tawang villagers do this partly to maintain age-old traditions but also to strive after virtue for their present and future lives. Some villagers in the Dirang or Kalaktang pay tribute at their respective local monasteries.

References

Amundsen I (2001) On Bhutanese & Tibetan Dzongs. J Bhutan Stud 5(W):8–41

Barua S, Ahmad SI (1995) Monpa: Kalaktang. In: Dutta P, Ahmad SI (eds) People of India, Arunachal Pradesh. Seagull Books, Calcutta, pp 232–236

Chakravarty LN (1953) Secret: notes on Indo-Tibetan relations. From: 1772–1951 (with special references to the McMahon Line). The Assam Government Press, Shilling

Directoral of Research (Archaeological Section) (without date) Taklung Dzong, West Kameng District, Arunachal Pradesh. Government of Arunachal Pradesh, Itanagar

Dorjee R (2006) The Monpas: their social, economic and religious institutions. In: Dutta S, Tripathy B (eds) Buddhism in North-east India. Indus Publishing Company, New Delhi, pp 194–203

Kawaguchi E (2012 [1909]) Three years in Tibet. General Books LLC™, Memphis

Narain Jha B (2006) Buddhism, Buddhist monasteries and the societies of the Monyul: a historical study. In: Dutta S, Tripathy B (eds) Buddhism in North-east India. Indus Publishing Company, New Delhi, pp 229–240

Sarkar N (1975a) Minor religious structures of the Monpas and Sherdukpens. Resarun 1:150–4

Sarkar N (1975b) Historical account of the introduction of Buddhism among the Monpas and Sherdukpens. Resarun 1:23–44

Sarkar N (1978) A historical account of Tawang Monastery. Res Arunachal 1951–1976:50–57

Sarkar N (1980) Buddhism among the Monpas & the Sherdukpens. Directorate of Research, Government of Arunachal Pradesh, Shillong

Sarkar N (1981) Tawang Monastery. Directorate of Research, Government of Arunachal Pradesh, Shillong

Tenpa L (2013) The life & activities of *Rgyal sras* Blo bzang bstan pa'i sgron me, a 16th century Tibeto-Mon monk from Tawang. Tibet J XXXVIII (3 & 4):55–72

Tenpa L (2014) The centenary of the McMahon Line (1914–2014) and the status of Monyul until 1951–2. Tibet J XXXIX(2):41–86

Thub bstan cos'phel (1988) Hin rdu btsan'dzul dmag gis mon khul du btsan'dzul byas pa'i don dngos. In: Nga phod ngag 'jigs med (ed) *Bod kyi lo rgyus rig gnas dpyad gzhi'i rgyu cha bdams bsgrigs* (Selected Research Materials for Tibetan History and Culture: 10, Mi rigs dpe skrun khang: 24–43, Beijing

Wakita M (2009) Ethnic dress as representation: a case study of the Monpas of Arunachal Pradesh, India. In: Proceedings of the Graduate School of Human Relations, Keio University, vol 68, pp 35–58 (in Japanese)

Yamaguchi Z (1987) Tibet: 1. Toyo Sensyo, Tokyo (in Japanese)

Ye shes'phrin las (1983) Mon yul gzhi rtsa'i gnas tshul. In: *Bod kyi rig gnas lo rgyus dpyad gzhi'i rgyu cha bdams bsgrigs*/'don thengs gnyis pa. Bod rang skyong ljongs chab srid gros tshogs rig gnas lo rgyus rgyu cha u yon lhan khang, Lhasa: 132–163

Chapter 4
Tibetan Buddhism, Bon, and Animism

Nuns pounding grain in Singsur Buddhist nunnery [sing zur a ne *dgon pa*], Tawang: they live with teachers, who are also nuns, in a dormitory known as a *shak* [shag] and study at a local village school.

K. Mizuno, L. Tenpa, *Himalayan Nature and Tibetan Buddhist Culture in Arunachal Pradesh, India*, International Perspectives in Geography 6,
DOI 10.1007/978-4-431-55492-9_4

Abstract People in the Monpa area have followed the Bon religion for many centuries. The Nyingma [rnying ma] sect of Tibetan Buddhism has existed in that area since the eighth century; the Gelug [dge lugs] sect arose in the fifteenth century. Tawang Monastery has long been the central institution of Tibetan Buddhism in the Monpa area. Elements of Tibetan Buddhism and Bon were combined to create a synthesized faith, which has exerted a strong impact on the cultures and societies of that region. In Bon religion, spirits and deities of the mountains have been worshipped as mountain gods for ages. Mountain gods are particular to each area and local people pray and give ritual offerings to their mountain gods in ceremonies. In the Monpa area, Tantric (Esoteric) Buddhism has also been a major influence. The focus of this chapter is on the current relationship between the inhabitants in the Monpa area and their faith. This chapter also examines the daily lives of people engaged in various religious practices.

Keywords Tibetan Buddhism • Monastery • Ritual • Tantric Buddhism • Fortune-telling

4.1 Monpa Area and Tibetan Buddhism

Tibetan Buddhism consists of four major sects: Nyingma, Kagyu [bka' rgyud], Sakya [sa skya], and Gelug. There are also two minor sects—Jonang [jo nang] and Bodong [bo dong]. The latter is particularly associated with Lumla in Tawang.

The Nyingma sect of Tibetan Buddhism, the Red Hat sect, is the successor of the religious tradition from the Tibetan Empire, and its devotees place their faith in Tantric Buddhism based on the earliest translations of Buddhist scriptures. The Nyingma sect follows Mahayana Buddhism, and this sect appeals to the local people's inclination toward miraculous factors. Through this dual composition, the school has achieved mass appeal.

The Gelug sect, which is also known as the Yellow Hat sect, underwent reformation in renouncing the excessive reliance on tantras observed in other sects and rejecting the decadence of the Unexcelled Yoga Tantra, the highest form of Tantric Buddhism. The central practice of the Gelug sect is ascetic training based on strict commandments after entering the priesthood. Gushi Khan of Mongol had conquered most parts of Tibet by 1642 and then supported the fifth Dalai Lama, who became the supreme political and religious authority of both Tibet and Tibetan Buddhism. The Tibetan central government was established with the Dalai Lama as its head, and the Gelug sect became the largest school in Tibetan Buddhism.

The spread of Tibetan Buddhism in the Monpa area is believed to have begun in the seventh century. However, it is likely that monastic institutions did not become established until the thirteenth or fourteenth centuries (Tenpa and Tempa 2013). The textual sources refer to Jang Kyine [byang bskyid gnas] Gompa, which is believed to have been built by the third Karmapa or followers of the Karmapa (Rgyal sras

Fig. 4.1 In a *gompa* [*dgon pa*] in the Monpa area, this statue of Je Tsongkhapa Lobsang Drakpa [rje tsong kha pa blo bzang grags pa] (1357–1419), founder of the Gelug [dge lugs] sect, wears a yellow hat, which indicates that the Gelug sect is on the observer's left (see the photograph on the *left*); this statue of Longchen Rabjampa Drime Ozer [klong chen rab 'byams pa dri med 'od zer] (1308–1364), one of the greatest scholars of the Nyingma sect, wears a red hat, which indicates that the Nyingma [rnying ma] sect is on the observer's right (see the photograph on the *right*). Statues of the two sects are often found together in a *gompa*. The layout of the statues here is that of the Gelug sect. The positions of the two statues are reversed in a *gompa* of the Nyingma sect

sprul sku 2009). A priest named Ugen Sangpo [u rgyan bzang po], the brother of the famous Nyingma master Pema Lingpa [padma gling pa] (1450–1521), traveled from Bhutan to Tawang and built three Nyingma monasteries: Urgyenling [u rgyan gling], Sangeling [sangs rgyas gling], and Tsorgeling [tshogs rgyas gling]. Because the area was close to Bhutan, the main stream of Tibetan Buddhism in the Monpa area became the Nyingma sect. However, the Gelug sect also arrived in the area in the sixteenth century when the second Dalai Lama requested Tenpe Dronme and Lobsang Khetsun [blo bzang mkhas btsun] to establish monasteries in the Monpa area in order to promote the Gelug tradition (Tenpa 2013). Those two individuals built the monasteries of Argyadung [ar yag gdung], Lhangateng [sla nga steng], Sanglamphe [gsang lam phel], and Tadung [rta gdung]. Taklung [stag lung] Gompa at Sanglam in the Kalaktang area was established by the Gelug priest Tenpe Dronme. Eventually, conflicts developed among the Tibetan Buddhist sects. A military force supporting the Gelug sect was sent from Tibet, and the Kagyu and Nyingma factions were largely eliminated. Tawang Monastery—Tawang Ganden Namgyal Lhatse [rta dbang dga' ldan rnam rgyal lha rtse]—was constructed after political consolidation was achieved in 1681. The Gelug sect can be considered the mainstream sect in the Monpa area today, but the Nyingma sect maintains a certain level of popularity, especially in the Dirang area.

It may be assumed that the Gelug and Nyingma traditions merged in the Monpa area. Many *gompa* in the Dirang area belong to the Gelug sect, but they often enshrine Buddha statues with a yellow hat indicating the Gelug sect and a red hat signifying the Nyingma sect (Fig. 4.1). Monks of the two sects often jointly participate in religious ceremonies, such as during the recitation of the Kangyur [bka' 'gyur], Tibetan translation of the Buddha's teaching in 108 volumes. On such occasions, the Gelug

Fig. 4.2 Khinmey [khyi nyal mes] Monastery, or Sangngak Choekhorling [gsang bsngags chos 'khor gling], of the Nyingma sect in Tawang

monks sit to the right and the Nyingma monks to the left in Gelug temples. In Nyingma temples, this order is reversed. Buddha statues are also configured in a similar manner (Fig. 4.1). Whereas monks of the Gelug sect follow strict disciplines and are not allowed to marry, those of the Nyingma and Kagyu sects, who practice or follow tantric traditions, may do so. However, according to Vinaya, the Buddhist monastic regulation, the monks from the latter two sects are not allowed to marry or have female consorts if they have taken full monastic vows. Those monks who have taken a full monastic vow are known as *gelong* [*dge slong*]. Contrasting with celibate monks such as *gelong*, monks who are married or intending to marry wear white shirts or grow their hair.

Sangngak Choekhorling [gsang bsngags chos 'khor gling] Monastery, also known as Khinmey [khyi nyan mes], belongs to the Nyingma sect and is located in Tawang (Fig. 4.2). At the time of the fieldwork in 2010, 95 monks were undertaking religious training at the monastery. The monastery receives cheese and butter as donations from pastoral people living in the surrounding area. The Thegtsepa [theg brtse pa] *rinpoche* of this monastery, visits the lake called "holy lake" once a year, and a nearby village holds a ceremony to welcome him. On this occasion, farmers donate cereal grains, such as corn and finger millet, and pastoralists donate cheese and butter.

Fig. 4.3 Part of the mandala mural in Khinmey Monastery: it depicts male deities (*heruka*) embracing female consorts (*dakini*)

Paintings of male deities embracing female consorts appear on all the walls of the main hall of Sangngak Choekhorling Monastery (Fig. 4.3). The Nyingma sect is the most traditional of the schools, and actual tantric training is undertaken in this sect; however, tantric training is practiced only symbolically through visualizing meditations upon initiation in the Gelug, Kagyu, and Sakya sects.

Tantric Buddhism is a form of late Esoteric Buddhism. Its highly mystic and symbolic doctrines were passed down over time through hereditary succession from a master to his apprentice. For both Hinduism and Buddhism, Tantrism involves the manifestation of ultimate truth in the fusion of male and female principles at the level of definitive existence. Tantrism is the ultimate completion and is defined as the state of blissfulness—*mahaasukha* (Kawasaki 1993). In Tantric Buddhism, ultimate existence is the state where the masculine form, *upaya*, representing compassion and skillful means, fuses with the feminine form, *prajna*, signifying unchanging and uncharacteristic emptiness. That is said to be enlightenment (Kawasaki 1993). A common feature in the Buddhist art of Tibet and Nepal is a *yab-yum* [yab yum], a mandala in which the main male deity, Khandroma [mkha"gro ma], embraces his female consort, Dakini, or lesser deities embrace their female counterparts (Figs. 4.3 and 4.4). In such iconographies, the body parts of each character are fused into one another to indicate the sexual union, the state of *yuganaddha* (Fig. 4.4). It should be noted that the *Tibetan Book of the Dead* warns repeatedly that sensual appetites have

Fig. 4.4 *Yab-yum* [yab yum] in Khardung Tashi Cholling Gompa [mkhar gdung bkra shis chos gling *dgon pa*] of the Tawang area. Ultimate existence is the state where the masculine form, *upaya*, representing compassion and skillful means, fuses itself into the feminine form, *prajna*, signifying unchanging and uncharacteristic emptiness. This figure is believed to represent Vajrayana Buddhist deities- Heruka Cakrasamvara and his consort Vajravarahi

to be excluded when such *yab-yum* are viewed: upon reaching the emptiness that lies beneath all beings and beyond lust, ultimate bliss is attained and enlightenment is achieved (Kawasaki 1993). Even today, the *Tibetan Book of the Dead* is a sutra that is chanted by a monk priest when a person dies in Tibet. On the mandala of the Unexcelled Yoga Tantra, common combinations of *yab-yum* as father-mother are Cakrasamvara and his consort Vajravarahi (Fig. 4.4), Hevajra and Nairatmya, and also Kalachakra and Vishvamata, the Mahayan Buddhist deities.

Wang [dbang], which means "empowerment" in Tibetan, was originally a ritual in which a qualified master poured water onto an apprentice's head for sanctification. It was a form of initiation in Esoteric Buddhism, but later became a different kind of ritual that consisted of a number of processes. Tibetan Buddhism has four phases of *wang*, the initiations for empowerment: vase empowerment; secret empowerment; knowledge-wisdom empowerment; and suchness empowerment. Other than vase empowerment, the other three types of empowerment initiation are kept secret. In Tibetan Esoteric Buddhism, an apprentice who has received only the first vase initiation may learn only as far as the initial process. The remaining three initiations are indispensable to begin learning the ultimate processes.

The Nyingma sect is divided into the Rabjung [rab 'byung] and Ngakpa [bsngags pa] schools. The Rabjung school was formed by the agreement of the *Khenlop Choesum* [*mkhan slob chos gsum*], which comprised Guru Padma Sambhawa (Guru Rinpoche), *Khenchen* [mkhan chen] Bodhisatva Shantirakshita, and the King of Tibet, Trisong Deutsen [khri srong lde'u btsan] in the eighth century. At Khinmey Monastery, only the *rinpoche* belongs to the Ngakpa school; all other monks belong to the Rabjung school. Ngakpa monks are allowed to have a partner and can marry, but monks of the Rabjung school must remain celibate and may not marry. Most recent young monks of the Nyingma sect have joined the Rabjung school, because Ngakpa monks have to perform ascetic training in mountain caves for 3–4 years. If Nyingma monks can control their own various desires, particularly sexual ones, they can join the Ngakpa school. At Palyul Changchub Dargeling [dpal yul byang chub dar rgyas gling] Nyingma Monastery, (Leung Gompa) in Dirang, all monks belongs to the Rabjung school. At Sangngak Choekhor Dargyeling [gsang bsngags chos 'khor dar rgyas gling] Monastery (Chillipam Gompa at Rupa), while the former *rinpoche* belonged to the Ngakpa school, his son, the present *Rinpoche*, belonged to the Rabjung school at the fieldwork in 2013. Of the 150 monks at that monastery, five belong to the Ngakpa school and the rest to the Rabjung school. If the five monks of the Ngakpa school marry, they must leave the monastery. To become a monk of the Rabjung school, the monks must promise to reject marriage and take monastic vows, known as *domba* [sdom pa]. That includes *getshul* [dge tshul] as the first stage and *gelong*. Monks belonging to the Rabjung school wear a red robe, *tsan* [gzan], whereas Ngakpa monks don a distinctive white robe.

The six monasteries of the Nyingma sect in Tibet are as follows: Dorjee Dak [rdo rje brag]; Mindroling [smin grol gling]; Shechen [ze chen]; Dzokchen [rdzogs chen]; Kathok [ka' thog]; and Palyul [dpal yul]. All of these monasteries were destroyed during the invasion by Chinese military in the 1950s; the Shechen *rinpoche* and monks escaped to Nepal, whereas the *rinpoche* and monks from other monasteries went to India. There is a replica of Palyul Monastery, called Palyul Namdroling [dpal yul rnam grol gling] Monastery, in Mysore, South India. Its branch monastery is called Palyul Changchub Dargeling Nyingma Monastery and was founded in Dirang. The head abbot of Palyul Namdroling Monastery also governs Palyul Monastery in Dirang. The head of the Palyul Monastery in Dirang is a *khenpo* [*mkhan po*], the abbot. The *khenpo* of the branch monastery is appointed from among the *khenpo* learned in Palyul Namdroling Monastery in Mysore. The main responsibility of the *khenpo* is to assist the studies of about 40 *gethruk* [dge 'phrug]—student monks at the monastery—for 8 years. Student monks may undertake further study at Palyul Namdroling Monastery for at least 9 years. After 17 years of study, a student can obtain the degree of *lopon* [slob dpon]. *Lopon* monks have to assist the *khenpo* or work as teachers to newly admitted monk students. After a further 5 years of study, the learned *lopon* can become a *khenpo*. *Khenpo* represents the highest academic degree offered in the monastic institutions of Nyingma, Sakya, and Kagyu. Such highest degree in Gelug and Bon schools, is called *geshe* [*dge shes*]. It is considered equivalent to a PhD awarded at modern universities. *Khenpo* translates as "abbot," whereas *geshe* means "virtuous friend" or "scholar." Both statuses are highly regarded, and monks with these credentials are considered learned and accomplished in Tibetan Buddhist traditions.

The initiation processes in Tantric Buddhism may seem eccentric to the non-educated Buddhist practitioners. However, in Tibetan Buddhism, this initiation is undertaken on the basis of a strong mutual trust between masters and disciples. This can be inferred from an interview with Geshe Rabten (*dge shes* rab brtan) undertaken by one of his apprentices, Alan Wallace. The interview was conducted to determine how Rabten had attained *geshe* status. Rabten describes his experiences and observations at a senior class in Madhyamika philosophy. The following is extracted from an interview cited by Wallace (1980: 94–95):

> Disciple: Who instructed you during these classes?
>
> Geshé: A master named Geshé Thutob. He lived in the same upstairs room as my former teacher; I made guru devotions to him when I visited him, often staying until eleven o'clock at night. Since the place to urinate was far away, he kept a pot in his room. Every morning I went up and insisted on taking out this pot to empty it. While doing so, I would place it on my head and offer a prayer. Then I would pour a little in my hand, drink it, and throw the rest away. This is not a Tibetan or a monkish custom. I was moved to do so by my deep faith in my guru, although there is not necessarily any relationship between drinking urine and receiving his blessing. There is one only if one has heartfelt faith in the guru.
>
> Disciple: Such a practice would probably sound very strange to most Westerners.
>
> Geshé: Yes, but this is of no great importance. Some will understand the reason behind it, and to others it will remain a mystery. Religious faith is known throughout the world; it simply has different ways of manifesting itself. Out of pure devotion, some disciples nurse their lama when he is sick, and by doing so gain great insights. There are a few instances of disciples washing their gurus when they were too ill to leave their bed to defecate; as they carried out the excrement, they experienced great clairvoyance, gaining awareness of the consciousness of other beings including even tiny insects. Such immediate, rather than gradual, insights are due to the disciple's great faith in his guru combined with the guru's blessing. To understand the profundity of this devotion, you need long experience in the Buddhadharma and a deep understanding of the stages of the path to enlightenment, as well as tantra.

The relationship between a master and his apprentice is far more important in Tibetan Buddhism than in Exoteric Buddhism. The reason that Tibetan Buddhism is sometimes called Lamaism is that the high priests, lamas, are respected as the living Buddha (Kelsang and Masaki 2000a, b).

In Tibetan Buddhism, the term *gompa* [dgon pa] in every part of the Monpa area refers either to a temple or a monastery. The term *gompa* denotes a range of temples from an austere temple where no lama is in residence to a large institution like Tawang Monastery, where a number of lamas are permanently based, which operates a school to develop future generations of monks. The latter larger institution is often called a *dratsang* [grwa tshang]. A *dratsang* is equivalent to a college. Certain monasteries, such as Sera, Drepung, and Namdoling, have two or three *dratsang*.

Fig. 4.5 The *Kangyur* [bka' 'gyur]—the 98 or 108 volumes of the translated words of the Buddha

One ritual involves chanting and reading the Kangyur [bka' 'gyur], the translated words of the Buddha (Figs. 4.5, and 4.7). Though the Kangyur is commonly accepted as consisting of 108 volumes, some editions comprise more or fewer volumes. Its simplified version is the *Bum* ['bum], which consists of 12–16 volumes. The variation in the number of volumes is due to the size of the written characters or the particular edition. A shorter version of the *Bum*, known as *Nithri* [nyi khri], consists of three volumes. The *Geytong* [brgyad stong] is a further simplified version and comprises only one volume: it consists of 630 folios or pages, which can be recited by three monks in a day (Fig. 4.7). Other variations include the *Dhomang* [mdo mang] and *Tsungdhue* [zung 'dus]. The *Dhomang* has more than 700 folios or pages, which five monks can complete chanting in a day. The *Tsungdhue* consists of more than 100 parts, each part made up of 1–50 folios or pages, and only parts relevant to an occasion are chanted.

During our fieldwork, we observed copies of the *Geytong*, *Dhomang*, and *Tsungdhue* at the home of a part-time monk. Of these, only the *Geytong* was printed on *boi shuku* [bod shog] (Figs. 3.7 and 4.6); the others were printed on modern paper. There were also smaller versions of the *Bum*: the *Bhoncheung* ['bum chung], with 29 folios (pages), and the *Sherab Nyingpo* [shes rab snying po], with just five pages. A 9-year-old nun who could not attend school because of a head injury was reading the *Bhoncheung* when we visited Singsur Buddhist nunnery as part of our field observation.

Fig. 4.6 *Geytong* [brgyad stong] is a sutra of Tibetan Buddhism that consists of just one 630-page volume. It is a digest of the *Kangyur* [bka' 'gyur]

Minakpa lama are part-time monks who ordinarily work as farmers in their villages but act as monks on particular occasions. They are also referred to as *Ngakpa lama*, but are simply called lama in the Tawang area. Some *gompa* are also supported by visiting full-time lamas from nearby monasteries. Meals for monks are offered by villages, and villagers provide an on-duty cook for them. Only central villages have a *gompa*, and the rituals vary depending on the *gompa*. The term "lama" originally meant a mentor priest in Tibet, but all full-time monks, including apprentices, are called lamas in the Monpa area, especially in the Dirang and Kalaktang areas, to distinguish them from part-time *Minakpa* or *Ngakpa lama*.

Approximately 20 (mostly part-time and a few full-time) monks recite sutras for Kangyur chant ritual in concert for 40 days from October to November every year at Khastung [mkhas stung] Gompa in the village of Dirang (Figs. 4.7 and 4.8). The villagers gather at the *gompa* and receive the blessing of the monks on the final day. The monks touch the heads of the villagers using written sutras. The monks also put *srungma* [*srung ma*] around a person's neck. The *srungma* is a knotted and blessed thin rope held by the head lama during the ceremony (see Sect. 7.4). It is thought that this *gompa* was founded by Tenpe Dronme at the end of the sixteenth century.

The *koenyer lama* of this *gompa* is chosen from among *Minakpa lama* to serve as its caretaker for a period of 3 years. He looks after visitors during such events as the Kangyur chant ritual throughout the year. To support this ritual, 160 rupees as

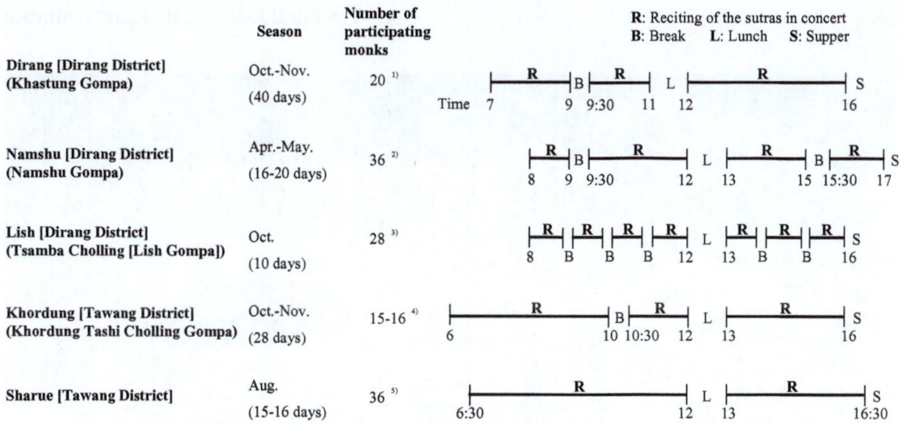

Ritual of *Kangyur*

The ritual in which the Kangyur, the ninety-eight volumes of Translation of the Word of the Buddha, is chanted by monks and performed at the central Gompa, Tibetan Buddhist temple, in each region. In the case of the Khastung Gompa in the Dirang village, mostly part-time and a few full-time monks recite sutras in concert for forty days from October to November. On the final day, the villagers gather at the Gompa, and receive the blessing from the monks: The monks caress the villagers' heads gently with sutras or put Rhungma, a necklace that is supposed to expel evils and is made of strings, around the villagers' necks.

1) Mostly Minakpa Lamas, part-time monks, around the Gompa with some Lamas, full-time monks. One hundred volumes of sutras are recited.
2) Entirely Minakpa Lamas. One hundred and nine volumes are recited.
3) Mostly Minakpa Lamas around the Gompa with some Lamas. Eight to nine volumes are recited every year.
4) Mostly Minakpa Lamas around the Gompa with some Lamas from the Tawang Monastery. One hundred volumes are recited.
5) Seventeen Lamas from the Jangdhaparie Monastery and 19 Minakpa Lamas around the Gompa. One hundred volumes are recited.

Fig. 4.7 Ritual of Kangyur [bka"gyur]

Fig. 4.8 Khastung Gompa [mkhas stung *dgon pa*] in Dirang [rdi rang] village

Fig. 4.9 Namshu Gompa [nam zhu *dgon pa*] in the village of Namshu [num shu]

well as corn or sheep are collected from each household in the villages of Dirang. The collected donations serves as wages for the monks, and butter, cheese, and other food items are collected from local pastoralists and served to the monks as meals. Another ritual in which the *Bum* ['bum], the shortened version of the Kangyur, is chanted is held twice a year for 2 days in May or June and November. In Namshu Gompa of the village of Namshu, approximately 35 part-time monks recite the sutras in concert for 16–20 days from April to May (Figs. 4.9 and 4.10). Villagers cook and serve the monks meals at the Kangyur chant ritual. It is believed that this Gompa was also founded by Tenpe Drome.

In Tsanba Choeling [btsan pa chos gling] Gompa, also known as Lish Gompa, in the village of Lish, approximately 28 monks (mostly *Minakpa lama* with a few lamas) recite eight to nine volumes of the sutras in concert for 10 days in October. A ritual during which the 12 volumes of the *Bum* are recited is also performed twice a year. For the Kangyur chant, 10 *brey* of corn is collected from every household. Two cooks are hired, and rice, cheese, and butter are consumed. It is believed that Lish Gompa was also established by Tenpe Dronme.

At Khardung Tashi Cholling [mkhar gdung bkra shis chos gling] Gompa in the Tawang region, 15 or 16 monks perform the Kangyur chant ritual for 28 days from October to November. Some of them are full-time monks from Tawang Monastery whereas others are part-time monks from the village. For the ritual, 500 rupees as well as cheese and butter are collected from every household. The *koenyer lama* of Khardung *Gompa* is appointed by succession.

Fig. 4.10 Ritual of the *Kangyur* [bka' 'gyur] chant in Namshu Gompa [nam shu *dgon pa*]

The ritual of the Kangyur chant is not only performed in *gompa*s. When we conducted the field observation in August 2010, 36 visiting monks chanted the Kangyur at the community center of the village of Seru in the Tawang region over a period of 15–16 days. We observed 21 part-time monks and 15 full-time monks reciting the approximately 100 volumes of the sutras in two separate rooms. The 15 lamas were from Jangdokpalri [zangs mdog dpal ri] Monastery in the village of Jang (Fig. 4.37). Some elderly part-time monks took part, but the majority were novice monks. The full-time monks were almost entirely apprentice monks sent from the monastery (Fig. 4.11). We observed a similar composition of monks performing the same ritual at other *gompa*s.

There are also privately owned *gompa*. Rimphokpa [rin 'phug pa] Gompa in the Dirang Dzong area is one such *gompa*. A recital of the 12 volumes of *Bum* in the private *gompa* can be performed on request in exchange for food or money.

The largest and most central *dratsang* in this area is Tawang Monastery. However, Palyul Changchub Dargeling Nyingma Monastery in the Dirang region is also becoming an important center for local people as well as nearby villagers (Fig. 4.12). A school for apprentice monks is operated in conjunction with this monastery.

Tibetan Buddhism was able to replace Bon largely because Buddhist monks understood the thinking of Bon believers to a considerable extent. It is also worth noting that Buddhist monks became more skilled in non-Buddhist practices, such as the worship of local spirits, receiving oracles, and fortune-telling, than did Bon believers (Snellgrove and Richardson 1980).

Fig. 4.11 Ritual of the *Kangyur* [bka' 'gyur] chant at the community center of Seru [bse ru] village: 21 part-time monks and 15 full-time monks recited the approximately 100 volumes of the sutras in two separate rooms. The photograph shows lamas from Jangdokpalri [zangs mdog dpal ri] Monastery in Jang [byang] village. Many of them are apprentice monks

The part-time monks living in some villages are called *tsespa lama* [rtsis pa bla ma]. Here, *rtsis* means "to calculate" or "to estimate," and *pa* signifies "one who calculates": "astrologer" is a common translation. These part-time monks in the Monpa area are consulted by the villagers on such matters as marriage, birth, and sickness, and they also conduct death rituals. On such occasions, these monks carry out animist and shamanist practices, and *tsespa lama* play a leading role in those practices. When a villager dies, the family invites the *tsespa lama* of the village to the home. He visits the house of the deceased, even if called during the night. The *tsespa lama* then gathers information about the deceased, such as the cause and exact time of death, the circumstances of the person's death, and their birthday and 12 signs of the Chinese zodiac (Fig. 4.13). Through full use of their horoscopes and books, *tsespa lama* determine matters related to reincarnation, such as where the deceased used to live in their previous incarnation and where he or she will be reborn; the *tsespa lama* explains to the family how they can revitalize the deceased. The monk also determines the details of the funeral. Funerals in the area usually involve water or ground burial, cremation, aerial sepulchers and so only. The *tsespa lama* informs the family of any specific form of funeral that they must not choose in a particular case. The family then selects and undertakes one of the remaining forms.

Fig. 4.12 A festival held at Palyul Changchub Dargeling Nyingma [dpal yul byang chub dar rgyas gling rnying ma'i *dgon pa*] Monastery

The *tsespa lama* advises the specific time and direction for sending the deceased out of the house. He also indicates one or two zodiac animals that must not touch the deceased and prohibits people having those zodiac signs to convey or touch the deceased.

A *tsespa lama* determines the background to a consultation by first using a horoscope called a *tsethang* [rtsis thang] (Fig. 4.14), and he then examines the astrological book called the *Kartsi* [dkar rtsis] (Fig. 4.15) to obtain relevant information. Finally, the *tsespa lama* consults the latest almanac, called the *lotho* [lo tho], to clarify the life history of the deceased from birth to death. In the case of a marriage, the *tsespa lama* asks for the couple's zodiac signs and performs a reading. A manuscript called a *baktsi* [bag rtsis] is used specifically for marriages (Fig. 4.16). Another manuscript called a *naktsi* [nag rtsis] is employed in the case of death. The process of fortune-telling starts from the horoscope and involves a number of examinations of manuscripts in proper order. One small mistake in any part of the process may result in a serious error, and so the *tsespa lama* performs the reading with great care. In the case of a villager's death, the process involving only the *lotho* and the *naktsi* takes about an hour. If more manuscripts are involved, it may last up to 9 days. The process for providing advice about marriage takes just 2–3 h if only the *lotho* and *baktsi* are used. However, it could take 30 h if more manuscripts are necessary.

Fig. 4.13 *Tsespa* [rtsis pa] *lama* gather information on the deceased such as the cause and exact time of death, the circumstances of their life, their birthday, and zodiac sign. By making full use of their horoscopes and books, *tsespa lama* tells matters related to reincarnation, such as where the deceased lived in their previous incarnation and where he or she will be reborn, and explain to the family how they can revitalize the deceased

The following experience of the author (Mizuno), which occurred during a field observation of August 2011 in a village, may provide a better grasp of the use of an almanac and horoscope in the Monpa area.

> A *tsespa lama* asked me about my zodiac sign, and he determined my birth year after consulting the 2011 *lotho*. When I replied that I already knew my own birth year, the *tsespa lama* asked doubtfully, "How could a Japanese person know his birth year just from his zodiac sign?" Many inhabitants of that area know their zodiac signs but not their birth year or age. There is a great gap in the sense of time between people of Monpa society, who use the Tibetan calendar, and the majority of outsiders, such as myself, who use the solar calendar. A serious hindrance in field observations in the Monpa area always arose when I asked someone which month something occurred. For example, I asked, "In which months do you carry out seeding and harvesting of corn?" The villagers always began counting off months on their fingers, became confused, and a serious debate often arose. Mr. Rinchin, who kindly took care of me in the village of Dirang, heard my age and claimed to have the same age. However, his passport indicated that he was several years younger than me.

According to Yamaguchi (1987), the Tibetan calendar is neither a solar nor lunar calendar but a variation of a lunar calendar mixed with elements from a solar calendar. Instead of the matrix of 12 zodiacal signs and 10 brotherly celestial stems as traditionally used in China, the Tibetan calendar features 12 signs and male/female combinations. It is a calendar system based on a particular tantra introduced to Tibet.

Fig. 4.14 *Tsethang* [rtsis thang or lo tho] is the horoscope referred to by a *tsespa* or *tsipa lama* (*rtsis pa bla ma* in Tibetan) in consultations

Fig. 4.15 A *tsespa lama* first consults the horoscope called *tsethang* [rtsis thang] (Fig. 4.14) and then examines the astrological book called the *Kartsi* [dkar rtsis] to obtain relevant information

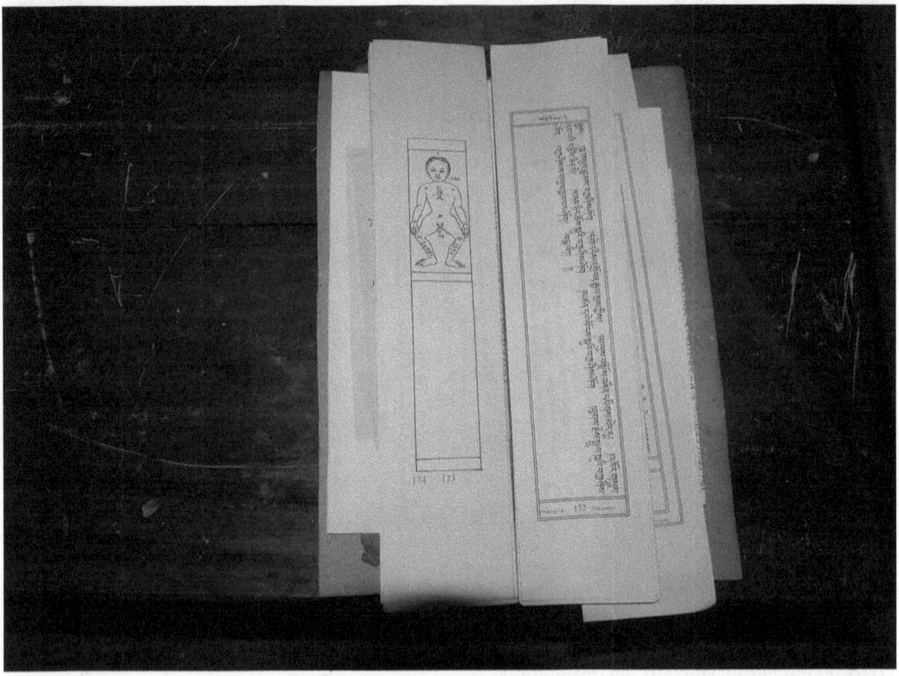

Fig. 4.16 This manuscript is used for births

It is locally believed that a *tsespa lama* is said to execute a calculation using 40,400 factors, and the process of learning the method is difficult. It is also said that fewer than 10 % of apprentices can successfully complete the learning process. A *tsespa lama* is not a hereditary job or title. A *tsespa lama* teaches a number of children who show interest in his work. Fees to *tsespa lama* are paid in cash, but most households pay 2 *brey* of cereal grains of barley or corn in the harvest seasons for his services.

Water burial is common and cremation is rare among the Monpa people. Carrying out a water burial involves cutting the body into 108 parts and throwing them into the river. Villagers believe that cremation pollutes the air and sky and that ground burial produces insects, but water burial feeds fish and contributes to nature and ecosystems. Most villagers in the area made similar comments in this regard, and it may be assumed that the views are widely held. However, in reality, cremation is an option but only for wealthy people owing to its expense. At a cremation, eight to nine monks recite sutras, which demands a large expenditure to cover the monks' fee and the cost of their meals. In contrast, only one monk is required for a water burial.

According to Norbu (2008), there are four ways to process a dead body. They are placement of the body in a mountain cave; water burial; ground burial, and cremation. With the first method, the body is placed in a wooden box, transported to a high mountain, and placed in a cave. This is used only for infants and the elderly. Ground burial is restricted to deaths from infectious diseases and confirmed felons (Norbu 2008); however, the Boot Monpa people practice ground burial (Barua 1995).

This is presumably related to the fact that they practice animism and shamanism. In Monpa societies in which Tibetan Buddhism is predominant, a ritual and feast are held on the 49th day of the period of mourning. Apart from the anniversary of death, that is the final ritual after a person's death. During the field observations, we observed the Kangyur chant ritual in the village of Namshu and photographed a villager setting out meals for the monks. We brought back prints of the photographs on my next visit to the village and was informed that the villager had died: the ritual of the 49th day of the mourning period was being held for him that very day.

Kawaguchi (1909) describes Tibetan funeral rites. According to him, the best burial method in Tibet was considered to be leaving a corpse exposed to vultures—sky burial. The second-best method was cremation, followed by water burial. Ground burial was considered the worst method. Kawaguchi states that Tibetans detested this method, and it was practiced only for those who have died from smallpox to prevent the disease being transmitted to birds and fish. Cremation was practiced only for the wealthy people because firewood was scarce and yak dung was inappropriate for use as a funeral pyre. Water burial was practiced at several places along large rivers. The corpse is cut up into smaller parts, such as the head, arms, and legs, to avoid it becoming lodged on banks and cliffs and to enable fish to easily consume the remains. Kawaguchi (1909) explained that Indian philosophy states that a human body consists of soil, water, fire, and wind, and there are therefore four separate ways to return it to the four elements, namely ground burial, water burial, cremation, and sky burial.

In the Monpa area, the locations available for water burial, usually called *dutai* [dur du] or *dukhroe* [dur khrod], are fixed. In several parts of the area, water burial is also called *dursha* [dur sa] and *rikili* (Fig. 4.17). At the beginning of a water burial, a monk recites a sutra to receive the consent of the river's sprits and to expel evil. Then, two young men drink alcohol and remove their clothes. They cut off the head and lay it in a safe, appropriate place. Then, they cut the body in half. Each of the two men handles either the upper or lower body beyond this point. They each cut up their part and one by one throw small pieces of the body into the river while counting. In the final phase, they cut the head in half and further cut up the halves to produce 108 pieces in total. After completing these phases, the men go into the river to wash away the blood and then return home. They are not allowed to enter their homes before washing their entire body with clean water at the entrance. The family of the deceased gives an amount of money to the two men. They drink at the beginning of the task of cutting up the corpse because it is considered unbearable without being moderately inebriated.

Every village has more than one part-time monk, who is called a *minakpa lama* or *ngakpa lama*. A *minakpa lama* normally works as a farmer and can marry but assumes the role of monk when necessary. The term *ngakpa* is considered originated from the *Ngakpa* school of the Nyingma sect. The *Ngakpa* school allow its monks to marry. *Ngakpa lama* originally signified a monk who could marry or a tantric practitioner of Tibetan Buddhism. However, those terms are used by villagers only to distinguish part-time from full-time monks. *Minakpa lama* and *ngakpa lama* are called "Lama" in a face to face conservation. The term *minakpa lama* is mainly used in the Kalaktang area, and *ngakpa lama* in the Dirang area. All full-time and part-time monks are called lama in the Tawang area; however, those belonging to monasteries

Fig. 4.17 *Dutai* or *dukhroe* [dur du or dur khrod], a site used to cut the body of a deceased person into 108 pieces as part of a water burial: every area has a riverside site for this purpose

are sometimes distinguished from others and referred to as *drapa* [grwa pa]—"monk." High priests such as *rinpoche* are called *kusho* or *kusho-la* [sku shogs *lags*], where the *la* [*lags*] syllable indicates respect. Every household in a village invites a nearby part-time monk to their home a few times a month and requests him to recite a Buddhist scripture—*choesak* [chos bsag]—and to pray. Households invite part-time monks for a specific purpose and hence a particular ritual is conducted by the monk. *Drolchoe mandal zhepa* [sgrol chos mandal zhes pa] is a ritual to pray for the recovery of a sick person, longevity, wealth, and prosperity, and it can also be used for exorcism. In the field observation stage of this study, we had an opportunity to witness this ritual (Fig. 4.18). The two brothers requested a monk to pray for their father, who was going to have a surgical operation in 4 days' time. On another occasion, we observed the ritual being practiced in another household. The monk was praying to a goddess named *Jetsun Dolma* [rje btsun sgrol ma]—Taradevi in Sanskrit.

Rituals like these typically take a whole day. At around 7 a.m., about four part-time monks start making *torma* [gtor ma] (religious offering made of dough) (see Sect. 3.2) (Fig. 4.19), set up incense sticks, melt butter for fuel lamps, and then align the lamps. The preparation is completed by 10 a.m. The monks eat breakfast prepared by the client family. A breakfast menu that we observed while undertaking fieldwork consisted of *gruntsum bokpai* [bra'o zan gro phye] (see Sect. 7.4) (Fig. 7.10), cucumber, tomato, and cheese. The actual reciting of sutras began at around 10:30 after breakfast. The reciting with a lunch break lasted until sunset. After the ritual, the monks were served supper.

Fig. 4.18 *Drolchoe Mandal Zipa* [sgrol chos mandala zhes pa] is a ritual to pray for the recovery of a sick person, longevity, wealth, or prosperity, or it may be used in an exorcism

Fig. 4.19 *Minakpa lama* [mi nag pa bla ma], a part-time monk, produces a *torma* [gtor ma]

Part-time monks, especially *tsespa lama*, offer traditional medical care to villagers. For example, they use an instrument called a *tsardham* [tsha dam] (Fig. 7.20) to heal sickness. Characters are engraved on its surface, and it resembles a personal seal. A monk places the engraved characters onto the affected part of a patient's body to ease pain and cure illness. This technique has been in use for a long time. Another instrument commonly used is a funnel-shaped device (Fig. 7.21). One of its ends is placed on the affected body part to suck out the cause of sickness. Nagahashi (1999) reported that witchcraft was employed in Ladakh and other parts of northern and northwestern India to suck out disease via a metal pipe at the time of his study. Nagahashi (1999) observed that common requests made of a shaman included cures for headache, neuralgia, chest pain, and stomachache; the suction of old blood either by means of a metal pipe or the shaman's mouth was often performed.

There is a strong relationship between the Monpa people and their religions, Tibetan Buddhism and Bon. An instrument called a *phurpa* [*phur pa*] like dagger, which is used to expel evil spirits, is a very familiar object in their households, and many people wear a small version of the instrument as a pendant. As another religious tool, wooden models of a virile member are often suspended from house roofs to provide protection from evil spirits and the evil eye (Fig. 4.20). Such a wooden model is called *meley* [*me ley*] in Tawang Monpa and *laktang* or *liktang* [*rlig thang*] in Dirang Monpa; however, the word must not be used in front of women.

Fig. 4.20 Wooden models of a virile member are often suspended from the roof of a house as protection from evil spirits and the evil eye

4.2 Relationships Between Tawang Monastery and Villagers

Tawang Monastery is the central institution of Tibetan Buddhism in the Monpa area. It is located on the hilltop overlooking the town of Tawang (Fig. 4.21).

Tawang Monastery owes its origin to the ruling group of the Gelug sect. During the sectarian power struggles among the various Tibetan Buddhist sects in the fifteenth to eighteenth centuries, the Gelugpa was opposed to the Karmapa, which is a subgroup of the Kagyupa, and also the Nyingmapa. Supporters of the Bhutanese Drukpa ['brug pa], a subgroup of the Kagyupa, were also hostile to supporters of the Gelugpa during that period. The Tibetan and Monpa people referred to the Bhutanese as Drukpa. The Drukpa supporters intended to subjugate Tawang as a whole to their traditions. While engaged in confrontations with the Drukpa, Merak [me rag] Lama Lodoe Gyatso [blo gros rgya mtsho] was struggling to found a temple for the Gelug sect. Conflict between them took place after the 1660s and lasted till 1680. Finally, with support from the fifth Dalai Lama, Ngawang Lobsang Gyatso (1617–1682), Merak Lama established a temple in Sanglamphel sometime after 1670 and built Tawang Monastery with the help of Tsona magistrate, Tsona Dzonpon [mtsho sna *rdzong dpon*] in 1681 (Aris 1980, 1988).

Fig. 4.21 Tawang Monastery is located on the hilltop overlooking the town of Tawang. It has the form of a fortress and used to be a base for collecting a tax called *khray* [khral] from the entire Monpa region (Monyul). The tax is still collected from Tawang Monpa locals

The etymology of "Tawang" signifies its genesis as the location of the monastery. In the Tibetan language, *rta* means "horse" and *wang* or *dbang* signifies "to choose" or "to bless." Both *wang* and *dbang* appear in some Tibetan texts between the seventeenth and early eighteenth century. The name refers only to the monastic complex or the particular mountain on which it is located. A number of the seventeenth and early eighteenth century texts mention La'og yulsum [la'og yul *gsum*] or Shar Nima Tshosum [shar nyi ma tsho gsum] as a subdivision of present-day Tawang. Only after the later eighteenth century texts, the term of Tawang [rta wang/dbang] is commonly used to refer to the modern Tawang district.

According to local oral traditions, Merak Lama in the late seventeenth century was contemplating where to build his monastery and often visited a cave to practice meditation. After mediating one day, he found that his horse had disappeared from outside of the cave. He was surprised since this had not happened before, and he followed the hoof tracks. They took him to the hill where the castle of Gyalpo Kala Wangpo [rgyal po ka la dbang po] was formerly situated. Therefore Merak Lama decided to build his monastery on the hilltop to which his horse had led him.

The Tawang people gradually began cooperating with Merak Lama. They accepted his proselytizing efforts and assisted with them. The influence of the Tibetan Buddhist Gelug sect began to spread, and penetrated areas of the Monpa and Sherdukpen as well as those of the Miji and Bugun (Fig. 4.22). The latter three areas had been particularly influenced by the Nyingma sect and Bon religion.

Fig. 4.22 People listen to a high monk preach the "teaching"- *sungchoe* [gsung chos] outside an auditorium called a *dhukhang* ['du khang] in Tawang Monastery

Although more than 300 years have passed since its construction, Tawang Monastery remains the central focus of Buddhism and people's religious lives in the Tawang and West Kameng districts, and it is central to the lives of other ethnic groups in Arunachal Pradesh. The monastery is located 3,050 m above sea level, where routes from Tibet, Bhutan, and East Kameng intersect. The monastic areas and monks' residential complexes are surrounded by walls with a total length of 610 m.

The tax system that supported Tawang Monastery was as follows. Every household in the Tawang area was obliged to pay a tax known locally as *khray* [khral]. The amount of *khray* that an average household had to pay was 20 *brey* [bre] of cereal grains, such as corn, finger millet, rice, buckwheat, and barley. As noted in Sect. 3.3, about 20 *brey* of cereal grain is equivalent to 12 kg in the International System of Units. However, this was only a rough standard, and the amount depended on the land productivity. A parcel of land ranging from 1 to 5 acres (4,000–20,000 m^2) was referred to as 1 *khreikang* [khral khang], and it was the area on which the obligation to pay 20 *brey* in taxes was based. A household that owned a large or highly productive land parcel was obliged to pay 2 *khray* (i.e. 40 *brey*). Fuel wood, rather than cereal grains, was also acceptable as tax. In the Tawang area, the tax collected in winter was called *ton thre* [ston khral] and consisted mainly of barley; tax collected in summer was called *yar thre* [d.yar khral] and was largely finger millet. According to Dhar (2005), it was considered that every *khreikang* was able to pay 20 *brey* of wheat during *dawa ngapa* [zla ba lnga pa], the 5th month in the Tibetan lunar calendar (from around June to July); the same amount of millet or barley was sufficient during *dawa gupa* [zla ba dgu pa], the 9th month in the Tibetan lunar calendar (about October to November). Biswal (2006) indicates that the amount of the first tax was 10 *brey* and the second 20 *brey*. We were able to confirm in the field observations of the present study that the tax payment was made only once a year. This indicates that there may have been some regional variations in the amount of *khray* due. Biswal (2006) believes that the *khray* system started in the days of Merak Lama. However, although this taxation system must be very old, it is difficult to date: it may have preceded the time of Merak Lama. Tibetan sources indicate that a number of lamas as well as secular or peripheral rulers also depended on taxation prior to the period of Merak Lama (Tenpa and Tempa 2013).

Dorjee Khandu [rdo rje mkha' 'gro], who was chief minister of Arunachal Pradesh between 2006 and 2011, was originally from the village of Gyangkhar near Gyangkhar Dzong (Fig. 2.5), and he well understood living conditions in that area. Khandu decided to reduce the tribute to Tawang Monastery to 25 % of the traditional amount, and compensated for the difference using the treasury of the state government. After this decision, the tribute that each household had to pay was decreased to 5 *brey* regardless of the size of the land cultivated by the household. For the owners of small pieces of land, the tribute to the monastery could be as low as 2–3 *brey*. This was in effect a governmental bailout plan for the villagers of that area. Even if brothers got married and their family's land parcel was split between them, a collective tribute of 5 *brey* paid by the brothers was accepted as fulfillment of their obligation.

Khandu went missing in a helicopter accident on April 30, 2011. We were conducting the field observation at that time and had to halt all planned interviews since many people, led by *tsorgen*, went into the mountains to search for him. It was estimated that 15,000 people joined the search on May 3. The helicopter was discovered, and his death was confirmed on May 4. When we were traveling on Sela Pass from Tawang to Dirang, we saw many mourners heading toward Tawang. Khandu was loved and respected by the locals. After the incident, Jarbon Gamlin, who had been the minister of tourism, became the next chief minister. Pema Khandu [padma mkha' 'gro], Dorjee Khandu's son, who had inherited his father's electoral base of support, became minister of tourism.

A form of land-lease system called *boima* [bogs ma] is common in the Tawang area. With this system, a farmer cultivates a plot of land owned by Tawang Monastery, and the harvested rice is divided between the monastery and the farmer. Norbu (2008) uses the term *buiwa* [phyul ba] rather than *boima* for this and indicates that the monastery receives a quarter of the harvest. Biswal (2006) also states that the monastery's share is one-quarter. The background to this system is the frequency of farmers' migration from Tawang to Dirang and Kalaktang and also to Bhutan owing to the severe taxes imposed by *dzongpon* sent from Tibet. The monastery acquired many parcels of land as it took over a number of ownerless properties, and some pieces of land were donated by villagers. This practice continues today.

In the village of Soma (so ma) near Tawang (Fig. 4.23), the *boima* system is extensively used. When we conducted interviews (Mizuno, May 2011) in Soma as

Fig. 4.23 Agricultural field of Soma village, where the *boima* [bogs ma] system is practiced: this is a locally common form of land lease in which a farmer uses a piece of land owned by Tawang Monastery. Harvested rice is divided between the farmer and the monastery

part of this study, we found that farmers and the monastery each take half of the harvest after setting aside seed rice for planting. The cost of labor for harvesting is also covered jointly by the farmers and the monastery on a 50–50 basis. In the harvesting season, around October, six to ten monks arrive in Soma from Tawang Monastery and stay at the village's community hall for a week. This is the week that starts just before and ends just after the harvest, and the monks' duty is to confirm the harvest yield. During that week, the villagers offer food, fuel wood, and sleeping mats to the monks and prepare their meals. After staying for 1 week, the monks transport the monastery's share back to Tawang Monastery. The farmers of this village pay barley as their *khray* tribute and rice for the *boima* system.

Other villages where the *boima* system operates to a notable extent in the Tawang area include Tengteng [steng steng], Saikharteng [sras mkhar steng], and Bekhar [sbas mkhar]. Interviews we conducted in the village of Aodung near Tawang (Fig. 3.1) revealed that two households supported that system. This implies that a number of households in other villages may also take advantage of the land-lease system.

During the *torgya* [*gtor rgyag*] (held once a year in January or February) and larger *dungyur* [*dung 'gyur*] (held every 3rd year) festivals, young women who do not normally wear traditional dress don traditional costumes, which consist of Monpa clothes. They gather at Tawang Monastery to celebrate the festival (Fig. 4.24).

Fig. 4.24 Young Monpa women in traditional costumes walk alongside a *mani* wall on their way to Tawang Monastery. They wear such clothing only for traditional ceremonies

4.3 Life and Education in a Tibetan Monastery

Until the later twentieth century, it has been common, or even mandatory, for all households in the Monpa area with more than three sons to send their second son to a Buddhist monastery (Dorjee 2006). However, only a few families still adhere to this tradition. For example, during our field observations in the village of Dirang Dzong, one household that supported this tradition sent the second son to a monastery in southern India and the third son to a monastery in Nepal. The second son visited the parental home in February 2011 and the third son in May of the same year. It had been 8–9 years since their previous visits home. Each stayed for about 2 weeks. We took a photograph of the second son during his homecoming and were able to show it to the third son in May. The latter had known his elder brother only as a small child, and it was the first time for him to see his brother as a grown-up youth. Another family in the village of Gunthung [dgun 'thung] near Thembang Dzong showed us a photograph of their second son who was living as a monk in a monastery in Taiwan. The first sons of both families were living in their family villages. Monks from distant monasteries occasionally visit the villages and invite boys of the appropriate age to join their institutions. The young monks mentioned here went to their respective monasteries with their childhood friends and grew up together.

Full-time monks, lamas, and apprentice monks live together in Tibetan Buddhist monasteries. Large monasteries usually have their own schools. These monasteries also have a residential hall for pupils, and a residential unit consisting of three to six rooms is called a *shak* [shag] or *drashak* [grwa shag] (Fig. 4.25). As part of our field observations, we visited a *shak* that accommodated one lama and six apprentices (Fig. 4.26). There was a kitchen on the ground floor and a room for two apprentices behind it. The central room upstairs had a fire pit and was used by two apprentices. The next room was divided in two by a curtain. One of the spaces contained a small sacellume, and the other was used by an apprentice. Two rooms were located on the other side of the floor. One was used by the lama and contained a television set and a radio-cassette recorder. The lama's office was attached to this private room, and also contained a personal computer. The bathroom and toilet were located outside the main building. At the time of our visit, two of the apprentices were away on a study tour to Banaras.

Altogether, Tawang Buddhist Monastery of the Gelug sect had 530 monks and apprentice monks housed in 75 residential units at the time of our field observation (Mizuno, August 2011). One teacher monk, *gegan* or *gergan* [dge rgan], and a number of apprentice monks, *genyen* [dge bsnyen] or *gethruk* [dge phrug], lived together in each *shak*. The age of the teacher monks ranged from 27 to 86 years, and that for the apprentice monks from 6 to 22 years. The number of *genyen* living in a *shak* was most commonly six to eight, but it could be as low as one and as high as 26 (Fig. 4.25). The age of boys enrolling at the school is 5–12 years. Apprentice monks, *genyen*, learn to read and write from a book called a *kakha* (alphabet book), read Buddhist textbooks, and perform devotional exercises. They also study English, Hindi, Buddhist philosophy, mathematics, social studies, and some other subjects. After completing the initial phase, a *genyen* is promoted to a *getshul* [dge tshul]. When a

Fig. 4.25 Residential hall of Tawang Monastery: a residential unit consisting of three to six rooms is known as a *shak* [shag] or *drashak* [grwa shag]. The monastery has 75 *shak*. In each *shak*, one teacher monk, *gegan* or *gergan* [dge rgan], and a number of apprentice monks, *genyen* [dge bsnyen] or *gethruk* [dge phrug], live together

getshul turns 21 and has sufficient knowledge of Buddhist scriptures, he is promoted to a *gelong [dge slong*], a fully ordained monk, or *bhikshu* in Sanskrit. A *gelong* is a lama who is qualified to conduct various religious rituals (Dhar 2005; Biswal 2006).

At the time of our field observations, there were no monks aged 23–26 at Tawang Monastery since *gelong* of that age had been sent to external institutions and monasteries for further study and training. Before the border with Tibet was closed, *gelong* were sent to different monasteries, particularly Drepung Monastery in Tibet, for more advanced training (Dhar 2005). They are now mostly sent to other monasteries within India, Nepal, or Bhutan.

Small kitchens were attached to each *shak*, but a communal kitchen was established in 2008. Each member of the monastery had received 13 *brey* of cereal grain each month until 2008. After the communal kitchen was set up, cereal grain became uniformly managed at the monastery's storehouse. This change resulted in an excess of cereal grain, and in August 2011 every monk and apprentice received surplus 13 *brey* of grain (Fig. 4.27).

A typical day in the monastery begins and ends as follows. Monks and apprentices get up at 4 a.m. and carry out devotional exercises in the auditorium called *dhukhang* ['du khang] (Fig. 4.28). This exercise is practiced according to a fixed schedule. It consists of two reciting of sutras (morning and early evening) for the first 50 days and then one worship every morning for the following 50 days.

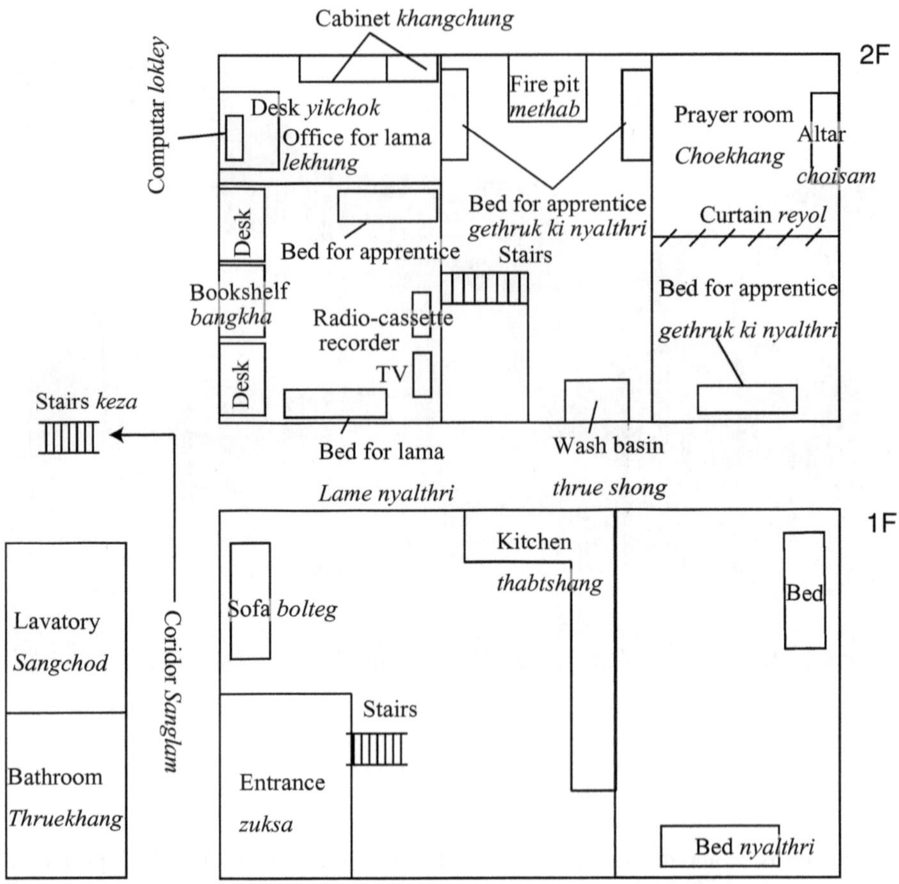

Fig. 4.26 Plan of a dormitory house, *shak* [*shag*], in Tawang Monastery of the Gelug [dge lugs] sect (Produced from field observations)

School in the monastery starts at 8 a.m. Pupils of the third to the most senior (10th) grade practice the catechism in each grade group in the plaza in front of the school (Fig. 4.29). In this procedure, the standing pupils ask a question in Tibetan and clap their hands; the seated pupils then answer. The handclap is said to aid in clarifying the mind. According to Kawaguchi (2012 [1909]: 305), the practice of the catechism begins with the following procedure:

> The catechized sits in a certain posture while the questioner stands up with a rosary in his left hand and walks towards him. He stretches out his hands with the palm of the left hand downwards and that of the right hand upwards and claps them together, uttering the words, *Chi! chi tawa choe chan* [*ji lta ba chos can*]. Here "Chi" means the heart of the Bodhisattva Manjushuri and its utterance is supposed to make the questioner one with Him, whose real body is knowledge. The rest of the utterance literally means "in that nature of the truth." The sense of the whole is: "We shall begin the discussion following the nature of Truth as it is manifested in the Universe." Then the discussion begins in earnest according to the rules of the logic of Nyaya (rule].

Fig. 4.27 Every monk and apprentice receives 13 *brey* [bre] of cereal grains as an extra ration at storage house of Tawang Monastery. The photograph shows apprentice monks transporting the extra rations to their *shak* [shag]

Fig. 4.28 A devotional exercise is being performed in the *dhukhang* ['du khang] auditorium in Tawang Monastery. Musical instruments are used in this regular morning exercise

Fig. 4.29 Pupils of the third to the most senior (10th) grade practice the catechism according to each grade group in the plaza in front of the school of Tawang Monasery. In this debate practice, standing pupils ask a question in Tibetan, clap their hands, and then the sitting pupils answer

What Kawaguchi observed in Tibet 100 years ago remains largely unchanged in Tawang today. The morning assembly is held at 8:45 a.m. (Fig. 4.30), and classes start at 9 a.m. Tuition is undertaken by seven teacher monks and four teachers who are not monks. The former teach Tibetan Buddhism, English, and Hindi, and the latter teach mathematics, science, geography, and social studies (Fig. 4.31).

Sometimes, 10 rupees is given to the juvenile apprentices as pocket money. They line up at the monastery gate, *kakaling* [ka ka gling], to receive it. They take great delight in buying candy with their pocket money at either of the two stores near the gate (Fig. 4.32).

The classes recommence at 1 p.m. after lunch and end at 2:30. If a worship service is scheduled, it is performed from 2:30 to 3:30. All the apprentices return to their *shak*, read sutras, recite them from memory, and finish their homework. Supper, which is eaten at their own *shak*, is provided to them in the communal kitchen in the early evening (Fig. 4.33). They retire to bed before 10 p.m.

Nunneries are frugal than monasteries. There are three nunneries in Tawang. The only nunnery supported by Tawang Monastery is Gyangong Ani Gompa [rgyang gong a ne *dgon pa*], where approximately 50 nuns live. The main income for the other two nunneries—Drama Dungchung Ani Gompa [brag dmar gdong chung a ne *dgon pa*] with 40 nuns and Singsur Ani Gompa [sing zur a ne *dgon pa*]—consists of donations from their local patrons, particularly nearby villagers.

Fig. 4.30 Morning assembly held at the school belonging to Tawang Monastery

Fig. 4.31 Class at the school belonging to Tawang Monastery

Fig. 4.32 Sometimes, juvenile apprentices of Tawang Monasery receive 10 rupees as pocket money. It is a great delight for them to buy candy at either of the two stores close to the gate—*kakaling* [ka ka gling]—of the monastery with this money

Fig. 4.33 Supper is provided to apprentice monks in the communal kitchen of Tawang Monastery in the early evening

Fig. 4.34 *Shak* [shag] and nuns of Singsur Ani Gompa [sing sur a ni *dgon pa*] of the Gelug [dge lugs] sect

Until 1985 at Singsur Ani Gompa of the Gelug sect, the resident nuns had requested donations from local villages once a year. At the time of our field observations in 2011, the nunnery had a small vegetable garden, which produced chilies, radishes, and potatoes for the nunnery's own use. At that time, 47 nuns and apprentice nuns lived in 35 *shak* (Fig. 4.34). They included the following: 15 nuns who had completed their ascetic training and were older than 70 years; 10 teacher nuns who were 40–60 years; 8 nuns who had finished their education and were older than 15 years; and 14 apprentice nuns who were in their early teenage. Since this nunnery does not have its own school, 8 of the 14 apprentice nuns attended a nearby junior high school, and the remaining 6 went to an elementary school. The nuns and former apprentice nuns who have completed their education move to a different *shak* so that they can live with new apprentice nuns and take care of them. The floor plan of a *shak* housing one teacher nun and two apprentice nuns at that nunnery appears in Fig. 4.35. The living conditions at Singsur Ani Gompa are more basic than those at Tawang Monastery and another monastery mentioned below. There are, for example, no appliances such as a television set, radio-cassette recorder, or personal computer in this *shak*. A fire pit and kitchen are located in one of the rooms of the *shak*, as shown in Fig. 4.35, and there is a bed on either side of the fire pit. One of the two beds is for a teacher nun, and the two apprentice nuns together use the other bed (Fig. 4.36).

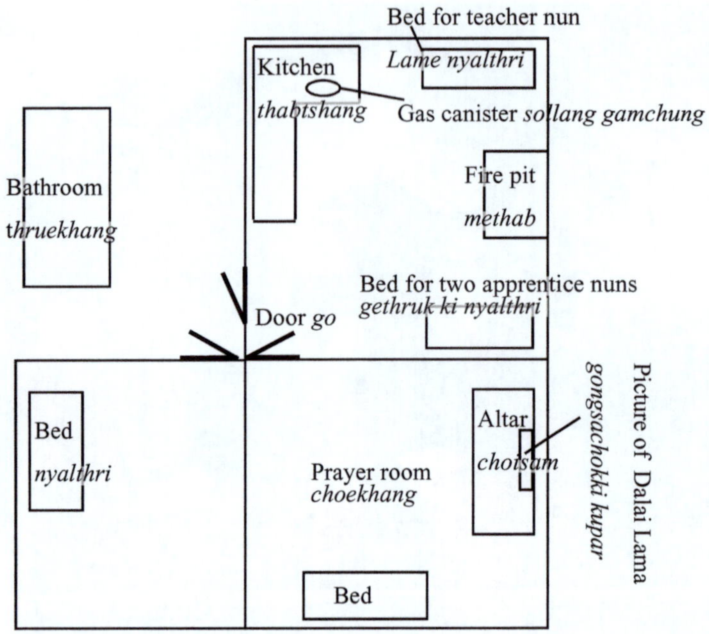

Fig. 4.35 Plan of a dormitory house, *shak* [*shag*], in Singsur Nunnery [sing zur a ne *dgon pa*] of the Gelug [dge lugs] sect (Produced from field observations)

Fig. 4.36 Interior of a *shak* [*shag*] in Singsur Ani *Gompa* [sing sur a ni *dgon pa*] of the Gelug [dge lugs] sect

The next room is a prayer room, where a bed and various items are placed haphazardly. Another bed is situated in the remaining room, which is treated almost like a storage room. Though the teachers at Tawang Monastery speak English, those at Singsur Ani Gompa do not, and they are also not fluent in Hindi.

Their daily routine is as follows. The nuns and apprentice nuns get up at 3 a.m. and read sutras. After washing at 4 a.m., they drink tea and eat breakfast. The nuns assign homework to the apprentices and visit their local patrons' houses to recite sutras. If the nuns have no other work to do, they perform domestic duties and teach the young apprentices. The apprentice nuns go to their elementary or junior high school, finish their homework after school, and receive supervision from the teacher nuns. The nuns and apprentice nuns take supper between 7 and 8 p.m., then read sutras and retire to bed at around 9 p.m.

The formal name of Jangdokpalri (zangs mdog dpal ri) Gompa of the Kagyu sect, which is located on the hill of the village of Jang [byang] (Fig. 2.5) in Tawang District, is Mon Palpung Jangchub Choekhorling (mon dpal spungs byang chub chos 'khor gling). The following descriptions were derived from field observations that we carried out in 2011. Approximately 60 monks and apprentice monks were then resident at the monastery (Fig. 4.37). The school within the monastery had six teachers, of whom two were monks. Thirty monks aged 15–30 years have finished their educational courses. There were also 30 apprentice monks aged 7–14 years.

Fig. 4.37 These apprentice monks, called *genyen* [dge bsnyen] or *geythuk* [dge phrug], belong to Jangdokpalri [zangs mdog dpal ri] Monastery of the Kagyu [bka' brgyud] sect in Jang village

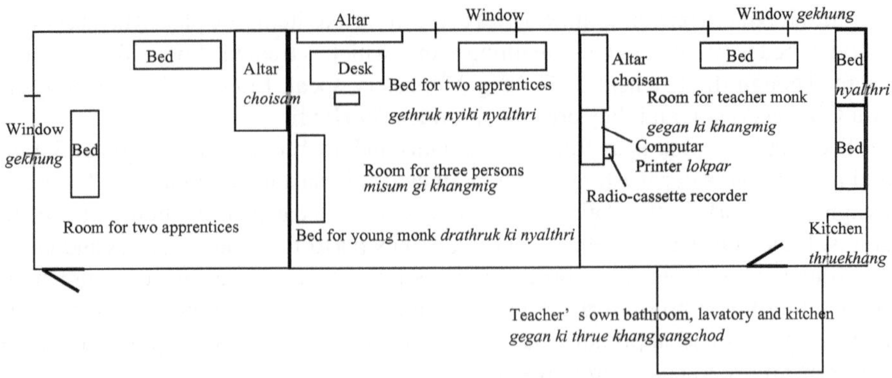

Fig. 4.38 Plan of a dormitory house, *shak* [*shag*], in Jangdokpalri [zang mdog dpal ri] Monastery of the Kagyu [bka' rgyud] sect (Produced from field observations)

The educational classes consisted of Tibetan Buddhism, English, mathematics, Hindi, science, geography, social studies, and art. In the art classes, the apprentice monks practiced sketching exercises. They all lived in 45 *shak*. Though Tawang Monastery and Singsur Ani Gompa have numerous independent *shak* for the monks and apprentice monks, at Mon Palpung Jangchub Choekhorling the one large two-story building is divided into 45 *shak*. All of the apprentice monks share the same toilet in the building. The teacher monks and apprentices live separately. Every teacher monk has his own room, whereas two to three apprentices share one room. A teacher's room has its own bathroom, toilet, and kitchen, and is equipped with a computer, printer, and radio-cassette recorder (Fig. 4.38).

The standard daily schedule at Mon Palpung Jangchub Choekhorling is as follows. The monks get up at 4:30 a.m., start reading sutras at 5 a.m., have breakfast at 7 a.m., and begin transcribing *sutras* at 7:30. There is a tea break at 9 a.m., and the morning meeting starts at 9:30. Morning classes at the school are held from 10 a.m. to noon. They have a lunch break at noon, and afternoon classes begin at 1. A tea break is held at 3–4 p.m.; then afternoon classes commence and continue until 6 in winter and 6:30 in summer. In these late-afternoon classes, elder apprentices practice dictation while a teacher monk talks and younger apprentices read sutras. After the classes, supper is served in the communal kitchen. Before going to bed, the monks study alone from 7 to 8:30 p.m. in winter and 9 p.m. in summer.

4.4 Bon and Animism: Shamanism and Local Societies

Bon is an ethnic religion with a long history in Tibet. It has developed its own highly sophisticated system of principles in relation to various indigenous aspects of the region. The Bon religion and the Nyingma sect of Tibetan Buddhism have influenced each other, which has contributed to the development of both religions. They share an inclination to collect material related to old legends and ceremonies as well as the method of meditation known as *dzogchen* [*rdzogs chen*]. Buddhism and Bon have never been mutually exclusive in their relations to each other. On the contrary, Buddhism assimilates any native beliefs that attract local people (Karmay 1988).

Bon has also absorbed the theories and practices of Buddhism to a great extent. Since the introduction of Buddhism from the seventh century, Bon has expanded its doctrines and practices over hundreds of years (Snellgrove and Richardson 1980). The history of Bon can be divided into three phases. The first phase was the period before the introduction of Buddhism. During that phase, Bon was a religion that involved the expulsion of evil spirits and worship of good spirits. The second phase, which lasted until the first half of the eighth century, was a period in which Bon rituals were consolidated with a central focus on ceremonies for the dead. The influence of Buddhism was still minimal during that phase. The third phase has lasted from the late eighth century to the present. After initial conflicts, Bon metamorphosed into a religion in which enlightenment as a mental facility could be pursued under the influence of Buddhism. Elements of shamanism also became part of the transformation process of Bon after the eleventh century (Tachikawa 2009).

Although the Monpa people began adopting Tibetan Buddhism in the twelfth century, they have retained faith in their traditional religion of Bon (Dhar 2005). In Bon rituals, the procedures and offerings are conducted by a shaman, whom the locals hold in awe (Dhar 2005).

The manner in which Bon remains an active religion in the Monpa area is illustrated in the following example. The village of Thembang consists of dwellings within and around Thembang Dzong. A Bon festival (*bompu* [*bon phul*] in the Dirang Monpa language) called *lhasushi* [*lha zhu gzi*] is to make offerings to the gods of the mountains, and is held once every 6 years in the village of Thembang. In 2011, it took place on February 20–23. On the 2nd day, a procession of people wearing traditional costumes left the western gate of Thembang Dzong (Fig. 4.39), and a ceremony and feast were held in the plaza (Fig. 4.40). The seating assignment had been arranged in advance for each of the four senior clans—*bapu*. People of junior clans—*gilla*—were seated behind the *bapu*, whom they serve. On the 3rd day, a procession of *bapu* people with white flags symbolizing their status left the eastern gate of the fortress. Each clan then conducted a ceremony and held a feast in the plaza (Fig. 4.41). Serving food was a task for the *gilla*, and the *bapu* were not supposed to participate in this. On the 4th day, a procession of *bapu* people with white flags left the eastern gate followed by *gilla* people bearing colorful flags to symbolize their status (Fig. 4.42). A ceremony and feast were then held in the plaza: this was the *lhasushi* for the *bapu*; the *lhasushi* for the *gilla* was performed on another day. The people of Attajepu clan (Table 2.1) did not participate in the *lhalushi* in 2011. This was because they would have had to prepare and serve their own feast since they had no *gilla*, and that would have been too humiliating for them.

Fig. 4.39 The 2nd day of the Bon festival of *lhasushi* [lha zhu gzhi] at Thembang Dzong [them bang *rdzong*]: people wearing the traditional Monpa costume parade from the western gate of Thembang Dzong

In the *lhasushi* ritual, offerings were made to the gods of five mountains (indicated below). We obtained the following story of the origins of the ritual from interviews (Mizuno, August 2011) with villagers conducted as part of this study:

> Two gods called *chachani* [cha cha gnyis] who were brothers came down to the land of Thembang. No human being was living in the area at that time. There were no other gods present either. The elder god asked his younger brother whether he was intending to become a human. The younger brother replied, "You should, my elder brother." After this conversation, however, the elder brother became a new god named Jowhodi [*jo bo di*], and the younger brother became the king of mankind. Many years passed and two brothers, who were descendants of the first king, became kings in turn. They killed their uncle, who said to them with his dying breath, "Since you have killed me, people will kill you." Years later, the two kings collected tax from that area, which is today known as Assam. There were language problems between the kings and people of Assam, and the result was that the kings were killed by the locals. The throne was unoccupied for a long time as the Boot Monpa subordinates searched for a new king.

This story leads to the oral tradition of the village of Thembang, mentioned in Sect. 2.12, although the reason for the empty throne is not the same. In the Thembang oral tradition, the Boot Monpa subordinates eventually find their new king. The god Jowhodi mentioned above is one of the five gods to whom offerings are made in the *lhasushi*. The other gods are as follows: Phulachung [phu lha chung] who oversees the senior clans, *bapu*; Tang [thang] who rules disasters, such as fire and flood and

Fig. 4.40 The 2nd day of a Bon festival in the Thembang [them bang] area: this festival is called *lhasushi* [*lha zhu gzhi*], and it features offerings made to gods of the mountains. On this day, the procession of people wearing traditional costumes started from the western gate of Thembang Dzong [them bang *rdzong*], and a ceremony and feast were held in the plaza. Members of four senior clans—called *bapu* [sba bu]—sat in front of the altar on which the offerings were placed. Members of the junior clans—called *gilla* [gyi la]—sat behind the *bapu*. The photograph shows the people of the Dirkhipa clan of *bapu* and those of the Nimsonga clan of *gilla* as well as a few Assamese

is also known as Salasily [sa lha sri gling]; Dumrihi [sdum ri hi]; and Amajomu [a ma jo mo] who oversees the junior clans. The mountains of Jowhodi, Phulachung, and Dumrihi are visible from Thembang, but those of Tang and Amajomu are not. The mountain of Amajomu is said to be in southeastern Tibet. The mountain of Dumrihi is located behind the Thembang Dzong and the old ruin of a *gompa* on its slopes is undergoing renovation (Fig. 4.41).

In the *lhasushi*, lamas recite sutras and pray to all the gods of the mountains. Shamans pray only to Tang and Salasily. A different kind of domestic animal is offered to each of the gods. A sheep is offered to Jowhodi; a *mithun*, or a *maincha* [ba men] (see Sect. 6.2), is offered to Phulachung; a red cow and a calf are offered to Tang; and a goat is offered to Amajomu. The animals are offered to the gods alive and are released after the ritual.

During our field observations, a shaman of the Boot Monpa conducted a ritual to invite the god of the local mountain and made an offering of a domestic animal, like goat (Fig. 4.43). The shaman who played the leading role in this ritual was called a

Fig. 4.41 The 3rd day of *lhasushi* [*lha zhu gzhi*]: the procession of *bapu* [sba bu] people with white flags symbolizing their status left the eastern gate of the fortress. Each clan then conducted a ceremony and held a feast in the plaza. The mountain in the background is Dumrihi

phrami [phywa mi]. He had an assistant called a *tsangmi* (tshang mi). They were surrounded by male dancers called *bropa* [bro pa] and female dancers called *bromo* [bro mo] (Fig. 4.44). Shamans cannot cut their hair, and they wear a distinctive cap with two extruding horns to hold their long hair. The shaman recited the history of the village using special language and his aide, the *tsangmi*, provided an interpretation for the villagers. After that session, the shaman visited every household. He picked up a piece of burning charcoal from the fireplace of each house, put it into his own basket, and then poured water onto the fire in the hearth. Each household recompensed him with butter and cheese. This part of the ritual is related to the special prayers by shamans for the god Tang who controls disasters such as fires.

The *bapu* have faith in Bon, Tibetan Buddhism, animism, and shamanism. The Boot Monpa have a long tradition of belief in animism and shamanism. The shaman in the *lhasushi* described above was invited from a Boot Monpa village. When a shaman dies, the *bapu* choose a successor by divination. The *bapu* give the shaman a sheep as remuneration for the first as well as the second occasion in which the shaman provides a service, and a yak for the third time and thereafter.

A festival that involves rituals similar to those of *lhasushi* is also practiced in the Dirang Dzong area. It is locally called *chisesoese* [*'chi srid gso srid*], and offerings are made to the gods of the mountains. The *chisesoese* is held for 2 days in August

Fig. 4.42 The 4th day of *lhasushi* [*lha zhu gzhi*]: the procession of *bapu* [sba bu] people with white flags symbolizing their status left the eastern gate of the fortress first, and the procession of *gilla* [gyi la] followed. Ceremonies and feasts were held in the plaza after the processions

and 1 day at the end of December. Events related to *chisesoese* are organized and performed by eight villages in the Dirang Dzong and Yewang areas on a joint basis. Eight *tsorgen* from each village take part as well as 12 *gomi* and 6 *chopon* (see Sect. 3.4) as inter-village staff for communal events and projects. Every year, villagers take turns acting as assistants. They visit households to collect donations for the festival. As was the case with the old taxation systems described in Sect. 3.4, the *tsorgen* instruct the *gomi* and *chopon* in the performance of specific tasks. The *tsorgen* play prominent roles not just in festivals and rituals but also in the construction and maintenance of bridges and roads as well as the management of schools.

For the *chisesoese*, each of the five senior clans commissions one shaman and the junior clans collectively ask for three shamans. These eight shamans are inhabitants of local villages. Being a shaman is a hereditary occupation, and the youngest son of a shaman normally takes over the position.

The shamans invite the gods of three mountains. They pray and make offerings to the gods from 5 to 9 p.m. Traditionally, one girl and one cow are offered to the god of a mountain named Attadungphu [a mtha' gdung phu]. Similarly, one girl and one cow are offered to the god of another mountain named Atabangle [a mtha' bang sles], and one sheep is offered to the goddess of another mountain named Jomula [jo mo la]. The offered girls must be unmarried and no older than 12 years in the ritual

Fig. 4.43 A shaman of the Boot Monpa [bud mon pa], *phrami* [*phywa mi*], wears a distinctive cap with two extruding horns as he conducts a ritual to invite the god of the mountain and make an offering of a domestic animal. The shaman is helped by an assistant called a *tsangmi* [*tshang mi*]. They are surrounded by male dancers called *bropa* [bro pa] and female dancers called *bromo* [bro mo]

offerings. They are released after the ritual. The local people claim that water pouring from a crack on a large rock near the top of Atabangle mountain hits another rock below and that this makes consecutive sounds like a drum roll. They believe that the sounds are representations of natural energy arising from the god. They also believe that there is a waterfall within Attadungphu mountain, which creates a thunderous roar.

The disparity between senior and junior clans can be observed to a certain extent during *chisesoese*. Traditionally, members of senior clans sit on chairs, and those of junior clans are seated on the floor. At one time, marriage between members of the two types of clans was not possible. Such customs are diminishing today (see Sect. 7.4); however, they still exist. Members of junior clans still offer labor to those of senior clans. Although it has decreased substantially, a remnant of the past pecking order becomes particularly evident during religious occasions.

Hosina [lho srid nag] is another Bon festival, and it involves offerings made to evil spirits for a few days in April or May every year at Thembang Dzong. In 2011, it was held from May 1 to 3, and we observed that (Figs. 4.45 and 4.46). The background to this festival is complex. One of the elements is the old taxation system. Another element is the dualistic custom of New Year celebrations in Thembang

Fig. 4.44 Male dancers called *bropa* [bro pa] and female dancers called *bromo* [bro mo] take part in the festival of *lhasushi* [lha zhu gzhi]. The white flags are the symbol of *bapu* [sba bu] senior clans, and the colorful flags represent *gilla* [gyi la] junior clans

village. Thembang celebrates not just the usual New Year, which is called *dawa dangpoi losar* [zla ba dang po'i lo gsar], but also the village's own New Year, which is known as *chunyeipe losar* [bcu nyis pa'i logsar]. The events are held, respectively, in the 1st and 12th month of the year according to the standard Tibetan lunar calendar. These two celebrations are related to *Hosina* as follows. The inhabitants of Thembang are *bapu* people, descendants of the kings who once ruled the area. The father or eldest son of each household of the village used to visit the Mazbat and Udalguri areas in Assam to collect tax before *dawa dangpoi losar*. This type of taxation was known as *posa*, and it is currently paid by the district administrative offices.

Posa has long been connected with *hosina*. Assam used to be governed by the kingdom of Ahom, whose residents are said to have migrated from northern Burma through the Patkai Mountains; they settled in and governed Assam from 1228 to 1826. During the rule of the Ahom dynasty, the relationship between the mountainous tribes of Arunachal Pradesh and the Assamese, who were under the rule of the Ahom kingdom, was not cordial. It has been recorded that the Ahom king, Pratap Singha (1603–1641) (Tenpa 2015), granted *posa* taxation privileges to various tribes of Arunachal Pradesh. This allowed certain tribes to collect taxes from a number of villages around the foothill regions of Assam on the condition that the tribes did not raid and disturb the Assam lowlands (Choudhury 1996).

Fig. 4.45 The Bon festival of *Hosina* [*lho srid nag*] in which offerings are made to evil spirits: it is said that boys from the village of Lish [rlis] were formerly sacrificed in this festival

The expectation of the Ahom kingdom was that the *posa* system would enable the mountainous tribes to settle peacefully and permanently. The privilege of *posa* was also offered to the Aka tribe and then to the kings of the Monpa and Sherdukpen (tribes known as Sathrajas in English sources). Accordingly, the mountainous tribes traveled to the lowlands to collect taxes from villages in the boundary zones. As determined by the mountainous tribes, the taxes consisted of such commodities as bunches of cotton yarn, fabric, clothes, handkerchiefs, and salt as well as domestic animals. However, the amount that the mountainous tribes were able to collect was fixed. The Sathrajas collected taxes from the Kuriapara *duar*[1] and Char *duar* in the districts of Udalguri and Sonitpur in Assam. In the nineteenth century, the British administration modified the system slightly. The British took control of northern Assam from the Ahom King Purandar Singha (1832–1838) in 1838, and they maintained the *posa* system. The payment method was changed from goods to cash. Because money had no value to the mountainous tribes, they initially rejected this

[1] A *duar* is a gateway or passage between Bhutan and India in the foothills of the eastern Himalayas. In Tibetan, it is called *lego* [las sgo]. There were approximately 20 *duar* located between Monyul and Bhutan in the north and Cooch Bihar, West Bengal, and Assam in the south. They are known as the Bengali *duar* and Assamese *duar*. Of these *duar*, Kuraipara *duar* and Char *duar* were among the Assamese *duar*; these *duar* have not served communication purposes since the mid-twentieth century.

Fig. 4.46 The Bon festival called *hosina* [lho srid nag] is held every year at Thembang Dzong. It was originally a festival to neutralize evil spirits sent from Assam. According to local tradition, a boy from the village of Lish used to be sacrificed in connection with *hosina*. In this sacrifice, three men wearing the traditional Miji animist costume and four men dressed as soldiers danced around a boy tied down to the ground in the small plaza in front of the community hall and killed him with their swords. Nowadays, a human figure is used to represent this. In this photograph, the human figure has already been removed and a *torma* [gtor ma] (a religious offerings made of dough) placed there instead

plan. However, they gradually accepted it, and the system was modernized in this way. Thus, the cleverly designed and maintained *posa* system introduced by Pratap Singha to deal with the northern mountainous tribes and passed on to the British was practiced until the 1940s (Choudhury 1996). The triangular relationship within this feudal system, which included the Ahom kingdom in Assam, the kings of Arunachal Pradesh, and hard-working peasants, was inherited by the British (Thakur 2006). It was also established by the Satrajas of Monpa in Tawang and the Sherdukpen in West Kameng, respectively, in 1844 and 1853 (Tenpa 2015).

The annual report for 1914–1915 prepared by Captain Nevill, Assam Government Agent, indicates that the following amounts of cash were paid to the Sathrajas through the *posa* system: 5,000 rupees to Tawang Monpa; 2,526 rupees to Rupa and Shergaon; 145 rupees to Tembang; 536 rupees to Aka Kovat Sun (Kopachors); 164 rupees to Akas Kutsun (Hazarikhowas); 3,631 rupees (1,020 rupees in Thakur 2006) to Daflas; and 1,124 rupees to Miri-Abors. Of the 5,000 rupees paid to Tawang Monpa, 500 rupees was taken by Tawang Monastery, and the remainder was forwarded to Lhasa. The amounts to Daflas and Miri were reduced to 2,440 and 935 rupees, respectively, between 1914 and 1923 (Reid 1997 [1942]). Thakur (2006) indicates that the amount paid to Daflas and Miri was 1,020 rupees during this period. *Posa* was also paid to the Bhutanese since they had extraterritorial rights over certain *duar* in Bengal and Assam until British-ruled India regulated those *duar* in 1865. Tenpa (2015) provides more details on this subject.

The local New Year festival, *chunyeipe losar*, was established to enable the masters of houses in Thembang to celebrate the New Year with their families. They returned to their village after the standard New Year, *dawa dangpoi losar*; the festival is believed to be associated with evil spirits that originated with the malice of the Assam people, who had suffered under the burden of paying *posa*. The *hosina* festival was originally established to neutralize these evil spirits by making offerings.

Traditionally with the *hosina*, a human body from which the internal organs had been removed was necessary as an offering to the evil spirits. An instruction paper—*sangchen gudrai rimo* [*gsang chen dgu dgra'i ri mo*]—specifying the requirements for the rituals in *hosina* depicts ornaments made from a human skull as well as the arm, breast, and hip bones (Fig. 4.47). The skins of elephants and tigers were also included among the offerings. A human body without organs is depicted in the

Fig. 4.47 Instruction paper specifying the requirements for the *Hosina* ritual [*lho srid nag*]—*Sangchen Gudrai Rimo* [*gsang chen dgu dgra'i ri mo*]—displays ornaments made of a human skull as well as bones of the arm, breast, and hip. Skins of elephants and tigers are also depicted as being included in the ritual. A human body without internal organs is depicted in the bottom-right corner of the *Sangchen Gudrai Rimo*

bottom-right corner of the *sangchen gudrai rimo*. It is said that in historical times, a human subject was killed and his body was used for the ritual. According to local tradition, a boy from the village of Lish was once sacrificed in Thembang for this purpose. However, we were unable to confirm this during our field observations. The procedures of the sacrifice revealed from an interview (Mizuno, May 2011) were described to us as follows:

> Three men wearing traditional Miji animist costumes start running, and two men in soldier costumes carrying large swords chase after them. At the small plaza in front of the community hall named *Mangbrang* [*mang grangs*], the three Miji animists disappear and two more men in soldier costumes join in. The four soldiers then dance around a boy tied down to the ground and finally kill him with their swords (Fig. 4.46).

According to the traditional account, a boy used to be sent for the ritual every year from Lish as a tax. At some point, the human sacrifice was replaced with a sheep from which the internal organs were removed. The sheep was normally killed and processed the day before the ritual, and *torma* were placed on its skin inside the community hall. In the small plaza in front of the hall, the blood of the sheep was injected into a human-shaped dough (*bokpai* [gro phye]) figure. In 2006, Tsona Gonpatse Rinpoche [mtsho sna dgon pa rtse rin po che], a respected Buddhist leader, prohibited the killing of a sheep for this yearly ritual, and a human-like figure made of *bokpai* has been used instead. We observed the following taking place. Red paint, symbolizing blood, and a raw egg were inserted into the head and chest of a human-shaped *bokpai*. First, the groin was cut with a knife, and the *bokpai* was then slashed from its head. The ritual was performed in the small plaza in front of the community hall, where a flagstaff called a *phanshing* [*'phan shing*] stood. At one time, delinquents were tied to the pole and whipped by two villagers 108 times.

On the day of *hosina*, part-time monks visit every household in Thembang and scatter corn and buckwheat to expel evil spirits. Their emptied *brey* container is thrown outside the house at the end of these visits. There is a local belief that if after being thrown, the *brey* stands with its base on the ground, that is a good omen; if it stands with its base uppermost, it is an ill omen.

The Boot Monpa people live in an area adjacent to that of the Dirang Monpa, and the Boot Monpa believe in shamanism and animism. Barua (1995: 221) outlines the nature of their belief as follows:

> The people believe in a number of spirits, the most important of which is *Jo-odi* [*jo o'di*], the hill spirit. They propitiate it once a year by sacrificing sheep and yaks. The sacred specialist is known as *Chaajee romu* [*phyag mdzad rol mo*], and he performs various rituals. He is also regarded as the witch doctor, and is engaged in curing diseases and protecting villagers from evil spirits. The liver of a fowl is examined to predict the future. A perfect liver is considered a good omen, white spots in it indicate otherwise. Cocks or hens are skinned and examined for prognostication.
>
> The Boot Monpas observe a number of festivals and ceremonies with socio-religious significance. The *Seksaphu* [*sreg sa phud*] festival is held sometime in January–February, in which the priest offers fish to the hill spirit and other spirits on behalf of the villagers. The *Chudung aphu* [*chu gdung sa phud*] festival is held in the months of October–November, and is associated with the harvesting of the crop. Yaks are sacrificed in the name of the hill spirit for the sake of a good harvest. The ceremony of *Samafen Phu* [*sa ma'phen phud*] is performed in the months of March–April to propitiate the various spirits for the germination of seeds.

Thus, the Tibetan Buddhism, Bon, animism and shamanism are inter-woven and fused, which represents a dynamic and intricate tapestry of religious faith, society, and culture in the Monyul (Monpa area).

Column 1

Development of Arunachal Pradesh and Effects of the Chinese Military Action in 1962

Arunachal Pradesh acquired an identity of its own for the first time as part of the North-East Frontier Tract organized by the British administration in 1914. It was renamed the North-East Frontier Agency (NEFA) by the government of India in 1946. It became the Union Territory of Arunachal Pradesh in 1972. It was ruled directly under the Ministry of Home Affairs of the central government. The central government restructured the political status of the region, and it became the 24th fully fledged state in the Union of India in 1987. However, the area was considered largely valueless and left derelict until 1990s. Historically, in 1914, during the negotiations to manage the conflict between Tibet and China, Sir Henry McMahon of the British administration persuaded Tibet to accept the border between Tibet and the Assam region of India, which was included in the Simla Accord (a treaty concerning the status of Tibet negotiated by representatives of the China, Tibet and the British India in Simla in 1913 and 1914). The McMahon Line (a line agreed to by Britain and Tibet as part of the Simla Accord, a treaty signed in 1914, and the effective boundary between China and India) then came into existence. It was drawn on the northern side of the Himalayan ridgeline in an area where the Tibetan population was distributed, but China did not recognize this boundary (Mehra 1971, 1972, 1974; Lamb 1966). The boundary dispute between China and India became a long-standing issue between the two countries. In 1962, the Chinese military launched a sudden attack in the border conflicts between China and India, and India suffered a crushing defeat. The Indian government then expanded paved roads to remote parts of Arunachal Pradesh and garrisoned military units to further enhance its effective control of the area.

Most major roads in this region are under control of the military, but they are open to inhabitants for daily use (Fig. 4.48). When constructing such roads in residential areas, the military relocated inhabitants to new settlements and paid them substantial indemnities via the state government. In nonresidential areas, the military simply grabbed up the necessary land. Stone material for road construction was bought up from clans and villages by the military. The military maintains and controls the roads and blocks them when strategically necessary. For the construction of less important state roads, the Arunachal Pradesh state government did not purchase the necessary land from clans and villages but hired locals to undertake the labor. Though existing settlements were relocated to permit the building of military roads, the state roads were designed and carefully constructed so that no relocation was required.

Fig. 4.48 For strategic reasons, most major roads in the state of Arunachal Pradesh are maintained by the Border Roads Organisation (BRO) of the central government of India

Major roads and strategically important roads in the Monpa area are built and maintained by the Border Roads Organisation (BRO), which is responsible for road development in India's border areas. The BRO cooperates with the General Reserve Engineer Force (GREF) and the Indian Army's corps of engineers. The Public Work Department (PWD) of the central government of India and the state government of Arunachal Pradesh are responsible for the construction of roads within towns or a district. Besides the above organizations like BRO, GREF and PWD, Pradhan Mantri Gram Sadak Yojana (PMGSY) is also involved in the constructions and maintenances of the road. PMGSY is a public project that is being undertaken by the Ministry of Rural Development of the Indian central government to connect villages in Arunachal Pradesh. The BRO is actively constructing roads in the Monpa area because this is a very important border region for India. The purpose of these roads is to promptly transport self-propelled artillery and tanks to every part of the national border in response to a possible Chinese military action.

A highway is currently being constructed from Tawang to Pasighat in Arunachal Pradesh, and it is necessary for both socioeconomic and strategic reasons. Whereas current roads pass through a wildlife sanctuary in Arunachal Pradesh, the new roads avoid such areas.

Women by the roadside crushing rocks with hammers to make pebbles are a common sight in Arunachal Pradesh (Fig. 4.49). The pebbles are used as construction

Fig. 4.49 Women crush rocks with a hammer to make pebbles by the roadside. The pebbles are used as construction material for roads and bridges

material for roads and bridges. A 50-kg bag of pebbles can be sold for about 30–200 rupees. Since the Indian government actively undertakes development in this state, the pebbles are in high demand, and the production of pebbles is a popular source of employment for local women. It is also notable that many migrating Nepalese workers have arrived in the state for road construction work.

Column 2

Land Ownership and Living Expenses

Most of the land in the Monpa region belongs to clans, especially in the Dirang area, or it is owned by the villages in the Tawang and Kalaktang areas. The urban areas in Itanagar, the state capital, and Bomdila, the district capital, are owned by the state, and residents are required to pay rent to the government. By contrast, urban areas in the Dirang and Tawang areas are owned privately.

Local people pay the government for electricity and water. For example, in the village of Lhou (Fig. 2.5) in the Tawang area, the electricity bill for an ordinary household is fixed at 150 rupees per month. A few households share a faucet connecting them to the public water supply system and collectively pay 50 rupees per month for the service. Households that use local water but not the public system do not have to pay a water bill. In towns, the electricity bill is determined by meters, which indicate actual consumption. The normal range of a monthly electricity bill is 190–400 rupees. In the village of Sanglem (Fig. 2.5) in the Kalaktang area, electricity is charged according to the number of sockets. This suits the way in which electricity is used in this remote village where electricity appliances are not abundant. The electricity bill is set to two levels: above poverty line (APL) and below poverty line (BPL; see Sect. 7.4). The monthly bill for one light bulb socket at the APL level is 80 rupees, whereas at the BPL level it is 45 rupees.

In a town, a license has to be paid to operate a business, and it must be renewed every year; however, the charge is small. Tax is mainly collected in the form of a consumption tax. The rate of consumption tax varies depending on the purchased goods. It is kept low for vital goods and daily commodities.

Wages paid to workers involved in development projects and salaries for military personnel are higher than in other states of India. Military personnel based in Arunachal Pradesh are paid by the state government in Assam, but they also receive a salary supplement from the central Indian government.

The state government of Arunachal Pradesh only owns some parcels of urban land. In the Dirang area, land belongs to the clans with a few privately owned exceptions. This means that the state government needs to acquire the rights to use land from the clans when constructing schools or public facilities.

Any matter related to a clan is determined by the clan's committee. Its conference members consist of *tsanpa* (btsan pa): *btsan* means a "clan," and *pa* is "person." The membership of *tsanpa* ranges from elders to educated young people. In the case of the Sharchokpa clan, the number of members is about 150, of whom 70–80 are *tsanpa*. No decision is made unless all the members present agree and put their signature to the decision. Once the document relating to the decision has been signed, no complaint can be made. This is the standard decision-making process of clans in the state.

The Indian central government proposed the building a hydroelectric power plant in the Dirang area in the 2007–2008. During our September 2010 field observations, we observed how some of the negotiations went through between the concerned parties. The government negotiated with clans in the two villages. We had the chance to witness the ongoing talks between *tsanpa* from the four local clans and officials in a governmental office. It came to light that the funding for the plan has been provided by the World Bank, a semi-private agency has been placed in charge of its management, and private companies were to have actual construction work. The power station was to be built in the Boot Monpa area, and the construction of a tunnel between the villages of Thembang and Namshu was planned to transport water for power generation. However, due to stiff opposition from the Chinese government, the World Bank has declined to fund and now it is expected that Indian government itself will meet the plan in the region.

The borders of pieces of land are normally streams and valleys. Such borders are called *tsampa* or *satsam* [mtshams pa or sa mtshams] in the Dirang area. When natural features are not available, a stone pillar—*tsampa lung* [mtshams pa lung] or *satsam doring* [sa mtshams rdo ring] in Tibetan—is erected and used as a landmark. Borders between the plots of land owned by clans are normally marked by these stone pillars in the Monpa area.

References

Aris M (1980) Notes on the history of the Mon-yul corridor. In: M Aris, Aung San Suu Kyi (eds) Tibetan studies in honor of Hugh Richardson, Vikas Publishing House, New Delhi, pp 9–20

Aris M (1988) Hidden treasures and secret lives: a study of Pemalingpa (1450–1521) and the sixth Dalai Lama (1683–1706). Indian Institute of Advanced Study, Motial Banarsidass, Delhi

Barua S (1995) Monpa: But, Dirang, Lish, Tawang. In: Dutta P, Ahmad SI (eds) People of India, Arunachal Pradesh. Seagull Books, Calcutta, pp 218–222, 228–231, 237–247

Biswal A (2006) Mystic Monpas of Tawang Himalaya. Indus Publishing Company, New Delhi

Choudhury SD (ed) (1996) Gazetteer of India, Arunachal Pradesh: East Kameng, West Kameng and Twang District. Government of Arunachal Pradesh, Shillong

Dhar B (2005) Arunachal Pradesh – the Monpas of Tawang in transition. Geophi Publishing House, Guwahati

Dorjee R (2006) The Monpas: their social, economic and religious institutions. In: Dutta S, Tripathy B (eds) Buddhism in North-east India. Indus Publishing Company, New Delhi, pp 194–203

Karmay S (1988) The arrow and the spindle: studies in history, myths, rituals, and beliefs in Tibet. Mandala Book Point, Kathmandu

Kawaguchi E (2012 [1909]) Three years in Tibet. General Books LLC™, Memphis

Kawasaki N (1993) (translated to Japanese) Tibetan book of the dead (Bardo Thodol) written by Padmasambhava. Chikuma Gakugei Bunko, Tokyo (in Japanese)

Kelsang T, Masaki A (2000a) Death practice in Tibet. Kadokawasyoten, Tokyo (in Japanese)

Kelsang T, Masaki A (2000b) Tibetan Buddhist tantrism. Chikuma Gakugeibunko, Tokyo (in Japanese)

Lamb A (1966) The McMahon Line: a study in the relations between India, China and Tibet, 1904 to 1914, 2 vols. Routledge & Kegan Paul, London

Mehra P (1971) The North-Eastern Frontier: a documentary study of the internecine rivalry between India, Tibet and China, 1914–54, 2 vols. Oxford University Press, London

Mehra P (1972) A forgotten chapter in the history of the Northeast Frontier: 1914–36. J Asian Stud 31(2):299–308

Mehra P (1974) The McMahon Line and after: a study of the triangular contest on India's North-Eastern Frontier between Britain, China and Tibet, 1904–1947. Macmillan Publishers, London

Nagahashi K (1999) Exploring Tibetan shaman. Kawade Shobo Shinsy, Tokyo (in Japanese)

Norbu T (2008) The Monpa of Tawang. Government of Arunachal Pradesh, Itanagar

Reid R (1997 [1942]) History of the frontier areas bordering on Assam from 1883–1941. Spectrum Publications, Guwahati

Rgyal sras sprul sku (2009 [1991]) Rta wang dgon pa'i lo rgyus mon yul gsal ba'i me long: the clear mirror of Monyul (Arunachal Pradesh); a history of Tawang Monastery. A myes rma chen bod kyi rig gzhung zhib'jug khang nas spar skrun zhus pa'o. Amnye Machen Institute, Dharamshala

Snellgrove D, Richardson H (1980) A cultural history of Tibet. Prajna Press, Boulder

Tachikawa M (2009) Bon religion and Mandala. In: National Museum of Ethnology (ed) Tibet: gods of Bon religion. Senri Bunka Zaidan, Osaka, pp 28–35 (in Japanese)

Tenpa L (2013) The life and activities of Rgyal sras Blo bzang bstan pa'i sgron me, a 16th century Tibeto-Mon monk from Tawang. Tibet J XXXVIII(3&4):55–72

Tenpa L (2015) Notes on the (Tibeto-) Monpa and Sherdukpen treaties with British India in the 19th century. Bulletin of Tibetology. Paper presented at the IATS UlaanBaatar, Mongolia, 21–27 July 2013 (in press)

Tenpa L, Tempa T (2013) A brief history of the establishment of Buddhism in Monyul: Tawang & West Kameng Districts, Arunachal Pradesh, India. Department of Karmic & Adhyamik, Government of Arunachal Pradesh Press, Itanagar

Thakur AK (2006) Polity of pre-colonial Arunachal Pradesh: interactions in trade, culture, state and Buddhism. In: Dutta S, Tripathy B (eds) Buddhism in North-east India. Indus Publishing Company, New Delhi, pp 241–263

Wallace BA (1980) The life and teaching of Geshé Rabten. George Allen & Unwin, London

Yamaguchi Z (1987) Tibet: 1. Toyo Sensyo, Tokyo (in Japanese)

Chapter 5
Forests in Monyul: Distribution and Management

A forest of *Quercas griffithii*, which is locally known as *soeba shing* [*gso pa shing*]—“forest for picking up fallen leaves”—in the Dirang [rdi rang] region. Here, farmers collect fallen leaves

K. Mizuno, L. Tenpa, *Himalayan Nature and Tibetan Buddhist Culture in Arunachal Pradesh, India*, International Perspectives in Geography 6,
DOI 10.1007/978-4-431-55492-9_5

Abstract Forests are important resources for the local people in mountainous Arunachal Pradesh. Inhabitants of the Dirang area in West Kameng categorize forests into three types: *soeba shing* [gso pa shing], those used for collecting fallen leaves; *borong* [bo rong], those used for gathering fuel wood; and *moon* [mun], those used for gathering timber and hunting. *Shing* means "tree". Fallen oak leaves are collected from *soeba shing* around the villages and spread over agricultural fields as fertilizer for barley and buckwheat, which are produced as subsidiary crops of maize, or as a mulch to inhibit weeds or prevent soil erosion during the rainy season. Each type of forest is managed according to its uses. However, this forest culture has undergone changes since the 1980s. Illegal logging has become common, and the price of timber has continued to rise. Local residents in these areas have only recently begun to practice forest conservation. The focus of this chapter is on the distribution of forests, their uses, and their management by the local inhabitants and the related changes that have occurred in the Monpa area.

Keywords Forest distribution • Fallen leaves • Oak trees • Timber use • Forest conservation

5.1 Distribution of Forests and Their Management

Large parts of the forested area in Arunachal Pradesh belong to clans, villages, and individuals. National parks comprise 4.79 % (2,468.24 km^2) of the forested area, and 13.70 % (7,059.75 km^2) of the forested area is located within wildlife sanctuaries. Protected and reserved forests constitute, respectively, 2.57 % (1,323.92 km^2) and 18.86 % (9,722.69 km^2) of the total forested area. The remaining forested area—60.08 % (309,653.39 km^2)—is privately or communally owned by clans, villages, and individuals. Within West Kameng and Tawang districts, the situation is more extreme: none of the forested areas is located within a national park, and only 7 % (317 km^2) is found within wildlife sanctuaries; none of the forests are protected, and reserved forest accounts for only 15 % (708 km^2) of the forested area. The area that is privately or communally owned by clans and individuals comprises 78 % (3,357 km^2) of the entire forested area (Dutta 2011). The protected forests are governmental plantations, and logging is possible in reserved forests if appropriate licenses can be obtained. Such licenses are not issued for national parks or wildlife sanctuaries.

A transition in forest type occurs at an altitude of approximately 2,400 m in the Dirang area in West Kameng. Below this line, evergreen broadleaf trees (e.g. *Quercus lanata*, *Quercus lamellosa*, and *Schima wallichii*) and deciduous broadleaf trees (e.g. *Quercus griffithii*, *Alnus nepalensis*, and *Juglans regia* var. *kamaonia*) are dominant. Above the line, these deciduous broadleaf trees are scarce. Instead, evergreen broadleaf trees, such as *Lithocarpus pachyphylla*, *Illicium griffithii*, and *Michelia oblonga*, as well as evergreen needleleaf trees, including *Juniperus indica*, prevail. In even higher zones above 2,800 m, evergreen needleleaf trees, such as *Abies spectabilis* and *Tsuga dumosa*, are more obviously dominant. The forest limit

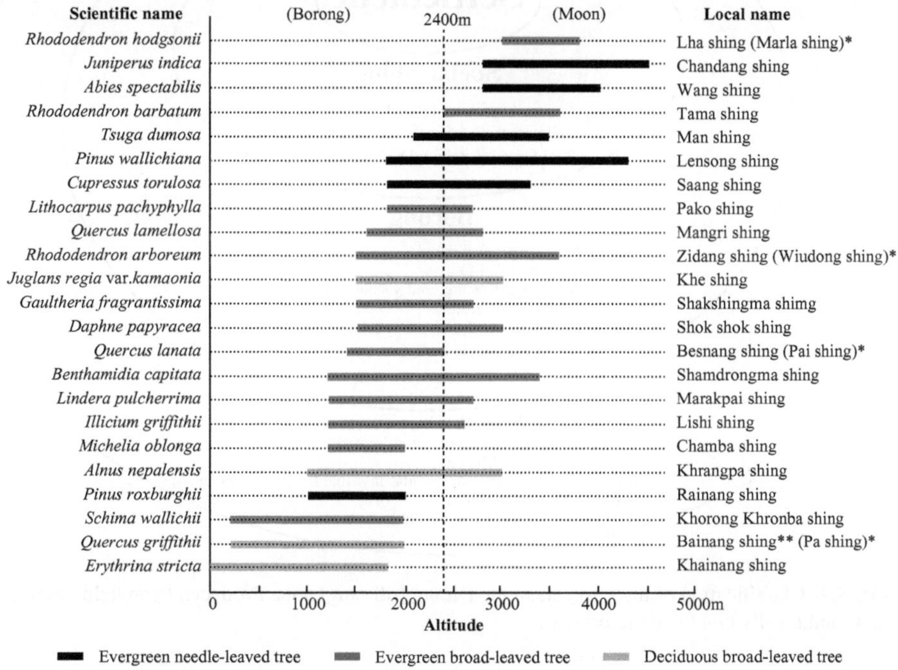

Fig. 5.1 Altitudinal distributions of tree species in the Dirang [rdi rang] and Tawang [rta wang/dbang] areas. Produced from field observations. The local names were obtained from interviews conducted in the area, and the scientific names are from Polunin and Stainton (1984) and other sources. *The local names in the Dirang area are identical to those in the Tawang area with four exceptions; their names in the Tawang area are indicated in parentheses. ***Bainang shing* [be nags shing] or *pa shing* [pa shing] is a generic name for several species of deciduous broadleaf trees belonging to the genus *Quercus*

is marked by *Abies spectabilis* fir trees, referred to locally as *wang shing* [wang shing]. These trees can be found in Sela Pass at an altitude of 3,900 m on the Tawang side (north side) and 4,050 m on the Dirang side (south side). Above this limit, the vegetation consists of grassland containing alpine plants.

We obtained the local names of trees through interviews with villagers as part of this study. The most noteworthy finding was that only four tree names differed between the Dirang and Tawang areas (Fig. 5.1). Given the marked linguistic differences between the two areas, this commonality in tree names is remarkable. It is to be noted that a tree is called *shing* in the Dirang Monpa language, while *sheng* in the Tawang Monpa language, and the written spelling is *shing* in Tibetan.

The altitude-dependent distribution of forests appears to influence the local people's access to and use of the forests (Fig. 5.2). The use of *soeba shing* [gso pa shing] in Dirang, the local name for forests used for collecting fallen leaves, is a good example of this relationship (Fig. 5.3). In the Monpa area, oak trees such as *Quercus griffithii* are known as *bainang shing*. Fallen oak leaves are collected from *soeba shing* around the villages and spread over agricultural fields as fertilizer for barley and buckwheat, which are produced as subsidiary crops of maize, or as a

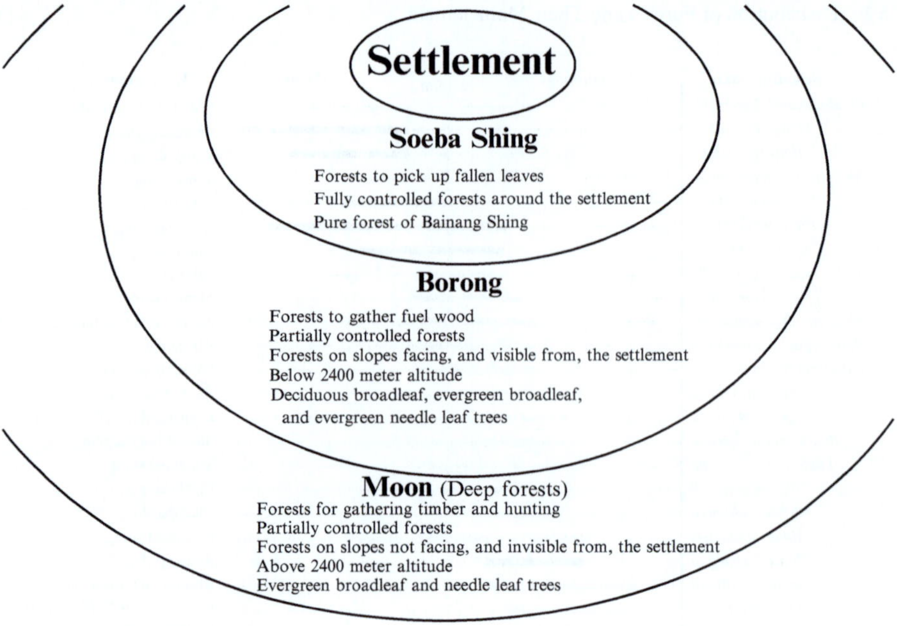

Fig. 5.2 Classification of three forests in the Dirang [rdi rang] area. Produced from field observations and locally conducted interviews

Fig. 5.3 A forested area called *soeba shing* [gso pa shing], used to collect fallen oak leaves, is often located close to a village's agricultural fields. The *soeba shing* is surrounded by *borong* [bo rong shing], which is a forest from which fuel wood can be gathered

Fig. 5.4 Villagers spread fallen oak leaves over agricultural fields of Thembang [them bang] village

mulch to inhibit weeds or prevent soil erosion during the rainy season (Fig. 5.4). This area receives high levels of precipitation, and the maize plants may topple over through soil erosion if the ground is not covered with a mulch of fallen leaves. A layer of fallen leaves, 3–5 cm thick, is strewn on the ground when the height of the maize reaches 30 cm, which is normally about 2 months after seeding (Fig. 5.5). Cow owners often mix fallen leaves with cow manure: fallen leaves, cow manure, and water are mixed together and kneaded by stomping. The mixture is left for 7–20 days before being spread on the ground as fertilizer.

Soeba shing are managed to produce pure forests of oak trees by weeding out the needleleaf trees, specifically *Pinus wallichiana*, locally referred to as *lensong shing* [len song shing]. As a result, land cover and land use in the Dirang area display particular patterns. Whereas forests around maize fields are dominated by oak trees, those around rice paddies are populated by *Pinus wallichiana.* This is because rice does not require fallen leaves as ground cover (Mizuno 2010).

Fallen oak leaves are gathered every year between January and February. For 2 weeks during this period, from 7 a.m. to 4 p.m., family members—often assisted by hired labor—gradually roll the fallen leaves downhill from the *soeba shing* (Fig. 5.6). A special rake-like tool made of pine—a *brak shing* [brag shing]—is used in this operation. Young male family members often use simple long sticks and apply force to move clumps of fallen leaves to the foot of a hill.

Fig. 5.5 Fallen oak leaves are spread over agricultural fields as fertilizer for barley and buckwheat, to inhibit weed growth and prevent soil erosion in the rainy season

Fig. 5.6 Fallen oak leaves are gathered between January and February every year. Family members gradually roll fallen leaves downhill from a *soeba shing* [gso pa shing] for 2 weeks during this period

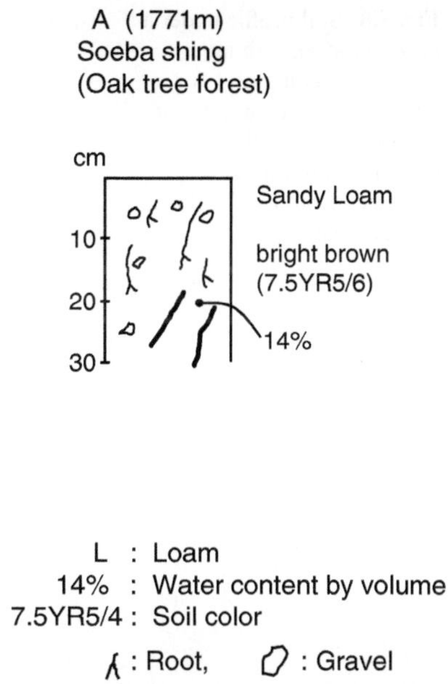

Fig. 5.7 Soil profile under a *soeba shing* [gso pa shing] forest, consisting of oak trees at an altitude of 1,771 m (location A in Fig. 2.5). Produced from local field observations

Fallen leaves are gathered only from an individual's own land. Fallen leaves from a *soeba shing* owned by another person may be acquired in exchange for cash or a distilled spirit called *arak*. The ground of a *soeba shing* is inevitably infertile since the litter on the ground is removed, and humus is absent. Thus, the surface of the soil is light not black in color (Fig. 5.7).

Most noteworthy in the management of *soeba shing* is that nobody, not even the owner, is allowed to cut down live oak trees. Only dead trees, fallen trees, and branches can be used. In the event of unlawful logging, an offender is reported to the village leader, the *tsorgen*, and made to pay the fine of a cow. Most agricultural households own a section of *soeba shing*.

Soeba shing are surrounded by another type of forested area called *borong* [bo rong], which is further encircled by another forest type called *moon* [mun] (Fig. 5.2). *Borong* is the local name for forests whose lumber is used for fuel wood. Such forests are usually located on slopes close to and facing villages. The altitude of the *borong* is normally below around 2,400 m. This type of forest consists of deciduous broadleaf trees, evergreen broadleaf trees, and evergreen needleleaf trees. Its soil is covered with some litters and has more humus than the soil of *soeba shing* (Fig. 5.8).

Moon is the local name for dense forests (Fig. 5.9). These forests are located on slopes that do not face villages. *Moon* are found at altitudes above 2,400 m and are at some distance from settlements. Such forests are used for gathering timber for construction and for hunting (Fig. 5.2). There is less anthropogenic impact on *moon* forests than on the other two forest types. *Moon* are largely primary forests; their ground surface is covered with litter, and the surface soil has a black color because of the well-developed humus. The soil of *moon* forests is argillaceous and the moisture content is higher than in *soeba shing* or *borong* (Figs. 5.10 and 5.11).

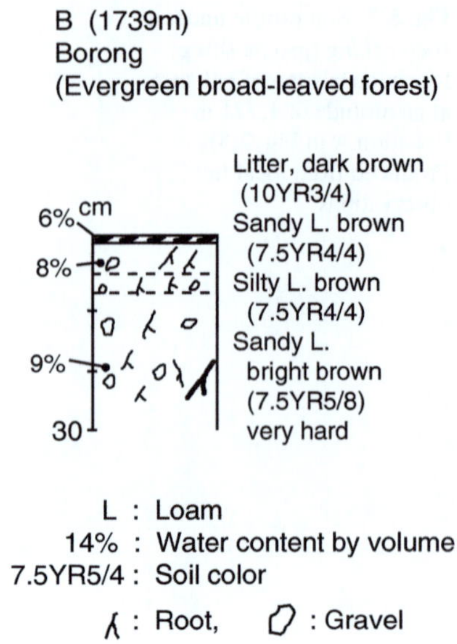

Fig. 5.8 Soil profile under a *borong* [bo rong shing] forest, consisting of evergreen broadleaf trees at an altitude of 1,739 m (location B in Fig. 2.5). Produced from local field observations

Fig. 5.9 A dense forest locally called *moon*. These forests are located at high altitudes and consist mostly of evergreen needleleaf trees

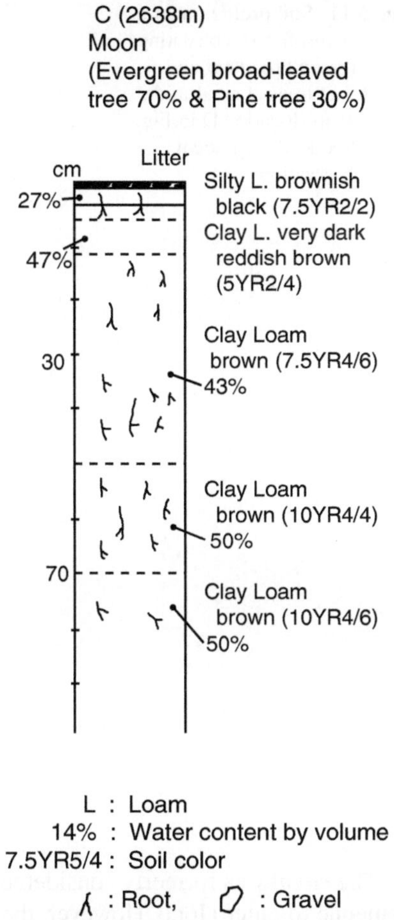

Fig. 5.10 Soil profile under a *moon* [mun shing] forest, consisting of evergreen broadleaf trees and pines at an altitude of 2,678 m (location C in Fig. 2.5). Produced from local field observations

The uses of *borong* and *moon* are, not surprisingly, related to who owns them. Some *borong* forests are communally owned by clans, whereas others are privately owned by individuals. *Moon* forests belong entirely to clans as a common property. In privately owned *borong* forests, their owners have the right to undertake logging operations and gather fuel wood. Individual, regardless of their clan, can use privately owned forests with obtaining the appropriate permission from their owners. Similarly, clan members have the right to acquire necessary resources from *borong* or *moon* forests that are owned by their clan. Permission is necessary from the clan head if a large quantity of timber is to be collected. Even members of other clans can acquire fuel wood if appropriate permission is received. If permission is granted, distilled spirits and a white scarf are presented to the forest owner as a sign of respect. If trees are cut down in a *borong* or *moon* forest without permission, the head of the clan that owns the forest can impose a fine of 5,000–15,000 rupees on the violator. The money received as a penalty is kept in a communal fund and used for the benefit of all members of the clan. For example, it may be used to cover the expense of rituals or as a solatium for the death of a clan member.

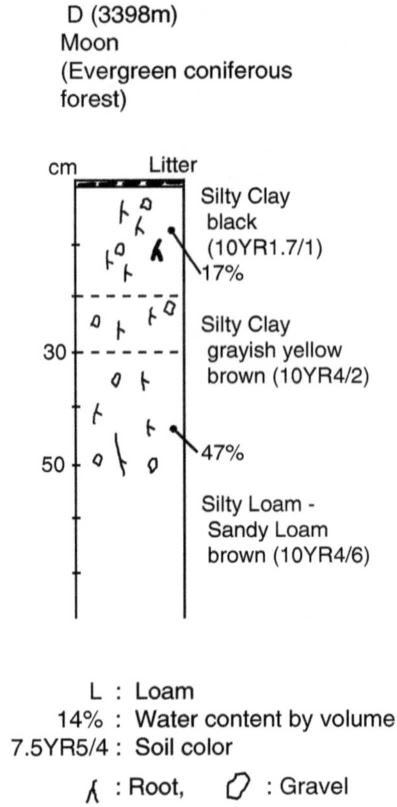

Fig. 5.11 Soil profile under a *moon* [*mun*] forest, consisting of evergreen broadleaf and needleleaf trees at an altitude of 3,398 m (location D in Fig. 2.5). Produced from local field observations

The *moon* was formerly considered deep, dark forests that were too daunting for someone to enter alone. However, the number of people using the resources available in *moon* forests has increased as a result of population growth, the expansion of road networks, and the prevalence of automobiles. Because of these changes, *moon* are now in a transition from primary forests consisting mainly of shade-tolerant trees to *borong* of secondary forests dominated by blue pines, *Pinus wallichiana*. As a result, *moon* are shrinking and *borong* are expanding; hence, *moon* forests become more distant from villages every year. Although *borong* are used to collect fuel wood, pines are not suitable for this purpose because they emit smoke when burned. Consequently, *lensong shing*, blue pines, remain untouched and are becoming abundant in many *borong* forests.

5.2 Use of Timber and Forest Changes

There is a clear relationship between the use of timber by the local inhabitants and changes in the forests of Arunachal Pradesh. The greater the utility of a particular type of timber, the more quickly it is exploited. A good example is *chamba shing* [sam pa

Fig. 5.12 Very hard timber can be obtained from the tree called *chamba shing* [sam pa shing] (*Michelia oblonga*). It is used to make furniture. Its price is currently soaring, and illegal logging of the tree is common

shing] (*Michelia oblonga*) (Fig. 5.12). Its wood is hard and suitable for carving, and it has gained popularity as a material for furniture making and building. The market price of one cubic foot of *chamba shing* (dimensions: 12 ft×12 in×1 in [365.8 cm×30.5 cm×2.5 cm]) was as high as 200 rupees in 2009, and rose to 500 rupees in 2010 (Table 5.1). The soaring price led to its extensive exploitation. Although *chamba shing* was formerly common in *moon* forests, it is now difficult to find this tree.

A number of trees other than *chamba shing* are used as construction materials. The following species are known locally to be suitable for such purposes: *man shing* [man shing] (*Tsuga dumosa*); *khe shing* [ke shing] (*Juglans regia* var. *kamaonia*); *wang shing* (*Abies spectabilis*); *khorong khoronba shing* [kho rong kho ron pa shing] (*Schima wallichii*); *kharangpa shing* [kha rang pa] (*Alnus nepalensis*); and *besnang shing* [*be snang shing*] (*Quercus lanata*). *Lensong shing* (*Pinus wallichiana*) from secondary forests is an oily and hence water-resistant tree species that is regularly used for the outer walls of buildings. Logging of these trees is increasing—at the same pace as the exploitation of *chamba shing*. Currently, there is significant demand for hard timber, including *chamba shing* and *khe shing*, both of which are used in furniture making.

As Table 5.1 indicates, the price of all timber used in construction is soaring. Many *chamba shing* trees have been cut down for commercial purposes and sold to such nearby cities as Tezpur and Guwahati as well as distant ones, such as Delhi and Mumbai, in addition to the states of Haryana, Punjab, and Gujarat.

In 1987, the Indian central government and state government tightened the commercial logging regulations for *chamba shing*, *lensong shing*, *wang shing*, and *man*

Table 5.1 Changes in the local price of one cubic foot (1 CFT*) of timber (rupees)

	Chamba shing	Lensong shing	Man shing	Wang shing	Khe shing
1987	100	60	60	60	No price
2009	200	200	200	200	300
2010	500	250	250	250	300
2011	500	300	300	300	400

Produced from locally conducted interviews
*1CFT is 12 ft. by 12 in. by 1 in. (365.8 cm × 30.5 cm × 2.5 cm)

shing. Licenses became necessary for the commercial exploitation of these timbers. However, licenses were not required to log these trees for private purposes. This constituted a legal loophole, and massive logging resulted along with the bribing of governmental officials. About 30 % of forests in the state are believed to have disappeared within 9 years of the change in logging regulations. The Supreme Court of India subsequently ordered a ban on the issuing of licenses and prohibited commercial logging entirely. That decision effected a halt in large-scale logging. However, illegal logging for commercial purposes on a small to medium scale continues today.

Though the government does not permit commercial or private logging, the actual owners of forests (i.e. mostly clans) allow logging for private purposes. This currently functions as a loophole for a certain level of commercial logging. Although clans prohibit the logging of *chamba shing*, other commercially valuable trees are cut down with clan approval. Logged timbers are covered and placed on trucks to be transported from the forests.

Gathering fuel wood is an important daily use of forests for villagers. As noted above, this task is performed mostly in *borong* forests. Trees used for this purpose are *besnang shing* (*Quercus lanata*), *bainang shing* [be nag shing] (*Quercus griffithii* and others), and *mangri shing* [man dri shing] (*Quercus lamellosa*). Charcoal made from *besnang shing* and *bainang shing* lasts longer than that made from other types of wood and is considered to have high utility. When used as fuel wood, *mangri shing* produces intense heat, which is considered an advantage. An increase in the consumption of fuel wood has encouraged the local people to look for wood not only in *borong* but also in *moon* forests. Large numbers of *pako shing* [spang kho shing] (*Lithocarpus pachyphylla*) have been cut and removed from *moon* forests. Above an altitude of 3,000 m, *tama shing* [stag ma shing] (*Rhododendron barbatum* and *Rhododendron campanulatum*) trees are indispensable as fuel wood for pastoralists. The characteristically large leaves of a similar tree, *lha shing* [stag ma shing or lar la shing] (*Rhododendron hodgsonii*), are used by pastoral people for wrapping butter and cheese (Figs. 5.13 and 6.11).

Several other useful resources are found in the forests. The berries of *lishi shing* [*li shi* or *li ka ra shing*] (*Illicium griffithii*) are collected between November and December; they are powdered and used as a spice (Fig. 5.14). The price of these berries is up to 25 rupees per kilogram when sold by villagers to local merchants. *Borong* forests at lower altitudes produce larger and hence more favorable *lishi shing* berries. The abundance of this tree in *moon* forests has proved to be very

Fig. 5.13 Leaves of *Rhododendron hodgsonii*, locally called *lha shing* [marla shing, stag ma shing, or lar la shing], are used by pastoralists to wrap butter and cheese

Fig. 5.14 When ground, the berries of *lishi shing* [li shi shing or li ka ra shing] (*Illicium griffithii*) are used as a type of spice, which is traded at high prices

valuable. *Shokshok shing* [*shog shog shing*] (*Daphne papyracea*) is an evergreen broadleaf tree and another useful resource since it is the raw material used to make the traditional paper of *boi shuku* [*bod shog*]. Strips of paper used for religious purposes—called *choephan* [*chos 'phan*] in the Dirang area and *shukphan* [*shog 'phan*] in the Tawang area—are mostly made from this tree (Fig. 3.7). *Khainag shing* [*khal nags shing*] (*Erythrina stricta*) is a deciduous broadleaf tree, and owing to its thorny character is often used in making fences around houses. This tree is easy to plant since a cutting grows when simply pushed into the ground. *Rainang shing* [*ras nags shing*] (*Pinus roxburghii*) is a type of pine, and its oil can be easily obtained by placing a container after making an incision on its trunk. The oil from this tree is sold to industrial firms. Leaves and branches of *saang shing* [bsangs shing] (*Cupressus torulosa*) and *chandang shing* [chang dwangs shing or shug pa shing] (*Juniperus indica*) are burned in stone fire pits—called *saang bum* (bsangs 'bum) in the Dirang area and *budpa* [bud pa] or *sangkhuk* [bsangs khug] in the Tawang area—to produce incense smoke for Tibetan Buddhist ceremonies (Fig. 5.15). Flowers of *zidang shing* [zi dwangs shing or stag ma shing] (*Rhododendron arboreum*) are used to make a drink of juice (Fig. 5.16).

Fig. 5.15 A stone-built fire pit called *saang bum* [bsangs 'bum] in the Dirang [rdi rang] area and *budpa* [bud pa] or *sangkhuk* [bsangs khug] in the Tawang area: the leaves and branches of *saang shing* [bsangs shing] (*Cupressus torulosa*) and *chandang shing* [chang dwangs shing or shug pa shing] (*Juniperus indica*) are burned in pits to produce incense smoke in Tibetan Buddhism ceremonies

Fig. 5.16 Flowers of *zidang shing* [zi dwangs shing or stag ma shing] (*Rhododendron arboreum*) are used to make juice to drink

5.3 Local Forest Conservation

The people of the village of Thembang were the first to realize the danger of excessive logging and overexploitation of medicinal herbs. Inhabitants of such towns as Dirang and its surrounding villages conducted logging to obtain fuel wood as well as for commercial purposes. Medicinal herbs were pulled up, including the roots: often the roots have pharmaceutical value. The situation was similar in Thembang. The problem was complex because the income from forest resources was indispensable to the local economy. To prevent environmental deterioration, it became necessary to develop alternative sources of income. After considering several potentially viable methods, the people of Thembang decided to collaborate with the World Wide Fund for Nature (WWF) to launch the Community Conserved Area (CCA) project. An office of the WWF was relocated from Itanagar (Fig. 1.1), the capital of the state to Dirang and began providing support for the project. The CCA project was launched in 2004, and active efforts began in 2007. It was funded by the Dorabji Tata Trust (http://www.dorabjitatatrust.org) via the WWF. In this project, three houses in Thembang were modified for use as guesthouses, with one room in each acting as a guestroom for tourists. The villages also began offering support and training for local trekking activities for tourists. Through these projects, the villagers attempted to compensate for the economic loss that resulted from stopping the exploitation of forest resources.

The Thembang Bapu CCA, which covers 312 km^2, was followed by the Pangchen Lhunpo Muchat CCA, covering 98 km^2, and the Pangchen Shoktsen Lakhar CCA, covering 102 km^2, in the Pangchenpa area (Dutta 2011). A number of villagers are currently working with WWF staff in implementing these projects. The WWF also conducts technical training for local inhabitants to enable them to conduct field surveys.

The use of yak dung as fuel has also been promoted. Dried yak dung was formerly used as fuel in the Monpa area, but it became replaced by fuel wood from forests because it is very smoky when burned and more labor intensive to collect. The use of yak dung thus declined, except in high-altitude areas above the forest line. To revitalize its use as an energy source, a device to streamline its production process was introduced (Fig. 5.17). This device uses burned, carbonized yak dung and converts it into a rod-shaped piece of fuel. The cost of producing this device is 16,000 rupees per unit, but the WWF provides the units to the CCA projects free of charge.

Fig. 5.17 This device was invented to streamline the process of fuel production from yak dung. Yak dung is inserted into the device, and burning and carbonization result in a piece of rod-shaped fuel (This photograph was provided by Mr. Pijush Kumar Dutta of the World Wide Fund for Nature)

References

Dutta PK (2011) Community based conservation in Western Arunachal Pradesh, national workshop on sharing of lessons and wise practices for sustaining Himalayan ecosystem. Gbpihed, Almora, 10–11 August

Mizuno K (2010) Forest distribution and human activities. In: Okumiya K (ed) Ecology of human life, aging and disease. Shouwado, Kyoto, pp 100–104 (in Japanese)

Polunin O, Stainton A (1984) Flowers of the Himalaya. Oxford University Press, New Delhi

Chapter 6
Yak Husbandry and Pastoral Communities

Land in the Dirang [rdi rang] region on which yaks and their hybrid varieties graze: the forests are cleared and converted to grazing land. Taxes known as *tsarin* [tswa rin] are paid to the agricultural clans that own the land

K. Mizuno, L. Tenpa, *Himalayan Nature and Tibetan Buddhist Culture in Arunachal Pradesh, India*, International Perspectives in Geography 6,
DOI 10.1007/978-4-431-55492-9_6

Abstract Yak husbandry is a crucial part of Monpa community life. These domestic animals of the Tibetan plateau and Himalayan Region, including highlands of Arunachal Pradesh have complex mating patterns. Each of a number of types of yak [*g.yag*] is used for a specific purpose. The relationship between the type of yak and altitude is especially important. Grazing land is produced by forest reduction. The traditional method is the application of aconite root extract to tree trunks after ring-barking, and the main method used today is simply to bark trees. Pastoral people migrate seasonally, but they typically allow yak grazing in areas around their settlements during the winter. Pastoralists never practice agriculture even in or around their winter settlements. In pastoral communities, yaks are much more than mere domestic animals: they are fundamental to the livelihood of Monpa communities. The focus of this chapter is on Monpa pastoral communities, and consideration is given to their relationship with agricultural people.

Keywords Grazing land • Hybrid • Pastoralists • Land rent • Yak

6.1 Pastoralists and Yak Grazing Land

Fir and hemlock trees are frequently found at an altitude above 3,000 m, and alpine roses are common understory plants in high-altitude areas in the Eastern Himalayas. Grazing lands are located in these needleleaf forest zones. Human activity has led to the reduction of forest areas to allow the development of grazing land. One method used to reduce forested areas has been the application of aconite root extract to tree trunks after ring-barking. Such a method is becoming less common since there has been a decrease in the number of local inhabitants with a detailed knowledge of the local vegetation. The main method used today is simply to bark trees. As a result of such activities, patches of grazing land are scattered throughout the forests (Fig. 6.1). Seasonal shelters of pastoral people can be found in these areas.

Pastoralists live in settlements at low altitudes and allow their herds to graze during winter (from October to March) (Figs. 6.2 and 6.3). They do not carry out agriculture—even in winter. The villagers of Senge Dzong and Nyukmadung of West Kameng (Fig. 2.5) are rare exceptions in that they practice both grazing and agriculture. The villagers are pastoral people but own plots of land, where they reside permanently. Typically, women are responsible for farming and men take charge of grazing. In Senge Dzong, where there are 86 households, 40 practice grazing and 46 families do not own domestic animals but conduct regular farmwork. In many of the 40 pastoral households, the husband and his brothers manage the grazing, and the wife remains in the village to undertake work on the farm. There are variations in this role sharing, and the segregation of duties can be flexible. For example, the primary *tsorgen* of Senge Dzong practiced agriculture and his younger brother was in charge of grazing in 2011.

The following record of pastoral activities in 2011 illustrates the pastoral lifestyle to a certain extent. An advance party departed the village of Senge Dzong in

Fig. 6.1 Nagagg is used as a rangeland for yaks. Rangelands are created through the clearing of natural forests

mid-June to take their yaks [*g.yag*] to grazing areas. The other party with highland cows, called *galang* [*ba glang*], and hybrids between cows and yaks, called *dzo* [mdzo] (males) and *dzomu* [mdzo mo] (females), left the village around July 20. The first party with their yaks returned to the village at the end of October. The second party with the *dzo* and *dzomu* had already returned by the end of September. Compared with the hybrids, yaks are resistant to cold but vulnerable to heat, and therefore they stay in the highland area longer.

Each pastoral household of the village of Senge Dzong has its own single piece of land for grazing. The primary *tsorgen* of the village claimed to have practiced grazing for 15 years—from his mid-teens to the end of his thirties—in a highland area. The parcel of land used for grazing is called a *lukguthang* [lug dgu thang]: *lug* means "sheep," *dgu* "nine," and *thang* "place." A *lukguthang* is thus a small area that can sustain only nine sheep.

Two groups of crops are grown in Senge Dzong. One is barley and buckwheat, which are cultivated around the village; the other is finger millet and maize, which are grown at lower altitudes within the village. Another source of income has become important in recent years: an increasing number of villagers undertake road construction work, and many of them are women. They manage daily agricultural tasks in the early morning and late afternoon; they engage in road construction work for 7 h from 7 a.m. to 3 p.m. for 30 days a month. The typical monthly income for

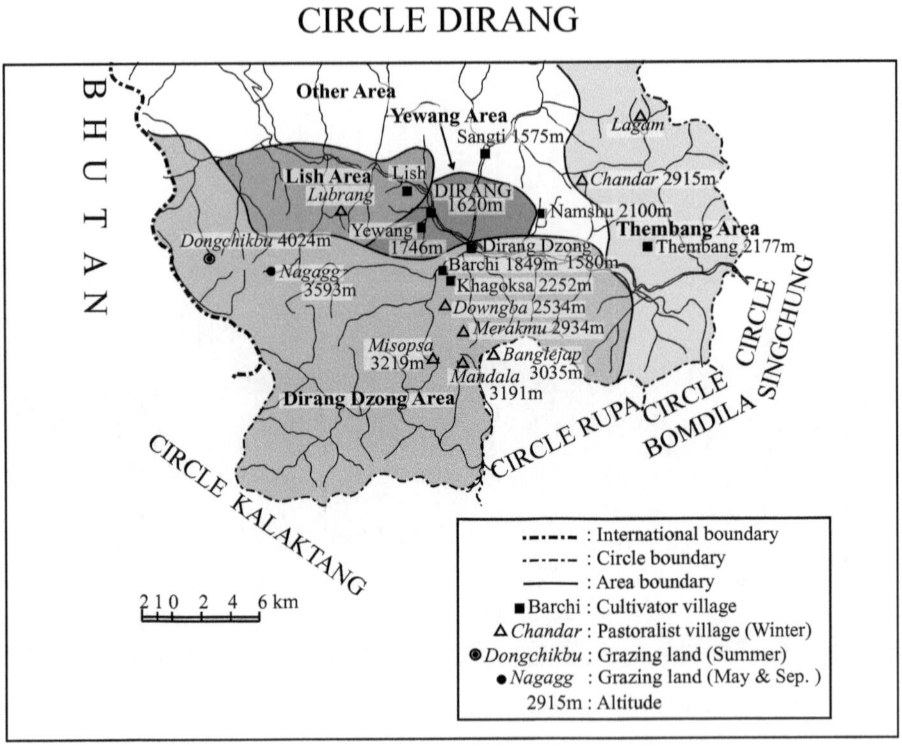

Fig. 6.2 Agricultural villages, pastoral villages, and rangelands in the Dirang [rdi rang] area. Produced from locally undertaken surveys and the *Census of India 2001 Administrative Atlas*

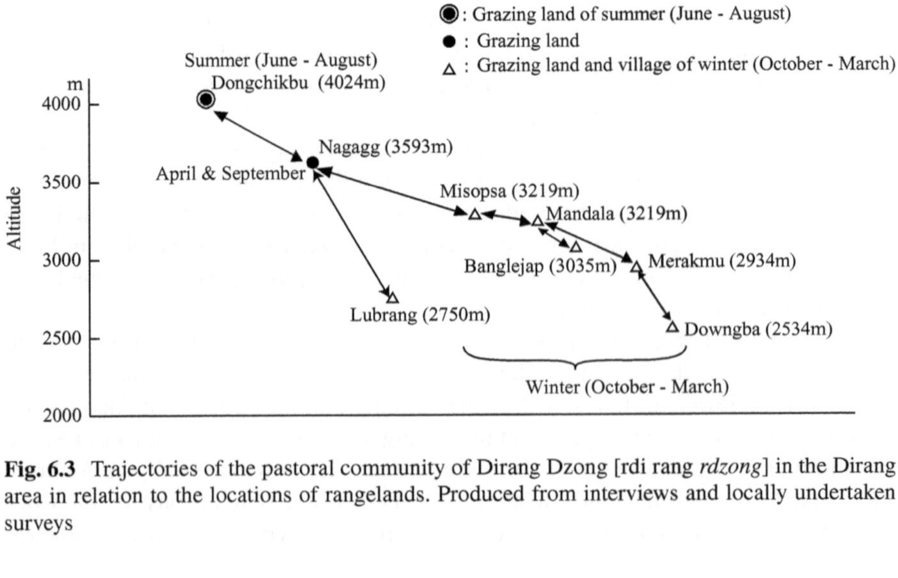

Fig. 6.3 Trajectories of the pastoral community of Dirang Dzong [rdi rang *rdzong*] in the Dirang area in relation to the locations of rangelands. Produced from interviews and locally undertaken surveys

a village woman for this type of construction work is 5,000 rupees. Since there are many steep slopes around the village, landslides are common; road construction laborers are therefore constantly in high demand.

Pastoralists in the Monyul (Monpa area) never form a clan. The village of Senge Dzong consisting of pastoralists is not an exception, and has no clan. In the village, pastoralists are known as *brokpa* or *drokpa* ['brog pa], and agricultural workers are known as *ungpa [lung pa]* or *yuel pa* [yul pa]. *Brokpa* and *ungpa* are considered ethnically different from the Dirang Monpa, and each has their own language (Table 1.1). *Brokpa* are said to be the descendants of migrants from Tibet, and their language, *brokkhe* ['brog skad], is much closer to the standard Tibetan language compared to the language of the agricultural people's *ungpa.* Senge Dzongpa, Nyukmadungpa, and pastoral people in Lubrang close to the border with Bhutan share a language and native costumes with the people of Merak and Saktang in Bhutan. The people of Mago, Thingbu, and Lukguthang in the northeast of the Tawang district and the people of Chandar [chan dar] and Lagan [la gan] in the Dirang area share a language and native costumes with the people of Tibet. The language is slightly different between these two groups.

The pastoralists of Mago, Thingbu, and Lukguthang own their own land. The number of people in a pastoral household in the Dirang Dzong area is between 3 and 11. Each household owns 20–300 domestic animals. Villages where the pastoralists stay during winter in the Dirang Dzong area include Downgba [dong ba] at an altitude of 2,534 m, Merakmu [me rag mu] at 1,934 m, Banglejap [spang la brgyab] at 3,035 m, Mandala [man da la] at 3,192 m, and Misopsa [mi sob sa] at 3,219 m (Fig. 6.2).

The movement of pastoral people over the course of a year is flexible. For example, a pastoral villager of the Downgba remains in the village in winter from December to April; he moves to Merakmu sometime between April and early May for a few days or weeks to let his livestock graze. The pastoralist then moves to an area between Mandala and Misopsa for a few days or few weeks, and he finally arrives at Dongchikbu [gdong gcig bu] near the national border with Bhutan (Fig. 6.3). The duration of stay at each location depends on the condition of the grass. The timing of travel changes accordingly. Dongchikbu is the grazing land for the summer. Pastoral people from the village of Downgba normally remain there for 3 months from June to August. They begin returning in early September and gradually arrive home in their winter village (Fig. 6.4).

The original home of the pastoralists in the areas of Dirang Dzong and Lish in the Dirang region was the village of Lubrang. Grazing land ran short at some point in the past as a result of increasing population pressure within that village; emigration to the villages of Downgba, Merakmu, Banglejap, Mandala, and Misopsa resulted. Lubrang is still the hub for pastoralists of the Dirang region, and it is therefore the distribution center for dairy products. Agricultural people visit this village to exchange their cereal grains for cheese and butter. The pastoralists of Lubrang leave their village in April, arrive in Nagagg [na ga zem bu] in May, stay at Dongchikbu, from June to September, and return to their home village by the end of October.

Fig. 6.4 As pastoral people travel, their *dzos* [mdzo] carry parcels and blankets called *lewu* [sle'u] or *lewa* [sle'u ba]. The blankets are made of yak [g.yag] wool and are water resistant

There is a diverse range of grasses on the high-altitude grazing lands used by the pastoralists in the Dirang region. They include genus of *Deyeuxia*, *Deschampsia*, *Poa*, *Hierochloe*, *Stipa*, *Agrostis*, *Calamagrostis*, *Helictotrichon*, and *Glyceria*. *Poaceae* and *Cyperaceae* grasses are very widely distributed (Chatterjee et al. 2003) and are used to feed livestock.

The village of Misopsa is located at an altitude of 3,219 m and consists of eight households of pastoral people. They return from their higher-altitude grazing lands and stay in the village until late March. They then move toward Nagagg in April.

At an altitude of 3,035 m, the village of Banglejap consists of six households. Children, their mothers, and senior people of three of those households remain in the village while all members of the other three households move to their grazing land for the summer. For example, during the fieldwork, we observed that in one of the three households, nine people left for the grazing land; one old woman, one mother, and her baby remained in the village during the summer (Fig. 6.5).

The state government has promoted a plan to encourage pastoralists to settle in the village of Merakmu to improve the children's educational standards. The government built 26 residential cottages as well as a school and clinic in Merakmu in

Fig. 6.5 In summer, only children, their mothers, and senior members of pastoral households remain in the village while all other members move to the grazing lands. The photograph shows a baby and his mother remaining in the village of Banglejap [spang la rgyab] at 3,035 m altitude. The mother was producing butter tea using a device known locally as a *jandong* [ja dong]

2007 (Fig. 6.6). Each cottage cost the government 20,000 rupees. However, the pastoral people did not choose to live there. The school is used by the children of migrant road construction workers from Nepal.

Dongba is located at an altitude of 2,543 m and is the lowest village inhabited by pastoralists in Dirang Dzong. It is in a bamboo forest zone, and clipped bamboo leaves are used as fodder during winter. Bamboo trees at high altitudes are called *shee* [bshed], and the local people believe that they bloom once in 50 years and then die. They are a useful resource and are used as material for roofs, fences, and baskets. At low altitudes, bamboo forests are called *rashu* [ra shu]. They do not die but are less useful than *shee*. A fruticose lichen, *Usnea pectinata*, which is locally called *bunpu* or *bunpa* ['bun 'bu], is used as a feedstuff for yaks. It is also used as a scrubbing brush for cleaning. When the pastoralists stay in shelters above an altitude of 3,000 m, *tama shing* [stag ma shing] (*Rhododendron barbatum* and *Rhododendron campanulatum*) trees are used as firewood. The surface of these trees is shiny and smooth. Its permeability is low, and it is suitable as a fuel wood in the rainy mountain regions.

Fig. 6.6 The government of Arunachal Pradesh promoted a plan to encourage pastoral people to settle in the village of Merakmu to improve their children's education. The government built 26 residential cottages as well as a school and clinic in 2007

Tree leaves are also used as fodder for yaks in winter settlements. Deciduous broadleaf trees, such as *Quercus wallichiana*, *Acer campbellii*, *Salix humboldtiana*, *Buddleja asiatica*, *Symplocos racemosa*, *Castanopsis* sp., *Berberis* sp., and *Ligustrum myrsinites*, are often used for this purpose (Chatterjee et al. 2003, 2006).

Pastoral communities in the Thembang area are notably different from those in Dirang Dzong. The presence of *tsorgen* is one of the main differences. Every village in the Thembang area has at least one village mayor, called a *tsorgen*, whereas villages in Dirang Dzong do not. Some villages in the Thembang area have several *tsorgen*.

At an altitude of 2,915 m, the village of Chandar in the Thembang area accommodates 16 households, which amount to over 60 people in winter. About 20 villagers, mainly seniors and children, remain in the village while about 40 people leave to manage the grazing in summer. They depart from the village sometime between June and July, and they return in late September. The houses in Chandar are larger than those in Dirang Dzong and have access to electricity.

Polyandry is traditionally practiced in the village of Lubrang. Normally, two men are married to one wife in this village. One of the husbands maintains the house, and the other manages the grazing of yaks. This is a customary practice in Tibet, and the husbands of the same wife are often brothers (Yamaguchi 1987).

6.2 Nature of Yaks

The yak is a common domestic animal in Tibetan culture and typically thrives in highland areas. A female yak is called a *bri* (*dri*) ['bri] (Fig. 6.7). Male and female highland cows are called, respectively, *galang* [ba glang] and *galangma* [ba glang ma]. A hybrid of a male yak and female cow is called a *galangma*, whereas a hybrid of a *galang* and *bri* is called a *dzo* [mdzo] if it is a male and a *dzomo* [mdzo mo] if it is a female (Fig. 6.8). Both *dzo* and *dzomo* resemble yaks and *bri*. Yaks and hybrids are distinguishable by the fur on their bellies, which is longer on yak and *bri* of female yak than on *dzo* and *dzomo*. These domestic animals are very important for the pastoral people of the highlands of Arunachal Pradesh.

There are complex mating patterns among the domestic animals of Arunachal Pradesh. Whereas *dzomo* are fertile, *dzo* are infertile. A hybrid between a *galang* and a *dzomo* is called a *khotu* [khu tu] (Table 6.1). A *khotu* produces only half the amount of milk that a *dzomo* does, and it is also of lesser quality. All newborn male *khotu* as well as most females are killed. The reason for killing most *khotu* is to minimize consumption of their mother's milk. A few female *khotu* are saved because the third hybrid between a male *galang* and female *khotu*—*gogyalang* [go rgya glang]—is considered to produce good-quality milk. The first and second hybrids

Fig. 6.7 A male yak [*g.yag*] appears on the right and a female yak—*bri* [*'bri*]—stands in the foreground

Fig. 6.8 *Dzo* [mdzo], a hybrid of a male yak and highland cow: *dzo* can carry greater burdens to lower altitudes than yaks and *galang* [ba glang]. The cargo they carry is often covered by a yak blanket, known as a *lewu* [sle'u] or *lewa* [sle'u ba]

Table 6.1 Types and mating of domestic animals of the pastoral communities

Male	Female
Yak	Bri
Dzo	Dzomo
Galang*	Galangma*
Jatsa	Jatsamin

Produced from locally conducted interviews

*Galang [ba glang] and galangma [ba glang ma] are types of highland cattle

1. Galang × Bri = Dzo & Dzomo (Brimdzo & Brimdzomo)
2. Yak × Galangma = Dzo & Dzomo (Bamdzo & Bamdzomo)
3. Galang × Dzomo = Khotu
4. Galang × Khotu = 1. Shingalang, 2. Tagalang, 3. Gogalang
5. Yak × Dzomo = Tuia
6. Yak × Tuia = Garyak & Garburi
7. Yak × Garburi = Yak & Bri
8. Cow × Galang or Bull × Galangma = Wagalang & Wagalangma
9. Mithun × Cow = Jatsa & Jatsamien

Fig. 6.9 Pastoral people produce butter and cheese from the milk of *bri* [*'bri*] and *dzomo* (mdzo mo)

between a male *galang* and female *khotu*—*singalang* [sing ba glang] and *tagalang* [rta ba glang], respectively—are believed to yield only low-quality milk. *Gogyalang* also produce good *galang*. Male *khotu* are infertile and are therefore killed at birth for leather and meat. Though the interval between breeding for a female yak, a *bri*, is 18–24 months, that of a *dzomo* is 15–18 months (Ramesha et al. 2009). To increase the number of *dzo* and *dzomo*, it is necessary to have a *galang* for breeding, and they are therefore highly valued in pastoral communities.

The amount of milk that an average *dzomo* produces is greater than that of a *bri*. Whereas a *bri* typically produces 3 liters of milk a day, a *dzomo* can produce 7 liters. More milk can be obtained from a *dzomo* because the *bri* feed their babies while a *dzomo* does not.

Milk from *bri*, *dzomo*, and *gogalang* are mixed in the same container. By stirring the milk with an appropriate tool (Fig. 6.9), the upper cream turns to butter and the remainder becomes cheese (Fig. 6.10). The butter and cheese are wrapped with leaves of *Rhododendron hodgsonii*, known locally as *lha shing* [stag ma shing or lar la shing] (Fig. 5.13), for storage and transportation (Fig. 6.11).

A *dzo* or *dzomo* as a hybrid between a *bri* and a *galang* is called a *brimdzo* ['bri mdzo] or *brimdzomo* ['bri mdzo mo] from more detail classification. A *dzo* or *dzomo* as a hybrid between a yak and a *galangma* is called a *bamdzo* [ba mdzo] or a *bamdzomo* [ba mdzo mo] in the same way. A hybrid between a yak and a *khotu* is called a *shingdzo* [shing mdzo] or a *shingdzomo* [shing mdzo mo]. *Bridzomo* yield the best milk and *shingdzomo* the worst (Table 6.1). A hybrid between a yak and a

Fig. 6.10 Butter produced by pastoralists

Fig. 6.11 Butter and cheese are wrapped with leaves of *Rhododendron hodgsonii*, locally called *lha shing* [stag ma shing or lar la shing], for storage or transportation

dzomo is called a *tuia* [thu'i]. Since its utility is low, all male and most female *tuia* are killed. A hybrid between a yak and a *tuia* can be either a *garyak* [dgar g.yag] or a *garburi* [dgar bu ri] of female garyak, which have physiology and ecology very similar to those of yak and bri. A hybrid between a yak and a *garburi* is almost identical to a yak or a *bri*.

A hybrid between a cow and a *galang* or a *galangma* is called a *wagalang* [wa ba glang] or a *wagalangma* [wa ba glang ma], respectively. The latter produces good milk.

Pastoralists favor *dzo* and *dzomo* and actively undertake breeding to acquire them. There are clear reasons for this preference. *Dzo* and *dzomo* are larger and stronger than yaks and highland cows. They also produce more milk and meat. They can carry more than a yak and a *galang* and can also convey loads to lower altitudes (Fig. 6.8).

A yak eats long grasses using its tongue, just as a cow does, and can also eat short grasses and rhizomes by means of its incisors and lips, like a sheep (Ramesha et al. 2009). Further, it can consume hard leaves and the stems of sedges in swamps. In winter, yaks remove layers of snow by using their forefeet to reach the wilted vegetation underneath. They employ a variety of methods to consume a wide range of plants and hence they have high adaptivity to many pasture areas (Ramesha et al. 2009).

Dzo and *dzomo* cannot live in high-altitude areas, where yaks and *bri* can survive, but are better able to adapt to altitude than cows; hence, they are used in mid-altitudinal areas. Temperature and humidity hinder yaks and *bri*, restricting them from low areas, and they normally remain at altitudes of 3,000–4,500 m. Pastoralists keep yaks as well as *dzo* and *dzomo* at altitudes above 2,500 m.

A *mithun* [ba men] (*bos frontalis*) is a partially domesticated animal that normally lives in forests. When it produces offspring, they are brought to a dwelling in a pastoral community and sheltered for 1–2 months before being returned to the forest. The owner of the forest sometimes gives salt to *mithun*. Young *mithun* often visit the homes of pastoralists as they become familiar with them. Forest owners cut off part of the ear of an infant *mithun* as a mark of ownership. *Mithun* are important assets for producing of good milk, and in animist communities they are offered from a man to his future wife's parents at the time of marriage. They are also often required in rituals. For example, Monpa people purchase *mithun* from the Miji people, who are animists, and use them as an offering to the gods of the mountains in Bon rituals, such as *lhasushi* (see Sect. 4.4).

As determined during the fieldwork of 2011, the price of a *mithun* was 15,000 rupees for a 1½-year-old animal and 50,000 rupees for a 6-year-old. A male hybrid between a *mithun* and a cow is called a *jatsa* [rgya rtsa], and a female hybrid is called a *jatsamin* [rgya rtsa men]. *Jatsamin* are regarded as producers of high-quality fatty milk, and the price of one could be up to 20,000 rupees for a 2-year-old. They are the most expensive domestic animals traded in the area. Other examples of prices we observed were as follows. A yak or a *bri* could cost 30,000 rupees for a 6-year-old and 7,000–10,000 rupees for a 1½ to 2-year-old. A *dzo* or a *dzomo* could cost 18,000–20,000 rupees for a 4-year-old animal. The meat of *dzo* and *dzomo* has become popular only within the last 20–30 years.

6.3 Use of Yaks and Their Social–Cultural Importance

Both yaks (and *bri* and *dzo*) and *dzomo* are similar in appearance. *Bapu*—senior clans in the Thembang area—and senior clans in the Dirang Dzong area traditionally consume only meat from yak, *bri*, sheep, deer, or fish. They never eat beef, pork, chevon, chicken, or the meat of *dzo* and *dzomo* (Table 2.1). *Gilla* (junior clans of the Thembang area), junior clans of the Dirang Dzong area, and Boot Monpa people consume most kinds of meat that are available. Overall, taboos on dietary habits are decreasing, especially among younger generations, with the exception of pork. That has always been absent from the local diet. When *bapu* require the meat of yaks, the *gilla* undertake the actual slaughtering. The *bapu* clans then receive the meat and offer any remaining organs to the *gilla* clans. The blood produced when beheading a yak is used to make sausages.

A yak can yield 70–190 kg of meat (Ramesha et al. 2009). The price of yak meat in the Dirang area was 70–80 rupees per kilogram in 2007. It increased to 120 rupees in 2010 and 150 rupees in 2011. A yak is often slaughtered on the road, and local inhabitants and travelers buy the meat (Fig. 6.12). Pastoralists sell butter and cheese by the roadside or take it to town to sell. The roadside prices of a kilogram of butter and cheese were 130 rupees and 80 rupees, respectively, in 2007; they were 180 rupees and 130–140 rupees, respectively, in towns. The roadside prices increased to 300 and 250 rupees, respectively, in 2010; unsurprisingly, the prices in

Fig. 6.12 A yak is often slaughtered on the roadside, and local inhabitants and travelers buy the meat

towns became even higher at 350 and 300 rupees, respectively. The roadside price of 1 kg of both butter and cheese was 300 rupees in 2011.

The shearing of yaks is undertaken once a year—normally between April and June, when they molt. Yak wool is naturally water resistant. It is used to produce ropes, hats, and blankets called *lewu* [sle'u] or *lewa* [sle'u ba]. *Lewu* are often used as a cover for harvested grain as well as for wrapping cheese and butter on *dzo* during transportation (Fig. 6.8). *Lewu* serve multiple purposes. They are sometimes stretched over the space between rocks to form a simple tent or temporary shelter. A more regular tent consisting of wooden poles and yak wool is called a *bura* [bu ra]. The tails of yaks are known as *tsampi* [mtshams pi] by the Dirang Monpa people and are used as dusters. They are called *ngama* [rnga ma] or *jukma* [mjug ma] in Tibetan. Since the tails of most yaks are black, exceptional white tails fetch higher prices. Hats made of yak hair are called *tsipe zhamu* [rtsid pa'i zhwa mo] in the Dirang area. They are often worn by elderly people as part of their traditional costume (Fig. 7.23).

Yaks are not just a source of milk, meat, and wool but are also considered a special form of livestock that are different from other domestic animals. This is evident in the *yakcham* [g.yag 'cham], the special yak dance that is performed at *losar* [lo gsar], the festival of the Tibetan New Year. In that dance, a life-size imitation of a yak made of yak fur or black clothing is maneuvered in a series of movements and dances by two people within the costume (Fig. 6.13). It is clear that yaks are still an integral part of community life in Monpa area.

Fig. 6.13 A life-size figure of a yak, made of yak's fur and black clothing, for the *yakcham* [g.yag 'cham] dance at *losar* [lo gsar], the Tibetan New Year: it is performed by two people within the yak figure

References

Chatterjee A, Basu A, Sarkar M, Das DN (2003) Comparative chemical evaluation of some high altitude tree fodders commonly fed to yak. J Hill Res 16(1):39–41

Chatterjee A, Ahmed FA, Bouragohain R, Pourouchottamane R, Ramesha KP (2006) Tannin and non-tannin phenolic compounds in some high altitude tree fodder of Arunachal Pradesh. Indian J Anim Sci 76(2):165–168

Ramesha KP, Manajit B, Kandeepan G, Chakravarty P (2009) Indigenous traditional knowledge of yak rearers. National Research Centre on Yak, Dirang

Yamaguchi Z (1987) Tibet: 1. Toyo Sensyo, Tokyo (in Japanese)

Chapter 7
Distribution of Farmland and Agricultural Communities

Agricultural people harvesting finger millet in the Tawang region: finger millet is used to produce a local alcoholic beverage as well as local food called *bokpai* [gro phye]. The latter is made by kneading cereal powder with hot water

K. Mizuno, L. Tenpa, *Himalayan Nature and Tibetan Buddhist Culture in Arunachal Pradesh, India*, International Perspectives in Geography 6,
DOI 10.1007/978-4-431-55492-9_7

Abstract Agriculture in the Monpa area can be classified into three zones in terms of altitude. The appropriate form of agriculture is practiced in each zone, and this results in their characteristic agricultural products. Fallen leaves from oak trees are traditionally applied to most cultivated land as fertilizer in the Monpa area. Most parts of the rangeland in this area belong to agricultural people practicing cultivation, and they receive land rent from pastoralists. A long relationship has existed between these two types of people. A clan system is dominant in the agricultural communities of the Dirang Monpa in the Dirang area. The structure of the dwellings and lifestyle of the agricultural people in the Dirang area differ from those in the Tawang area. Rapid developments are taking place in the Monpa area. These have led to the disappearance of many traditional aspects of life and the introduction of modern lifestyles in its agricultural communities. The main focus of this chapter is on these changes to the agricultural communities in the Monpa area.

Keywords Clan • Fallen leaves • Maize • Lifestyle • Loan system

7.1 Three Agricultural Zones in the Dirang Area

In the Monpa area, the introduction of the use of the plow for agriculture is believed to have taken place in the seventeenth century (Narain Jha 2006). It occurred at about the same time as Tawang Monastery was built. The plow used in the Monpa area is considered to have had a wooden edge and to have been better than those used in Brahmaputra valley, Assam (Narain Jha 2006).

Determined by altitude, there are three recognizable agricultural zones in the Dirang area. They are *yaer uang* [yer lung], which is below an altitude of approximately 1,790 m; *tsanbu uang* [gtsang po'i lung], which is between 1,790 and 2,000 m; and *ya uang* [yar lung], which is between 2,000 and 2,300 m. Since cereal cultivation is not feasible between 2,300 and 2,500 m, such cash crops as apples, walnuts, oranges, and kiwifruit are produced. Areas above 2,500 m are used as rangeland (Fig. 7.1). Since the air temperature of farmland at higher altitudes is lower than at lower elevations, a crop takes longer to grow in highland areas; therefore, the seasons of seeding and harvesting differ among the zones. This section presents observations at representative villages of each zone.

7.1.1 *Agriculture in* Yaer Uang

As noted above, *yaer uang* is below an altitude of approximately 1,790 m. Maize, finger millet, paddy field rice, and barley are produced in the village of Yewang near Dirang (Fig. 6.2), where the altitude is 1,750 m (Fig. 7.1). For maize, seeding is between June and August, and harvesting takes place in November. Soybeans and pinto beans are seeded at the same time as maize, and they are harvested 1 or 2 days after harvesting the maize. Cultivating beans is a practice that improves nitrogen fixation in the soil.

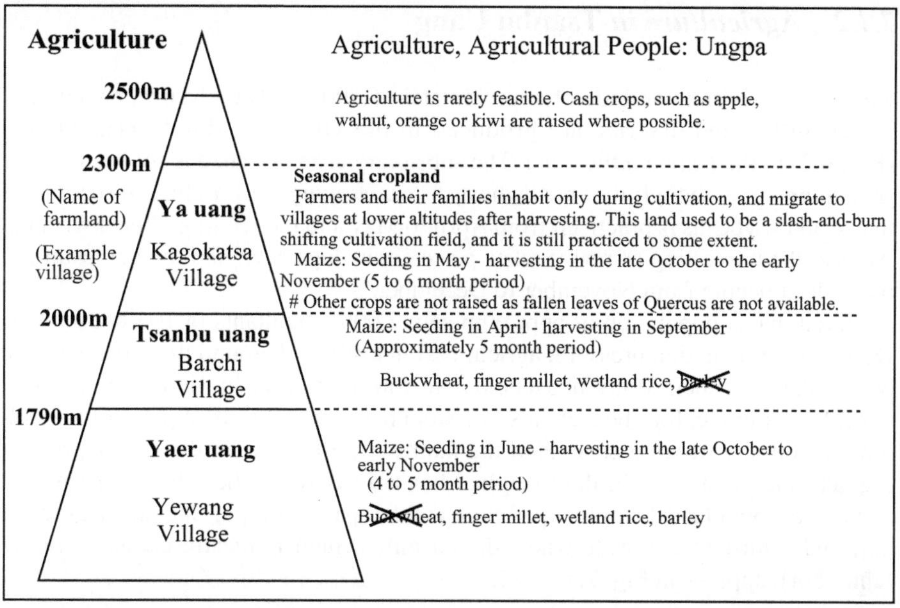

Fig. 7.1 Altitudinal distributions of types of agriculture in the Dirang [rdi rang] area. Produced from field observations and locally conducted interviews

As noted in Sect. 5.1, fallen leaves from oak trees are spread upon agricultural fields. This helps to restrain weeds, acts as a fertilizer for soil, and reduces soil erosion (Figs. 5.3 and 5.4). In the 1990s, the state government of Arunachal Pradesh once promoted the introduction of chemical fertilizer, which resulted in an increase in maize production but also in an exuberance of weeds. Chemical fertilizers are rarely used in the Monpa area today. The stems of maize are left in the field for 15 days after harvesting and are then burnt. The resultant ash is plowed into the soil when seeding soybeans. Owners of cows let their animals remain in agricultural fields after harvesting to use the dung as fertilizer. The cows are free-ranging during the day and are yoked up at night. They are taken out of the agricultural fields and kept in the mountains before the harvesting season.

There are 65 households in Yewang. The village is associated with the three small villages of Rakthi [rag khri], Rungkhung [rung khung], and Nahgrijapa [nags ri brgyab pa], which are located close by. Altogether, the area has 2,000–2,500 inhabitants. The community consists of a number of clans: the Serthipa [ser Ti pa]; Phaichulupa [pha'i phyi lugs pa]; Changphupa [chang 'phur pa]; Grangthungpa [grangs 'thung pa]; Damlapa [dam la pa]; Jamtsenpa [byams sems pa]; and Rakthipa [rag khri pa]. These clans use a loan system to lend money to other villagers. Each of the Serthipa, Phaichulupa, and Jamtsenpa clans has its own loan system. The Changphupa and Rakthipa clans have a joint loan system, and the Grangthungpa and Damlapa clans operate a similar system of cooperation.

7.1.2 *Agriculture in* Tsanbu Uang

Tsanbu uang is at an altitude of approximately 1,790–2,000 m. Maize, buckwheat, finger millet, and dry rice are produced in the village of Barchi near Dirang (Fig. 6.2) in the *tsanbu uang* zone. The subsidiary crop of maize is soybeans in the *yaer uang* zone, though it is buckwheat in *tsanbu uang*. In Barchi, the seeding of maize and soybeans is practiced from April to May and the seeding of buckwheat in August; the harvesting of maize takes place a few days later, and the harvesting of buckwheat occurs from November to December.

Barchi has a population of approximately 2,000 inhabitants in 105 households. Despite the population pressure, agricultural fields in and around the village cannot be expanded further. Many households are forced to farm in the higher *ya uang* zone. For example, four households in Barchi have dwellings on a piece of agricultural land in the *ya uang* zone. Half the family members live in the *ya uang* dwelling, and the others live in their original house in Barchi. The village community consists of four clans: Dungkharpa [gdung mkhar pa]; Serthipa; Bagiapa [ba brgyad pa]; and Jamtsenpa. A waterwheel-driven mill, which is locally called a *chukor* [chu'khor], appears in Fig. 7.2.

Fig. 7.2 A waterwheel-driven mill, locally called a *chukor* [chu 'khor]

7.1.3 Agriculture in Ya Uang

Ya uang is at an altitude of approximately 2,000–2,300 m. The agricultural fields of *ya uang* (Fig. 7.1) are seasonal. Farmers are resident only during the cultivation period and return to their villages at lower elevation after harvesting. Slash-and-burn and shifting cultivation used to be common in this zone, and are still practiced in some fields. Oaks are largely absent at this altitude, and therefore oak leaves are not readily available for agricultural use. Owing to the lack of appropriate fertilizer, maize is mainly suitable for producing with additional crops of buckwheat in *ya uang* (Fig. 7.3).

At an altitude of 2,250 m, Kagokatsa [bkag sgo kha tsa] is a typical village of the *ya uang* zone. Its inhabitants move from their original village in the *tsanbu uang* zone to this village in April or May and return to their original village in November. Seeding of maize takes place in May, and it is harvested in October. Seeding of buckwheat occurs at the beginning of September, and it is harvested in November. As noted above, oaks are scarce in this zone. Thus, using the few nearby oak trees, farmers are able to spread oak leaves on very few fields to produce buckwheat. The farmers clearly recognize the necessity of nutrients for buckwheat production. The period from seeding to harvesting is only 3 months. Dried stems of maize are used as fertilizer for buckwheat in addition to the scarce oak leaves. Weeding is undertaken in fields where maize is produced only when plowing; however, in buckwheat fields, weeding occurs at the time of seeding as well as one month afterward so that the weeds do not grow taller than the crop.

Fig. 7.3 Barley is the second seasonal crop after maize. Fallen oak leaves and the dried stems of corn are used as fertilizer for buckwheat

The dominant type of buckwheat is called *gruntsun* [bra'o zan](*Fagopyrum esculentum*), and it has a gray color. Rare ripe bitter buckwheat (Tartary buckwheat: *Fagopyrum tataricum*), which is locally called *brashum* [bra'o shom], is also produced in some areas. A local staple food called *bokpai* [gro phye] (Fig. 7.10), which is made especially from *brashum*, is considered good for pains like gout as well as for pregnant women and postpartum mothers. When cultivating *ya uang*, rent is paid to the clan that owns the piece of land. The standard rate of the rent is approximately 20 rupees per *dome* ['dom]. A *dome* is a land unit used locally in the Monpa area: it is the area that can be plowed by two cows in 1 day. Kawaguchi (2012 [1909]) mentions a similar rental system in Tibet. Thus, it may be that the *dome* land unit used in the Monpa area is or was similar to that employed in Tibet.

7.2 Rice Cultivation

Rice cultivation is practiced in lowland areas of the Dirang region. Villages in which rice cropping is undertaken include Lish, Chug, and Yewang near Dirang. Rice cultivation is generally also practiced in the Boot Monpa area. Rice transplanting is undertaken in June or July, and harvesting occurs in October or November (Fig. 7.4).

Fig. 7.4 Rice transplanting in Sangti [sangs rdi] village. Households work cooperatively in a system called *branpa* [bran pa]

Fig. 7.5 Terraced rice fields in the village of Nangro [nang ro]. Water management is achieved by adjusting the flow of descending water channels

Rice seedlings are grown in seed plots in fields used for pumpkins and beans from March or April.

Rice paddies are terraced in the village of Nangrijap in the Dirang area at an altitude of 1,674 m. The edge of each level of the terraced rice fields is banked, allowing water entering via channels to accumulate (Fig. 7.5). When abundantly available, water is provided to all levels. If availability is limited, water is provided to one level a day.

Cooperation among households is a characteristic of rice cultivation. Such mutual support is called *branpa* [bran pa], and it is the driving force behind labor-intensive work, such as rice transplanting.

Slightly sticky red rice is produced locally. However, the quantity of rice produced is limited. It is served only on special occasions, such as festivals and receptions. Local people normally eat low-gluten Punjabi rice. The locally produced red rice cannot be purchased using money but has to be obtained through barter exchange. This red rice is called *mondeb* [mon dep or mon 'bras], derived from *mon*, the region, and *deb*, which means "rice" in almost all villages in the Monyul region. This type of rice is also known as *khutsalu* [khu tsa lu], which originates from *khu* ("rice") and *tsa lu* ("red") in the Dirang area. In the Tawang area, this rice is called *mondeb* and *rachalu* [ra cha lu], which derives from *chalu* ("red") and *ra* ("rice"). In Tibet, this type of rice is generally known as *mondre* [mon 'bras]—"rice of Mon".

7.3 Cash Crops

Though the production of traditional crops, such as finger millet, maize, and buckwheat, is decreasing, the production of such cash crops as chilies, tomatoes, potatoes, and apples is clearly increasing in the Kalaktang region. The amount of cash crop production tends to be higher around towns. The production of such cash crops as tomatoes and cabbages accounts for nearly 80 % of agricultural production in the area of the Sherdukpen people, which is adjacent to the Kalaktang area. In the village of Rupa, agricultural production is almost entirely of cash crops, and it largely consists of tomatoes (Fig. 7.6). The planted areas of cash crops constitute around 20 % of the total agricultural land in the Dirang and Tawang areas. The cash crops produced in the Dirang area are potatoes, cabbages, tomatoes, apples, kiwifruit, and chilies. Namshu villagers produce and sell their chilies seasonally earlier than other villages. Whereas the standard market price of chilies is 30–60 rupees per kilogram, that of Namshu is 150–180 rupees per kilogram. Potatoes are mainly grown in the high-altitude Tawang area.

Fig. 7.6 Tomato field in the village of Rupa: such cash crops as tomatoes and cabbages are commonly grown in the area of the Sherdukpen [sher gtug spen] people

7.4 Agricultural Communities

Agricultural people are called *ungpa* [lung pa] and also *yuepa* [yul pa] or *zhingpa* [zhing pa]. The clan system is dominant in the agricultural communities of the Dirang Monpa in the Dirang area. Many decisions are made by clans rather than villages. The clans are called *khangrus* [khang rus] or *khang* by the Dirang Monpa, and *mirih* or *mirus* [mi rigs or mi rus] by the Tawang Monpa. Many people share a same name, and hence their clan names are indicated when their names are written down. Bonds within the clans are very tight. Marriages within the same clan are prohibited, and a wife joins her husband's clan after marriage. Their children also belong to the father's clan. Until around 1990, junior clan members were not permitted to enter the houses of senior clan members, and senior clan members did not eat or drink tea served in the houses of junior clan members. Marriage between senior and junior clan members was also uncommon before around 1990. In communities of the Tawang Monpa and the Kalaktang Monpa, the clan systems no longer function, and villages have become the decision-making units. It would appear that one of the reasons for the strong functioning of the clan system among Dirang Monpa communities is that their clans have royal origins (see Sect. 2.22).

The senior clans are associated with specific subordinate junior clans. The relationships in the Dirang Dzong and Thembang regions (Fig. 6.2) are summarized in Table 2.1. Five senior clans formerly governed the Dirang Dzong area. Each of the senior clans—*khangtangphu* [rkang thang phug], meaning "the large bone"—had a junior clan known as a *gilla*. The *gilla* served the senior clan members. Even today, where circumstances allow, the *gilla* still serve their corresponding *khangtangphu* on request. However, the extent of such obligations on the *gilla* in the Dirang Dzong area is less than in the Thembang area.

The mechanisms of the clan system in the Thembang area are evident in the land ownership structure. In that area, the senior clan people—called *bapu*—are descendants of royalty. The descendants of the servants—*gilla*—are junior clan members. The *gilla* use land parcels belonging to the senior clan members and offer their labor rather than rent to the senior clans. This relationship has survived for a very long time, and it is still in existence despite the increasing population. The requests of *bapu* become compelling duties for *gilla* in the Thembang area. By contrast, the Rachompa [ra bchom pa] people of the Boot Monpa do not own any land but live in an area of Dirkhipa owned by *bapu*. They do not offer any labor to the senior clans, but they do pay a land rent. Whereas the *gilla* were originally servants of *bapu*, the Boot Monpas were soldiers who served the royalty. With the exception of the Rachompa, the Boot Monpa people own land.

Aspects of the clan system are apparent in the Dirang area. The Dungkharpa clan in the Dirang Dzong area and the Jamtsenpa clan in the Yewang area support the local inhabitants; they operate, respectively, the Dungkharpa Welfare Association and the Jamtsenpa Welfare Association. These are non-governmental organizations that contribute to communal welfare, such as building dwellings for indigent households and informing the local administration of difficulties that villagers may be experiencing. For an example, in 2010, inhabitants of the villages in the Dirang area were dissatisfied with a village leader—the *tsorgen*—and a letter of appeal signed by many villagers was brought to the two associations run by the senior clans. A more formal letter of appeal with the signatures of the chairpersons of the associations was then sent to the Arunachal Pradesh state government. About 3 months later, an official order was given to the village that it should replace the *tsorgen*. The same year, the village of Lubrang requested the state government via the associations to approve an increase in the number of its *tsorgen*. A young, new *tsorgen* was then appointed as an addition to the existing one. The functions of the associations are clearly based on the local clan system, and this system is an effective, active mechanism for expressing the needs of communities in the Dirang area.

Most of the rangelands used by the pastoral people—*brokpa*—belong to clans of agricultural people. Every December, the *brokpa* pay rent to the clans that own the land. Such a rent is locally called *tsarin*, derived from *rin* ("due") and *tswa* ("grass"). In the Dirang Dzong area, four of the eight local clans—the Serthipa, Jamtsenpa, Dungkharpa, and Bagipa—own land in grazing areas. The areas surrounding Nagagg belong to the Jamtsenpa clan; the areas peripheral to Mandala are owned by the Dungkharpa; the areas below Nagagg belong to the Bagipa; and the areas adjacent to the international border with Bhutan are owned by the Serthipa and the Dungkharpa. The pastoralists pay *tsarin* to all the owners of the rangelands on their travel routes.

An example of *tsarin* is as follows. The Dungkharpa clan receives 6 kg of cheese and the same amount of butter from each of 16 pastoral households every year; the clan also obtains at least one yak or 7,000 rupees every 3 years. The agricultural clans and pastoral people used to exchange a form of written informed consent called *tansha* [brtan sha]. This normally permitted a group of pastoral people to practice 1 year of grazing in the land of an agricultural clan in return for one yak. However, the custom often caused conflict since *tansha* were exchanged with too many groups of pastoralists, who then experienced inconvenience in the actual grazing land. As a result, the offering of a yak from discontent pastoral groups to agricultural clans reduced to only once every few years. Some wealthy pastoralists still offer a yak to agricultural clans just to maintain a good relationship.

The other agricultural clans in the Dirang Dzong area also receive a similar level of *tsarin* to that obtained by the Dungkharpa clan from pastoral households. The cheese and butter obtained as part of *tsarin* are cooked and prepared as food served to the monks at the Kangyur in *gompa*s (see Sect. 4.1). The yaks that are received are divided among the clan members.

The pastoral village of Lubrang is located on land owned by agricultural people living in the village of Lish (Fig. 6.2). Some Lish villagers go to Lubrang in October to collect the land rent. The 42 households in Lubrang as well as the 13 households in the annexed village at a lower elevation known as Rama Camp collectively pay 45 kg of butter and one sheep every year. Every household in Lubrang and Rama Camp has to pay part of the rent, which is determined by the number of livestock it owns.

Cash obtained as part of *tsarin* is administered by the loan system of each clan. As noted above, each of the Serthipa, Jamtsenpa, and Dungkharpa has its own loan system. With this system, money is lent to villagers upon request. The interest rate is 3 % per month for a clan member and 5 % per month for an individual from a different clan. Maximum allowable 70 % of the income derived from the interest is kept as a fund for future loans, and about 30 % is used to assist sick people and indigent households as well as for other necessary expenditures. Any residual income is shared among the clan members. The Jamtsenpa clan was the pioneer of this loan system, but the fund of the Serthipa has recently become the largest of the three.

Among the four local clans, the Sharchokpa own more than 70 % of the land in the Thembang area. Grazing land used by the pastoral people belongs mostly to the Sharchokpa, with a smaller area owned by the Khochilu clan (Table 2.1). The village of Namshu once attempted to encourage the *bapu* descendants of royalty to become residents, and it allowed some of the Khochilu to immigrate from Thembang. This explained the support for the Khochilu by the village of Namshu when the Sharchokpa and Khochilu quarreled over land ownership in the Thembang area. The way in which *tsarin* was paid subsequently changed in the area. All the *tsarin* from the village of Chandar was formerly received by the Sharchokpa, but it is now paid to the Sharchokpa in the village of Thembang for two consecutive years and to the Khochilu every 3rd year. Villagers of the Thembang visit Chandar to receive *tsarin* in August or September every year.

The Dirkhipa clan are the descendants of the fourth son of the unknown named king who was several generations after Wongme Palder (see Fig. 2.1), and they have only a small amount of land; its members have mostly become merchants, technicians, and politicians. Interestingly, the accumulation of the income from these occupations has enabled the Dirkhipa to develop a larger asset base than any of the other three royal clans. The wealth of the Dirkhipa has gradually grown while the other three clans were contesting land ownership. When speaking of the Dirkhipa, local people say that they have managed to "catch fish in troubled waters."

Every agricultural village has at least one, but sometimes several, chiefs called *tsorgen*. They are chosen from among the villagers. *Tsorgen* are well acquainted with matters regarding borders between villages and local history. The most important quality of a *tsorgen* is to be knowledgeable and informed. Their main role is to possess a vision for the development of their villages and to promote social work, such as the renovation of roads and bridges. They coordinate the demands and opinions of villagers and negotiate with the local government. A *tsorgen* has a life-long tenure and is supported by assistants called *gomi*. There is a maximum of ten *gomi* per village. Under the supervision of *tsorgen*, *gomi* are in charge of practical welfare and administrative duties in the villages.

Cheese, butter, and yak meat obtained from pastoral people are cooked in various ways by agricultural villagers. Cheese, butter, rice, and cooked rice are called *chura*, *moh*, *koo*, and *tochan*, respectively, in the Dirang area. Cooked rice is strongly flavored using cheese. Yak meat is sometimes simply fried and served as is, but it is more often stirred into the cheese-flavored rice. Buckwheat noodles, which are

Fig. 7.7 *Grutsun phutang* [bra'o zan phu thang]—buckwheat noodles—and a bowl of yak broth used as a dipping sauce for the noodles

called *grutsun phutang* [bra'o zan phu thang] in the Dirang area, are dipped in yak soup just before eating (Fig. 7.7). Yak meat is also an ingredient of the local version of dumplings called *momo* [mog mog]. Chili is an indispensable ingredient in dishes in the Monpa area. For pregnant women, it is boiled and dried in the sun to lessen its heat (Fig. 7.8). Chili is called *tsolu* [so lo] in the Dirang area and *solu* in the Tawang area. Its market price was 130 rupees per kilogram in 2010 when we undertook the fieldwork. Fermented soybeans, which are relatively common in South, Southeast, and East Asia, are also produced and eaten in the Monpa area (Fig. 7.9). The fermented dish is called *lebechora* [le be phyur ra] in the Dirang area, *lebezetpan* [le be zan pag] in the Thembang area, and *graib chura* [greb phyur ra] or *shable chuerue* [shab li phyur rul] in the Tawang area.

Agricultural people drink butter tea, straight tea, and milk tea. Pastoralists basically drink only a local butter tea called *shuja* [gsol ja]. Local milk tea is sweet owing to the addition of sugar, but butter tea is salty since butter and salt are mixed using an instrument called a *jandong* [*ja dong*] (Fig. 6.5). Butter tea is crucial in rituals, such as Kangyur, to relieve the sore throats of monks who recite sutras over a prolonged period. Tea leaves are called *jari* [ja ril] in the Dirang area. They used to be imported from Tibet, but local villagers now buy Indian tea leaves in markets. Though jars are generally called *therbu* ['ther bu], brass jars in kitchens are referred to as *chuthar* [chu thar]. The locals believe that *chuthar* resonate with one another under a full moon.

Fig. 7.8 Chilies are sometimes boiled and dried in the sun. This reduces the chilies' heat and is done for pregnant women

Fig. 7.9 Fermented soybeans made in the Monpa area

Fig. 7.10 *Grutsun bokpai* [bra'o zan gro phye] of buckwheat: a local staple food made by kneading cereal powder with hot water is called *bokpai* [gro phye] by the Dirang Monpa and *tsan* [zan] by the Tawang Monpa

A local staple food made by kneading cereal powder with hot water is called *bokpai* [gro phye] by the Dirang Monpa and *tsan* [zan] by the Tawang Monpa. It is frequently eaten, and there are variations in the particular ingredients (Fig. 7.10). They are *phinang bokpai* [phye nang gro phye], which is made of maize powder, *ramphong* [ram phong] or *kongpu bokpai* [kong pu gro phye], which is made of finger millet, and *grutsun bokpai* [bra'o zan gro phye], which is made of buckwheat. A local confection made from either barley flour or naked barley flour is called *tsampa* [rtsam pa] or *pai* [pag] and is eaten as a snack.

Alcoholic drinks are called *yu* [yu] in the Dirang area. A number of alcoholic beverages are consumed, including the fine-tasting fermented alcohl *thangyu* [thang yu], which is made from maize; *kongpuyu* [kong pu yu], which is made from finger millet; *bong kongpuyu* [bong kong pu yu], which is made from barley and finger millet; and *khuyu* [khu yu], which is made from rice. *Bong kongpuyu* has a strong flavor and is particularly popular. Alcoholic drinks may be referred to as *chang* [*chang*], but distilled spirits are called *arak* [a rag] (Fig. 7.11); fermented alcohol (no distilled spirit) is called *bang chang* (sbang chang) in the Monyul region and Tibet. Although *Bang chang* is mainly used for personal consumption, arak is often used for the gift to other persons or the commercial products.

Barley is also used to make *torma* [gtor ma], which is cone-shaped dough used as offering to deities in a religious ceremony. The *torma* is produced by kneading barley flour, hot water, and butter (Fig. 4.19). In Monpa area, Tibetan Buddhism has

Fig. 7.11 Distilled alcoholic drinks are made by a simple method. Cereal grains are boiled in a large container, and a smaller container is placed within it, as indicated in the *upper photograph*. Vaporized alcohol is then cooled down by another vessel containing water, and distilled spirit accumulates in the small container as seen in the *lower photograph*

been mixed with Bon (see Sect. 4.1). Thus, torma is used for the Buddhist deities and gods of nature. After the ceremony, the *torma* is normally divided into pieces and given to the villager as a blessed food.

Pastoral and agricultural people have developed an inter-community relationship that lasts over generations at the level of each household. This bond between households is called *neatsang* [gnas tshang]. As part of the *neatsang* relationship, every pastoral household visits a particular agricultural household in October or November just before the snow season. The pastoral people bring cheese and butter to exchange for maize, rice, chilies, salt, and dried vegetables. The agricultural people of the villages of Thembang, Namshu, and Sangti in Thembang area have a *neatsang* relationship with the pastoral people of the villages of Mago and Lukguthang. The agricultural households of the Dirang Dzong area used to have a *neatsang* relationship with the pastoral households of the Merak and Sakteng of Bhutan. Some Sakteng later migrated to the Dirang Dzong area, and the *neatsang* relationship has been maintained between them and the agricultural households of that area.

There are two types of cylindrical measuring cups, locally called *brey*, used to measure cereals in such areas as Thembang. The two types are slightly different sizes. When lending cereal to someone, a smaller *brey*, 14 cm in diameter and 8 cm high, is used. The reimbursement is measured with a larger *brey*, 14.5 cm in diameter and 9 cm high (Fig. 7.12). The difference in size functions as the interest on the loan. When hiring labor for agricultural work, such as harvesting in the Dirang area, ten small *brey* of cereal is the standard daily wage. If the hired workers prefer cash, a daily wage of 150 rupees (as of 2011) is offered. The wage is fixed for all community members, and if raised the new wage applies to everyone.

India operates a rationing system and Arunachal Pradesh is no exception, although it is more privileged than other states. Local people are subject to two types of rationing depending on their income: above poverty line (APL) and below

Fig. 7.12 Two types of cylindrical measuring cups—locally called *brey* [bre]—are used for cereals in some areas, such as Thembang [them bang]. The two types have slightly different sizes. When lending cereal to someone, a smaller *brey* (14 cm diameter, 8 cm high) is used; when receiving reimbursement, a larger *brey* (14.5 cm diameter, 9 cm high) is used. The size difference functions as the interest

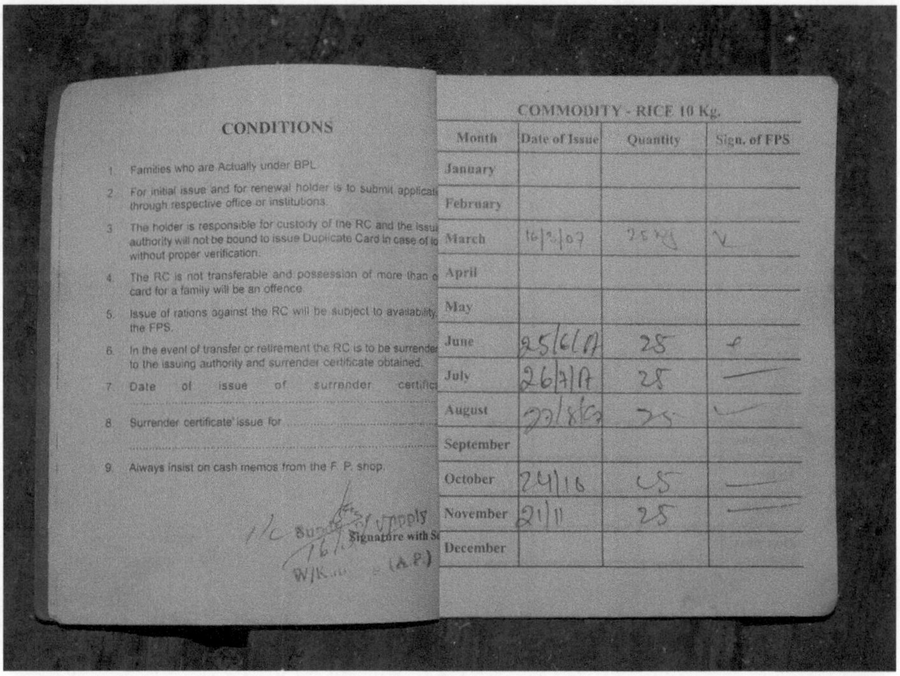

Fig. 7.13 India has a rationing system, and the state of Arunachal Pradesh is no exception. Two types of rationing apply to individuals depending on their income: above poverty line (APL) and below poverty line (BPL). A ration card indicates which of the two types applies to each household. The photograph shows a BPL card

poverty line (BPL). A ration card is issued indicating which of the two types applies to each household (Fig. 7.13). BPL is further divided into two subtypes. One BPL subtype enables a household to buy 1 kg of rice for 3 rupees; with the other subtype, the price is set at 6 rupees in 2011. In both cases, a household can buy 20–30 kg of rice a month. APL households can buy up to 50 kg of rice at a price of 8 rupees per kilogram. More can be bought by large households with an APL ration card. The market price without the ration card is 14 rupees per kilogram. The ration card also enables a household to purchase sugar and salt at reduced prices. Although available dates are limited, heating oil, cooking oil, and flour may be bought using the ration card at government-operated stores. Many people visit these stores when oil and flour become available to the ration cardholders (Fig. 7.14). The *tsorgen* in each village evaluates the type of card that a household receives, and he passes his recommendation on to the local government. At one time, many people illegally possessed more than one card; however, a recent tightening of regulations by the government has made such abuses of the system very difficult.

Wealthy agricultural households normally have several juveniles as servants. In the case of a *bapu* household in Thembang village, we observed that many

Fig. 7.14 A long line in front of a government-operated store, which is part of the national rationing system

juvenile servants were from the Assam region. Some households have juveniles from the Boot Monpa as servants. These juvenile servants typically do not attend school.

7.5 Dwellings and Lifestyle

The structure of the dwellings and lifestyle of the agricultural people in the Dirang area differ from those in the Tawang area. A dwelling in the Dirang area has a large room just inside the entrance, which is used as a bedroom or kitchen (Fig. 7.15). The dwellings in the Tawang area are duplexes, which have a balcony or an earth floor inside the entrance and a door to the actual rooms beyond (Fig. 7.16). Although the toilet is located outside the main building of the houses in the Dirang area, it is attached to the central dwelling unit in the Tawang area. In both areas, the entrances are deliberately kept small to keep out zombie-like specters locally called *rolangen* [ro lang mkhan]. Since *rolangen* are believed to be stiffened and cannot crouch, they are unable to enter a house through a small entrance. Steps are placed at the entrance to make the floor uneven and force *rolangen* to stumble (Fig. 7.16). Cereals are

Fig. 7.15 A dwelling in the Dirang Dzong (rdi rang *rdzong*) area: it has a large sitting room just inside the entrance

Fig. 7.16 A dwelling in the Tawang [rta wang/dbang] area with a balcony or earth floor inside the entrance. A door leads to the actual rooms. The entrances are deliberately made small and have steps to prevent the supposed entrance of zombie-like specters known locally as *rolangen* [ro lang mkhan]

Fig. 7.17 Cereals are stored above the kitchen. This arrangement ensures that smoke from the kitchen reaches the stored cereals and eliminates maize weevils

stored above the kitchen. This arrangement allows smoke from the kitchen to reach the stored cereals and eliminate maize weevils (Fig. 7.17). In the Dirang area, large houses normally have a prayer room called a *choisam* [mchod gshom] (Fig. 7.18). The kitchen is called *thabtsang* [thab tshang] in the Dirang area (Fig. 7.19). Kitchens contain pans, ladles, and other cooking tools, such as steamers to make *momo* dumplings.

Fallen oak leaves are used as toilet paper in the Tawang area. Used leaves are stored with the fecal waste under the toilet. When the waste container becomes full, its contents are removed and spread over agricultural fields as fertilizer. Alternatively, maize cores are used as toilet paper in the Dirang area.

Rapid developments are taking place in the Monpa area. These have led to the disappearance of many traditional aspects of life and the introduction of modern lifestyles. For example, traditional medicine—as practiced by monks and shamans who live in the villages (Figs 7.20 and 7.21)—is being replaced by modern medicine. *Amchi* [em chi or sman pa] are medical staff who have trained at the central institution for Tibetan medicine—Mentsee Khang [sman rtsis khang] in Dharamsala—and they diagnose illnesses through various methods and prescribe herbs or other medicinal products. There is a branch hospital of the Mentsee Khang in Tawang. Traditionally, the *tsirig dratsang* [rtsis'drwa tshang] of the College of Astrology (Chakpori Mentsee Khang) [lcags po ri sman rtsis khang] of the Institute of Tibetan Medicine and Astrology at Chakpori, also assists people who require advices in medical or health related issues. In Monyul, a local Monpa *amchi*

Fig. 7.18 Large houses normally have a prayer room

Fig. 7.19 A kitchen in a dwelling of the Tawang [rta wang/ dbang] area: kitchens usually contain pans, ladles, and other cooking tools, such as steamers for making *momo* [*mog mog*] dumplings

Fig. 7.20 The medical tool known as a *tsardham* [tsha dam]: the surface on which the characters are engraved is pressed against a diseased body part

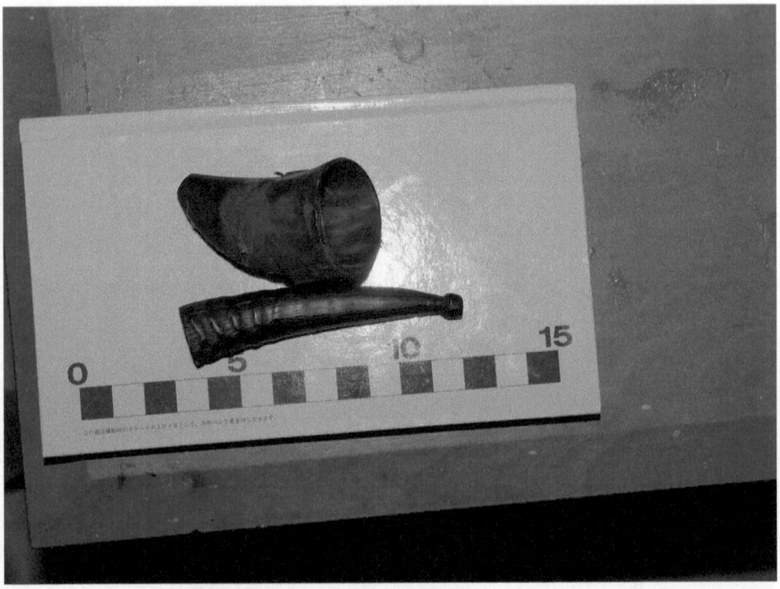

Fig. 7.21 A medical tool: priests press this tool against a diseased body part to extract sick elements

Fig. 7.22 Weaving a *toedung* [stod 'thung], the traditional jacket for Monpa women

provides medical care to villagers. During a medical interview, the *amchi* asks every patient where his or her home village is located. This is because there are numerous ethnic groups other than the Monpa living in Tawang, and the *amchi* has to consider their respective physical environments, eating habits, and daily activities when evaluating their symptoms (Nagaoka 2011).

Traditional forms of clothing are also becoming less common (Fig. 4.24). Young people wear them only on special occasions, such as the local New Year or other festivals. As a consequence of this change in apparel, the traditional craftsmanship required to manufacture such native clothing (Figs. 7.22 and 7.23) is declining. Mobile phones, cars, and motorcycles are increasingly widespread, and the daily lives of locals are becoming more governed by the pressures of time. Monpa speak their own mother tongue, but the Tibetan language is used almost only by monks and the relatively few native speakers. Additionally, the Assamese language, which used to be commonly spoken, has been largely replaced by Hindi and English in schools (Fig. 7.24). Immigration has also increased. Assamese (Fig. 7.25) and Nepalese (Fig. 7.26) individuals come to Arunachal Pradesh as seasonal workers, and some have married Monpa people and settled locally. Developments in the Monpa area have fused tradition with modernity as well as indigeneity with foreignness, and this has resulted in local communities and societies undergoing a great metamorphosis.

Fig. 7.23 Elderly Monpa ladies wear *shingka* or *shinka* [shing kha], the traditional one-piece dress. The woman on the left is also wearing a *tsipe zhamu* [rtsid pa'i zhwa mo], a traditional hat made of yak wool

Fig. 7.24 A class at the primary school in the village of Thembang [them bang]

Fig. 7.25 Monpa and Assamese children in the village of Zemithang [bye ma'i thang]

Fig. 7.26 Nepalese children in the village of Merakmu [me rag mu]: the original inhabitants of this village are pastoralists, but many seasonal workers from Nepal currently live there

References

Kawaguchi E (2012 [1909]) Three years in Tibet. General Books LLC™, Memphis

Nagaoka K (2011) Poisoning through the nails: narratives of poison at the border of Northeast India. Asian Afr Area Stud 11:64–68 (in Japanese)

Narain Jha B (2006) Buddhism, Buddhist monasteries and the societies of the Monyul: a historical study. In: Dutta S, Tripathy B (eds) Buddhism in North-East India. Indus Publishing Company, New Delhi, pp 229–240

Glossary of Monpa Languages

The phonetic transcriptions of the terms are listed alphabetically—for both the languages of Tawang (Dakpa) and Dirang and Kalaktang (Tshangla)

Term	Tawang	Dirang & Kalaktang	Tibetan
abhi tsatimu	-	-	abi tsati mu
agyal	-	-	a rgyal
ahuja	-	-	a hu ja
ahlung	-	-	a lung
alcohol	chang	yu	chang/ yu
amchi/ emchi	-	-	em chi/ sman pa
amojomu	-	-	a ma jo mo
aodung	-	-	a'u/o gdung
arak	-	-	arak
argyadung	-	-	ar yag gdung
ata devo dorjee	-	-	a tha bde bo rdo rje
attabangle		-	a mtha' bang les
attadungphu		-	a mtha' gdung phu
attajepu		-	a ta rje bu
atung kyi	-	-	a thung skyid
babu	-	-	sba bu
badzo	-	-	ba mdzo
badzomo	-	-	ba mdzo mo
bah gyetore	-	-	dbas rgyal to re
bainang shing	bainang sheng	-	be nags shing
baitang	-	-	bey Tram
baktsi	-	-	bag rtsis
bamen (mithun)	-	-	ba men
bang chang	-	-	sbang chang
banglejap	banglagyab	-	bang la brgyab

(continued)

K. Mizuno, L. Tenpa, *Himalayan Nature and Tibetan Buddhist Culture in Arunachal Pradesh, India*, International Perspectives in Geography 6,
DOI 10.1007/978-4-431-55492-9

Term	Tawang	Dirang & Kalaktang	Tibetan
bede tsan	-	-	sbas sde btsan
bekhar	-	-	sbas mkhar
besirijan	-	-	be si ri byan
besnang shing	besnang sheng	-	be snang shing
betang	-	-	dngul Trang
bodong	-	-	bo dong
bokpai	-	-	gro phye
boi shuku	-	-	bod shog gu
boima	-	-	bogs ma
bomdila	-	-	'bum 'di la
bon	-	-	bon
bong kongpuyu		-	bong kong pu yu
bonpo	-	-	bon po
bonpu	-	-	bon phul
boot	-	-	bud
boot monpa	-	-	bud mon pa
bootpa	-	-	bud pa
borong	-	-	bo rong
borong shing	borong sheng	-	bo rong shing
brack shing	brack sheng	-	brag shing
brami	-	-	brag mi
branpa	-	-	bran pa/ skyid sdug
brashum	-	-	bra'o shom
brey	-	-	bre
bri	dri	-	'bri
bridzo	-	-	'bri mdzo
bridzomo	-	-	'bri mdzo mo
brokkhe	brokke	-	'brog skad
brokpa	-	-	'brog pa
bromo	-	-	bro mo
bropa	-	-	bro pa
buckwheat	drawa	gruntsun	bra bo/ bra'o zan
budpa	-	-	bud pa
buiwa	-	-	phul ba
bum	-	-	'bum
bum chung	-	-	'bum chung
bumla	-	-	'bum la
bunpa	-	-	'bun 'bu
bura	-	-	bu ra
butter	mar	mo/ moam	mar
chachani	-	-	cha cha gnyis
chakpori mentsee khang	-	-	lcags po ri sman rtsis khang

(continued)

Term	Tawang	Dirang & Kalaktang	Tibetan
chaksam	-	-	lcags zam
chamba shing	chamba sheng	-	sam pa shing
chandar	-	-	chan dar/ phyan dar
chang	-	chang/ yu	
changdang shing	changdang sheng	-	chang dwangs shing/ shug pa shing
cheese	chura	chura	phyu ra
'chisesoese	-	-	'chi srid gso srid
chumsonga	-	-	chum song ga
chura	-	-	phyu ra
cow	ba	wa shui/ soi (in Lishpa/ Chugpa)	ba/ ba phyugs
cooked rice	toh/ topcha chang/ 'bras btsos pa	tochang/ toshang hoi/ mama (in Lishpa & Chugpa)	lto/ lto chas/ lto chang/ 'bras btsos pa
changphupa	-	-	chang 'phur pa
chebu thonglek tsan	-	-	gces bum thong legs btsan
chicken	khat	goga/ gogadama wakhi/ hokhi (in Lishpa & Chugpa)	bya/ bya mo
choesak	-	-	chos bsag
choisam	-	-	mchod gshom
chokro lekra	-	-	cog ro legs sgra
chongleywha		-	chong las ba
chopon	-	-	bcu dpon
chu dzong	-	-	chu rdzong
chug	-	-	phyug
chugpa	-	-	phyug pa
chukor	-	-	chu 'khor
chunyeipe losar	-	-	bcu nyis pa'i lo gsar
chuthar	-	-	chu thar/ cho skyogs
dakpa	-	-	dags pa
dakpanang	-	-	dags pa nang
dakpanangpa	-	-	dags pa nang pa
dakpa tshogey	-	-	dags pa tsho brgyad
dalai lama	-	-	ta' La'i bla ma
damlapa	-	-	dam la pa
daryab dzong	-	-	mda' g.yab rdzong
dawa dangpoi losar	-	-	zla ba dang po'i lo gsar
dawa gupa	-	-	zla ba dgu pa

(continued)

Term	Tawang	Dirang & Kalaktang	Tibetan
dawa ngapa	-	-	zla ba lnga pa
deb (rice)	-		deb/ mon deb/ mon 'bras
dechin	-	-	bde chen
ding	-	-	lding
dhomang	-	-	mdo mang
dongba	-	-	dong ba
dongchikbu	-	-	gdong gcig bu
dorna	-	-	dor ma
dome	dom	-	'dom
dompa	-	-	sdom pa
dor zomba	-	-	rdor 'dzom pa
dorjee	-	-	rdo rje 'chang
dorjee dak	-	-	rdo rje brag
dorjee khandu	-	-	rdo rje mkhan 'gro
dorjee sonam	-	-	rdo rje bsod nams
dirang	-	-	rdi rang
dirang dzong	-	-	rdi rang rdzong
dirkhipa	-	-	bde skyid pa
dirkhi dzong	-	-	bde skyid rdzong
drapa	-	-	grwa pa
drama dungchung ani gompa	-	-	brag dmar gdong chung a ne dgon pa
drangnang	-	-	sbrang nang
drangnangpa	-	-	sbrang nang pa
drangnang tshodruk	-	-	sbrang nang tsho 'drug
drashak	-	-	grwa shag
dratsang	-	-	grwa tshang
drepung	-	-	'bras spung
drigum tsanpo	-	-	'gri gum btsan po
drokpa	-	-	'brog pa
drolchoe mandal zhepa	-	-	sgrol chos maN Dal zhes pa
drukpa	-	-	'brug pa
dukhang	-	-	'du khang
dumrihi	-	-	sdum ri hi(?)
dungkharpa	-	-	gdung mkhar pa
dungyur	-	-	gdung 'gyur
dursha		-	dur sa
dutai/ dukhroe	-	-	dur du/ dur khrod
dzo	-	-	mdzo
dzokchen	-	-	rdzogs chen

(continued)

Term	Tawang	Dirang & Kalaktang	Tibetan
dzomu	dzomo	-	mdzo mo
dzong	-	-	rdzong
dzongdoe		-	rdzong sdo
dzongpon	-	dzongpen	rdzong dpon
family	mathtsan	pheisonga	ma tshang/ nang mi/ mi tshang
galang	balang	-	ba glang
galangma	balangma	-	ba glang ma
ganden	-	-	dga' ldan
ganden namgyal lhatse	-	-	dga' ldan rnam rgyal lha rtse
gapde tsan	-	-	'gab sde btsan
garburi	-	-	dgar bu ri
garyak	-	-	dgar g.yag
gedun gyatso	-	-	dge 'dun rgya mtsho
gegen/ gergan	-	-	dge rgan
gelong	-	-	dge slong
gelug	-	-	dge lugs
genyen	-	-	dge bsnyen
geshi	-	-	dge shes
gethruk	-	-	dge 'phrug
getshul	-	-	dge tshul
geytong	-	-	brgyad stong
gilla	-	-	gyi la
gogyalang	-	-	go rgya glang
gomkang	-	-	sgom khang
gompa	-	-	dgon pa
gombu	-	-	mgon po
gongkar gyal	-	-	gong dkar rgyal
gopa	-	-	sgo pa
gorsam chorten	-	-	gor zam mchod rten
graib chura	-	-	greb phyur ra
grangthugpa	-	-	grangs 'thung pa
grutsun	-	-	bra'o zan
grutsun phutang	-	-	bra'o zan phu thang
gungle gyal	-	-	dgung las rgyal
gungthung	-	-	gung thung
gyada	-	-	rgya mda'
gyalpo kala wangpo	-	-	rgyal po ka la dbang po
gyangong ani gompa	-	-	rgyang gong a ne dgon pa
gyangkhar	-	-	gyang mkhar
gyangkhar dzong	-	-	rgyang mkhar rdzong

(continued)

Term	Tawang	Dirang & Kalaktang	Tibetan
gyepo	-	-	rgan po
gyero	-	-	rgyal ro
horse	ta	kuruta sita/ tek/ sita (in Lishpa & Chugpa)	rta/ chibs rta
hosina		-	lho srid nag
jam tsering	-	-	byams pa tshe ring
jamo kharmo	-	-	bya mo dkar mo
jamtsenpa	-	-	byams sems pa
jandong	-	-	ja dong
jang	-	-	byang
jang kyine	-	-	byang bskyid gnas
jangdokpalri gompa	-	-	zangs mdog dpal ri dgon pa
jangri	-	-	byang ri
jaril	-	-	ja ril
jatsa	-	-	rgya rtsa
jatsamin	-	-	rgya rtsa men
jetsun dolma	-	-	rje btsun sgrol ma
jhalane	-	-	ja las gnas
jirigaon	-	-	byi ri sgang
jomula	-	-	jo mo la
jonang	-	-	jo nang
jora dzong	-	-	sbyor ra rdzong
josepa ahdir	-	-	jo sras pa a 'dir
jowhodi	-	-	jo bo 'di
kakaling	-	-	ka ka gling
kalaktang	-	-	kha legs steng
kagokatsa	-	-	bkag sgo kha tsa
kagyu	-	-	bka' rgyug
kangyur	-	-	bka' 'gyur
karkhang darding	-	-	kar khang dar lding
kartsi	-	-	dkar rtsis
kathok	-	-	ka' thog
khada	-	-	kha btags
khagei		-	kha 'gas
khainag shing	khainag sheng	-	khal nags shing
khandroma	-	-	mkha' 'gro ma
khang		-	krang
khangrus		-	khang rus
khangthangphu		-	krang thang phug
kharangpa shing	kharangpa sheng	-	kha rang pa shing

(continued)

Term	Tawang	Dirang & Kalaktang	Tibetan
khardung tashi cholling	-	-	mkhar gdung bkra shis chos gling
khastung	-	-	mkhas stung
khastung gompa	-	-	mkhas stung dgon pa
khe shing	khe sheng	-	ke shing
khenchen	-	-	mkhan chen
khenlop choesum	-	-	mkhan slob chos gsum
khenpo	-	-	mkhan po
khenpo nyima tashi	-	-	mkhan po nyi ma bkra shis
khinmey	-	-	khyi nyan mes
khobleytang	-	-	'khob legs steng
khocilu	-	-	kho phyi lu
khoina	-	-	khu'i sna/ khos sna
khoinapa	-	-	khu'i sna pa/ khos sna pa
khoitum	-	-	khul gtam
khorongba shing	khorongba sheng	-	kho rong pa shing
khorwong	-	-	'khor dbang
khotu	-	-	khu tu
khreikhang	-	-	khral khang
khrey/ khray	-	-	khral
khu/ koo (rice)		-	khu
khu tsa lu (red color rice)		-	khu tsa lu
khuyu		-	khu yu
kishong depa	-	-	skyid shongs sde pa
koenyer lama	-	-	dkon gnyer bla ma
kyakyen la	-	-	skyag gyen la
konchok	-	-	dkon mchog
kongpu bokpai	-	-	kong pu gro phye
kongpuyu		-	kong pu yu
kungo lanyer	-	-	sku ngo bla gnyer
kusho-la	-	-	sku shogs lags
lagan	-	-	la gan
laktang	-	liktang	rlig thang
lama	-	-	bla ma
lama gombu	-	-	bla ma mgon po
lama pradhar	-	-	bla ma spre'u thar
lamawang	-	-	bla ma dbang
lang darma	-	-	glang dar ma
laningru/ leinauri	-	-	la nin 'bru/ le'i na'u ri/ btsod
la-nyer	-	-	bla gnyer
la'og yulsum	-	-	la 'og yul gsum

(continued)

Term	Tawang	Dirang & Kalaktang	Tibetan
lebechora	-	-	le be phyur ra
lebezetpan	-	-	le be zan pag
lebo tshoshi	-	-	legs po tsho bzhi
lego gyadar	-	-	las sgo rgya dar
lekig gombu	-	-	legs skyes mgon po
lensong shing	lensong sheng	-	len song shing
lewu	-	-	sle'u
lewa	-	-	sle'u wa
lhangaten	-	-	sla nga steng
lhalung palkyi dorjee	-	-	lha lung dpal kyi rdo rje
lhasa	-	-	lha sa
lhase tsangma	-	-	lha sras gtsang ma
lhashing	lhasheng	-	lar la shing/ stag ma shing/ mar la shing
lhasushi	-	-	lha zhu gzi
lhosonga	-	-	lho song ga
lhou	-	-	lha'u
lhou tsho	-	-	lha'u tsho
lish	-	-	rlis
lishi shing	lishi sheng	-	li shi shing/ li ka shing
lishpa	-	-	rlis pa
lish-chug	-	-	rlis phyug
lish-chugpa	-	-	rlis phyug pa
lobsang tenpa	-	-	blo bzang bstan pa
lobsang gyatso	-	-	blo bzang rgya mtsho
lodoe gyatso	-	-	blo gros rgya mtsho
lopon	-	-	slob dpon
losar	-	-	lo gsar
lotho	-	-	lo tho
lubrang	-	-	lu 'brang/ lug grangs
lukguthang	-	-	lug dgu thang
lumla	-	-	klung la
mago	-	-	dmag sgo/ smag sgo
mago-thingbupa	-	-	dmag sgo thing bu pa
magthing luksum	-	-	dmag sgo thing bu lug dgu thang gsum
maize	oshum/ boma	pinthang/ pinang	a shom
maincha	-	-	ba men
mangri shing	mangri sheng	-	man dri shing
manshing	man sheng	-	man shing
mandala	-	-	ma da la
mandala phudung	-	-	man da la phu gdung

(continued)

Term	Tawang	Dirang & Kalaktang	Tibetan
mangbrang	-	-	mang grangs
manpat	-	-	sman pag
mentsee khang	-	-	sman rtsis khang
meley	-	-	me ley
merak	-	-	me rag
merakpa	-	-	me rag pa
merakmu	-	-	me rag mu
millet	kowp/ khra	khongpu	khre/ kong bu/ mon chag
minakpa lama	-	-	mi nags pa bla ma
mindroling	-	-	smin grol gling
mirih	-		mi rigs
mirus	-		mi rus
misopsa	-	-	mi sob sa
mo/ moh (butter)		-	mo/ mar
mo mo	-	-	mog mog
mon	-	-	mon
mon dep (rice)	-	-	mon dep/ mon 'bras
mon dre (rice)	-		mon 'bras
monkee	monskat	monskat	mon skad
monlam chenmo	-	-	smon lam chen mo
monpa	-	-	mon pa
monyul	-	-	mon yul
moon	-	-	mun
moon shing	moon sheng	-	mun shing
mugye chang	-	-	mu ge 'chag
mukto	mokto	-	mog tog
naetsang	-	-	gnas tshang
nagagg	-	-	na ga zem bu
nahgrijapa	-	-	nags ri brgyab pa
naktsi	-	-	nag rtsis
namkoe de	-	-	gnam bskos sde
namshu	-	-	nam shu
namshu gompa	-	-	nam shu dgon pa
nangro	-	-	nang ro
nekteng	-	-	nags steng
netan	-	-	gnas brtan
ngawang	-	-	ngag dbang
ngyangben		-	myang bon
ngyerpa	-	-	gnyer pa
nimsonga	-	-	nyi song ga
nithri	-	-	nyi khri
noodle	thug pa	thug pa	thug pa
ngakpa	-	-	bsngags pa

(continued)

Term	Tawang	Dirang & Kalaktang	Tibetan
nub kameng (West Kameng)	-	-	nub ka meng
nyatri tsanpo	-	-	ngnya' khri btsan po
nyei rithang	-	-	nyal ri thang
nyei shatra	-	-	nyal sha khra
nyingma	-	-	rnying ma
nyingmapa	-	-	rnying ma pa
nyukmadung	-	-	smyug ma gdung
nyukmadungpa	-	-	smyug ma gdung pa
oedsung	-	-	'od zung
ongla tsho	-	-	'ong la tsho
pa shing	pa sheng	-	pa shing
paddy	nu/ nun	ra/raha	yu/ 'bras sog
pai	pa	-	pag
pako shing	pako sheng	-	spang kho shing
palyul	-	-	dpal yul
palyul changchub dargeling	-	-	dpal yul byang chub dar rgyas gling
palyul namdroling	-	-	dpal yul rnam grol gling
pangchen	-	-	spang chen
pangchenpa	-	-	spang chen pa
pangchen dingdruk	-	-	spang chen lding drug
passang tsering	-	-	pa sangs tshe ring
pelkyi dar	-	-	dpal kyi dar
pelden kyi	-	-	dpal 'dren skyid
pema lingpa	-	-	padma gling pa
pema gombu	-	-	padma mgon po
pema khandu	-	-	padma mkha' 'gro
phaichulupa	-	-	pha'i phyi lugs pa
phaijanken		-	khyim rgyag mkhan
phanshing	phansheng	-	'phan shing
phok	-	-	phogs
phinang bokpai	-	-	phye nang gro phye
phrami	-	-	phywa mi
phulachung	-	-	phu lha chung
phundung thongtse	-	-	phu gdung mthong brtse
phurpa	-	-	phur pa
phurpa tsering	-	-	phur pa tshe ring
pos	-	-	spos
pungur	-	-	spung 'gyur
rabjung	-	-	rab 'byung
rachompa	-	-	ra bchom pa

(continued)

Term	Tawang	Dirang & Kalaktang	Tibetan
rachulu(red color rice)	-		ra cha lu/ ra tsa lu
rahung	-	-	ra hung
rahungpa	-	-	ra hung pa
rakth	-	-	rag
rakthipa	-	-	rag khri pa
ramphong	-	-	ram phong
rashu	-	-	ra shu
rainang shing	rainang sheng	-	ras nags shing
redhipongman	-	-	re rdi spong ma
rice	dep/ depu	khu	'bras/ deb
rikili		-	ri ki li
rin 'phug pa	-	-	rin 'phug pa
rinchin tsering	-	-	rin chen tshe ring
rolangen	-	-	ro lang mkhan
rongbanghen		-	rong bang mkhan
rongnang	-	-	rong nang
rongnang toemey	-	-	rong nang stod smed
rongnangpa	-	-	rong nang pa
rungkhung	-	-	rung khung
saikharteng	-	-	sras mkhar steng
sakteng	-	-	sag steng
sakya	-	-	sa skya
salasily	-	-	sa lha sri gling
salari	-	-	sa la ri
samyas	-	-	bsam yas
samyas gyalgo	-	-	sam yas rgyal sgo
saangbum		-	bsangs 'bum
saangkhuk	-		bsangs khug
saang shing	saang sheng	-	bsangs shing
sang tashi	-	-	sangs rgyas brka shis
sangchen gudrai rimo	-	-	gsang chen dgu dgra'i ri mo
sangde tsan	-	-	gsang sde btsan
sangeling	-	-	sangs rgyas gling
sanglam	-	-	gsang lam
sanglamphel	-	-	gsang lam phel
sangngak	-	-	gsang bsngags
sangngak choekhorling	-	-	gsang bsngags chos 'khor gling
sangngak choekhor dargyeling	-	-	gsang bsngags chos 'khor dar rgyas gling
sangti	-	-	sangs rdi

(continued)

Term	Tawang	Dirang & Kalaktang	Tibetan
seng[yab]	-	-	seng g.yab rdzong
sengedzong	-	-	seng ge rdzong
sengedzongpa	-	-	seng ge rdzong pa
sela	-	-	ze la
sera	-	-	se ra
serthipa	-	-	ser Ti pa
seru	-	-	bse ru
seru tsho	-	-	bse ru tsho
shable chuerue	-	-	shab li phyur rul
shak	-	-	shag
shakthi	-	-	shag thil
shaok	-	-	sha 'og
shar nima tshotsum	-	-	shar nyi ma tsho gsum
sharchokpa	-	-	shar phyogs pa
sharmu	-	-	shar mo
shartsho	-	-	shar tsho
shechen	-	-	ze chen
shee	-	-	bshed
sherab nyingpo	-	-	shes rab snying po
sheng	-	shing	shing
sherdukpen	-	-	gsher stug spen
shokshoku shing	shokshoku sheng	-	shog shog bu shing
shing	sheng	-	shing
shingka/ shinka	-	-	shing ka
shingdzo	-	-	shing mdzo
shingdzomo	-	-	shing mdzo mo
shukphan	shogphan	-	shog 'phan
shuja	-	-	gsol ja
shyo	-	-	zhol
singalang	-	-	sing ba glang
singsur ane gompa	-	-	sing zur a ne dgon pa
sinpo	-	-	srin po
soeba shing	soeba sheng	-	gso pa shing
soma	-	-	so ma
songa sha boianken		-	song ba sha bor mkhan
songtsan gampo	-	-	srong btsan sgam po
sungchoe	-	-	gsung chos
sungma	-	-	srung ma
surbi	-	-	zhur bi

(continued)

Term	Tawang	Dirang & Kalaktang	Tibetan
ta dzong	-	-	lta dzong
tadung	-	-	rta gdung
tagalang	-	-	rta ba glang
taicha	-	-	lta lcag
taklung dzong	-	-	stag lung rdzong
taklung gompa	-	-	stag lung dgon pa
tama shing	tama sheng	-	stag ma shing/ lar la shing
tang	-	-	thang
tansha	-	-	brtan sha
tawang	-	-	rta dbang/ rta wang
tashi phuntsok	-	-	bkra shis phun tshogs
tengshot	-	-	steng zhol
tengteng	-	-	steng steng
tenpe dronme	-	-	bstan pa'i sgron me
thabtsang	-	-	thab tshang
thangtong gyalpo	-	-	thang stong rgyal po
thangyu		-	thang yu
thapkay	-	-	thabs mkhas
thegtsepa	-	-	theg brtse pa
thembang	-	-	them bang/ them spang
thembang dzong	-	-	them bang rdzong
therbu	-	-	'ther bu
thonglek pel	-	-	mthong legs dpal
thubten choephel	-	-	thub bstan chos 'phel
thumi	-	-	mthu mi
tochan/ tochang		-	lto/ lto chas
toedung	-	-	stod 'thung
ton thre	-	-	ston khral
torgya	-	-	gtor rgyag
torma	-	-	gtor ma
trimi lha'i wangchuk	-	-	khri mi lha'i dbang phyug
tri ralpachen	-	-	khri ral pa chen
trisong deutsan	-	-	khri srong lde'u btsan
triten pel	-	-	khri brten dpal
tsalu (red)		-	tsa lu
tsampa	-	-	rtsam pa
tsampa/ satsam		-	mtshams pa/ sa mtshams
tsampa lung		-	mtshams pa lung/ sa mtshams rdo ring
tsampi	-	-	mtshams pi (rnga ma/ mjug ma)
tsan	-	-	btsan
tsan	zan	-	gzan

(continued)

Term	Tawang	Dirang & Kalaktang	Tibetan
tsan	-	-	zan
tsangbu uang	-	-	gtsang po'i lung
tsangong gyal	-	-	btsan gong rgyal
tsangmi	-	-	tshang mi
tsangporong	-	-	gtsang po rong
tsanpa	-	-	btsan pa
tsanpa choeling	-	-	btsan pa chos gling
tsardham	-	-	tsha dam
tsarin	-	-	rtswa rin
tsespa/ tsispa lama	-	-	rtsis pa bla ma
tsering	-	-	tshe ring
tsethang	-	-	btse thang
tsethang	-	-	rtsis thang
tshapa	-	-	tshwa pa
tsewang norbu	-	-	tshe dbang nor bu
tshangla	-	-	tshang la
tshosum	-	-	tsho gsum
tshosumpa	-	-	tsho gsum pa
tshuk	-	-	tshugs
tshukhang	-	-	tshugs khang
tshukpa	-	-	tshugs pa
tsipe zhamu	-	-	rtsid pa'i zhwa mo
tsirig dratsang	-	-	rtsisrigs drwa tshang
tsogeling	-	-	tshogs rgyas gling
tsolo	-	-	so lo
tsona	-	-	mtsho sna
tsona dzong	-	-	mtso sna rdzong
tsona dzongpon	-	tsona dzongpen	mtsho sna rdzong dpon
tsona gompatse rinpoche	-	-	mtsho sna dgon pa rtse rin po che
tsorgen	-	-	gtso rgan
tsungdhue	-	-	zung 'dus
tumbthru drema lhagang	-	-	ldum thru sgrol ma lha khang
tuia	-	-	thu'i
ueru	-	-	dbu ru
ula	-	-	'u lag
ungpa	-	-	lung pa
ungashing cheewn		-	'ung ga shing mkhan
urgen sangpo	-	-	u rgyan bzang po
urgen tsering	-	-	u rgyan tshe ring
urgyenling	-	-	u rgyan gling
village	yui	dung	yul/ dung/ gdung

(continued)

Term	Tawang	Dirang & Kalaktang	Tibetan
wagalang	-	-	wa ba glang
wagalangma	-	-	wa ba glang ma
wang	-	-	dbang
wangshing	wang sheng	-	wang shing
wongmaikhar	-	-	dbang ma'i mkhar
wongme palder	-	-	dbang ma'i dpal dar
ya uang		ya uang	ya lung
yabyum	-	-	yab yum
yaer uang		yaer uang	yer lung
yak	-	-	g.yag
yakcham	-	-	g.yag 'cham
yang palmo	-	-	g.yang dpal mo
yar thre	-	-	d.yar khral
yarlung	-	-	yar lung
yeshi tsering	-	-	ye shis tshe ring
yeshi wangmu	-	-	ye shis dbang mo
yewang	-	-	yid dbang
yuepa/ yuelpa	-	-	yul pa
yueru	-	-	g.yu ru
yumten	-	-	yum brtan
zemithang	-	-	bye ma'i thang
zhingpa	-	-	zhing pa
zidang shing	zidang sheng	-	zi dwangs shing/ stag ma shing

A hyphen (-) indicates that the words are the same; a blank space signifies that the term does not exist in the other language. The equivalent Tibetan transliteration appears in the final column

Index

K. Mizuno, L. Tenpa, *Himalayan Nature and Tibetan Buddhist Culture in Arunachal Pradesh, India*, International Perspectives in Geography 6,
DOI 10.1007/978-4-431-55492-9

Hydrological, Socioeconomic and Ecological Impacts of the North Atlantic Oscillation in the Mediterranean Region

ADVANCES IN GLOBAL CHANGE RESEARCH

VOLUME 46

For other titles published in this series, go to
www.springer.com/series/5588

Sergio M. Vicente-Serrano · Ricardo M. Trigo
Editors

Hydrological, Socioeconomic and Ecological Impacts of the North Atlantic Oscillation in the Mediterranean Region

Springer

Editors
Sergio M. Vicente-Serrano
Spanish National Research Council (CSIC)
Pyrenean Institute of Ecology
Avda. Montañana 1005
50059 Zaragoza
Spain
svicen@ipe.csic.es

Ricardo M. Trigo
Campo Grande
Faculdade de Ciencias
Instituto Dom Luiz, Universidade de Lisboa
Edificio C8
Piso 3
1749-016 Lisbon
Portugal
rmtrigo@fc.ul.pt

ISSN 1574-0919
ISBN 978-94-007-1371-0 e-ISBN 978-94-007-1372-7
DOI 10.1007/978-94-007-1372-7
Springer Dordrecht Heidelberg London New York

Library of Congress Control Number: 2011929639

Printed on acid-free paper

Springer is part of Springer Science+Business Media (www.springer.com)

Contents

Contributors

Marta Angulo-Martínez Estación Experimental de Aula Dei, Spanish National Research Council (CSIC), Zaragoza, Spain, mangulo@eead.csic.es

Cesar Azorin-Molina Instituto Pirenaico de Ecología, Spanish National Research Council (CSIC), Zaragoza, Spain, cazorin@ipe.csic.es

Carles Barriocanal Departament de Geografia, Universitat de Girona, Girona, Spain, carles.barriocanal@udg.edu

Santiago Beguería Estación Experimental de Aula Dei, Spanish National Research Council (CSIC), Zaragoza, Spain, sbegueria@eead.csic.es

Martin Beniston C3i-Climate Change and Climate Impacts Group, University of Geneva, Geneva, Switzerland, martin.beniston@unige.ch

Jesús Julio Camarero ARAID-Instituto Pirenaico de Ecología, Spanish National Research Council (CSIC), Zaragoza, Spain, jjcamarero@ipe.csic.es

Anna Dalla Marta Department of Plant, Soil and Environmental Science, University of Florence, Florence, Italy, anna.dallamarta@unifi.it

Uri Dayan Department of Geography, The Hebrew University of Jerusalem, Jerusalem, Israel, msudayan@mscc.huji.ac.il

Ahmed El Kenawy Instituto Pirenaico de Ecología, Spanish National Research Council (CSIC), Zaragoza, Spain, kenawy@ipe.csic.es

Oscar Gordo Departamento de Zoología y Antropología Física, Universidad Complutense de Madrid, Madrid, Spain, ogordo@bio.ucm.es

Célia Gouveia Faculty of Sciences, Instituto Dom Luiz (IDL), University of Lisbon, Lisbon, Portugal; EST, Polytechnic Institute of Setubal, Setúbal, Portugal, cmgouveia@fc.ul.pt

Daniele Grifoni Institute of Biometeorology, National Research Council, Florence, Italy, d.grifoni@ibimet.cnr.it

Ercan Kahya Istanbul Technical University, Istanbul, Turkey, kahyae@itu.edu.tr

Vicente Lara-Fanego Department of Physics, University of Jaén, Jaén, Spain, vlara@ujaen.es

Juan I. López-Moreno Instituto Pirenaico de Ecología, Spanish National Research Council (CSIC), Zaragoza, Spain, nlopez@ipe.csic.es

Jorge Lorenzo-Lacruz Instituto Pirenaico de Ecología, Spanish National Research Council (CSIC), Zaragoza, Spain, jlorenzo@ipe.csic.es

Marco Mancini Department of Plant, Soil and Environmental Science, University of Florence, Florence, Italy, marco.mancini@unifi.it

Francesc Maynou Institut de Ciències del Mar, Spanish National Research Council (CSIC), Barcelona, Spain, maynouf@icm.csic.es

Enrique Morán-Tejeda Instituto Pirenaico de Ecología, Spanish National Research Council (CSIC), Zaragoza, Spain, emoran@ipe.csic.es

Simone Orlandini Department of Plant, Soil and Environmental Science, University of Florence, Florence, Italy, simone.orlandini@unifi.it

Tim J. Osborn Climatic Research Unit, School of Environmental Sciences, University of East Anglia, Norwich, UK, t.osborn@uea.ac.uk

Edmond Pasho Instituto Pirenaico de Ecología, Spanish National Research Council (CSIC), Zaragoza, Spain, mondipasho@gmail.com

David Pozo-Vazquez Department of Physics, University of Jaén, Jaén, Spain, dpozo@ujaen.es

David Robson Institut Català d'Ornitologia, Barcelona, Spain, drbn@yahoo.com

Jose Antonio Ruiz-Arias Department of Physics, University of Jaén, Jaén, Spain, jararias@ujaen.es

Francisco Javier Santos-Alamillos Department of Physics, University of Jaén, Jaén, Spain, fsantos@ujaen.es

Joaquín Tovar-Pescador Department of Physics, University of Jaén, Jaén, Spain, jtovar@ujaen.es

Ricardo M. Trigo Instituto Dom Luiz, Universidade de Lisboa, Campo Grande, Faculdade de Ciencias, Edificio C8, Piso 3, 1749-016 Lisbon, Portugal, rmtrigo@fc.ul.pt

Sergio M. Vicente-Serrano Instituto Pirenaico de Ecología, Spanish National Research Council (CSIC), Zaragoza, Spain, svicen@ipe.csic.es

Javier Zabalza Instituto Pirenaico de Ecología, Spanish National Research Council (CSIC), Zaragoza, Spain, jzabalza@ipe.csic.es

José Luís Zêzere Centre for Geographical Studies, Institute of Geography and Spatial Planning, University of Lisbon, Lisboa, Portugal, zezere@campus.ul.pt

Introduction

Sergio M. Vicente-Serrano and Ricardo M. Trigo

This book is a collection of the main contributions in a thematic workshop devoted to the hydrological, socioeconomic, and ecological impacts of the NAO in the Mediterranean area that was held in Zaragoza (Spain), in May 2010, in the framework of the European Science Foundation (ESF) Mediterranean Climate Variability and Predictability (MedCLIVAR) program (http://www.medclivar.eu/).

According to the latest IPCC report, the Mediterranean basin represents one of the most important "hot spots" of climate change in the world, with recent trends towards a hotter and drier climate being related to changes in atmospheric circulation patterns. Previous work has shown that the interannual variability of Mediterranean climate is mostly associated to changes in certain relevant atmospheric circulation patterns (Dünkeloh and Jacobeit, 2003; Zorita et al., 1992; Xoplaki et al., 2003, 2004; Pauling et al., 2006; Trigo et al., 2006). Such changes can have significant impacts in the climate of this region but also on the natural environment and several socioeconomic activities. Among these patterns, the North Atlantic Oscillation (NAO) is the only one which shows a clear signal throughout the whole year, although with stronger intensity and extension during winter due to the stronger meridional gradients (Lamb and Peppler, 1987; Hurrell et al., 2003). During this season changes in the NAO phase lead to shifts in the location of the centers of action and in the associated storm tracks (Trigo, 2006).

The NAO is responsible for most of the climatic variability in the North Atlantic, modifying direction and intensity of the westerlies, the track of the polar depressions and the location of the anticyclones (Hurrell, 1995; Wanner et al., 2001). During the positive phases, the Azores subtropical high is reinforced, leading to sunny and dry weather in the Mediterranean region (Trigo et al., 2002). On the contrary, during the negative winters, cyclones move southward increasing precipitation on the western Mediterranean (Hurrell and Van Loon, 1997; García-Herrera et al., 2001; Moses et al., 1987). Cloudiness, temperature and solar radiation are also highly modulated by the NAO index (Trigo et al., 2002).

S.M. Vicente-Serrano (✉)
Instituto Pirenaico de Ecología, Spanish National Research Council (CSIC), Zaragoza, Spain
e-mail: svicen@ipe.csic.es

S.M. Vicente-Serrano, R.M. Trigo (eds.), *Hydrological, Socioeconomic and Ecological Impacts of the North Atlantic Oscillation in the Mediterranean Region*, Advances in Global Change Research 46, DOI 10.1007/978-94-007-1372-7_1,

The NAO behavior is characterized by a remarkable interannual variability, which is evident in long instrumental (Jones et al., 1997) but equally in paleo records (Cook et al., 2002; Luterbacher et al., 2002). In the instrumental period, the NAO has shown important decadal variability and a decreasing trend between 1940 and 1970. Nevertheless, the most unusual period was observed between the 1970s and 1990s when the NAO showed an increasing tendency towards positive phases (Osborn et al., 1999), coincident with severe drought conditions in the Mediterranean (López-Moreno and Vicente-Serrano, 2008; Sousa et al., 2011) and a reinforced NAO influence on climate (Vicente-Serrano and López-Moreno, 2008a). High variability has been recorded during the first decade of the twenty-first century, including extreme seasonal values of the NAO index. As usual, extreme winter NAO values are at the root of large climate impacts on different natural hazards (e.g. floods, droughts, landslides) but also several important socio-economic areas such as agriculture, renewable energies production, water resources. Thus the winter 2001 (2005) was characterised by several monthly negative (positive) values of NAO inducing an abnormally wet (dry) year in western Mediterranean (García-Herrera et al., 2007). However, the recent winter 2010 was characterised by the most negative NAO winter value ever recorded, and it has caused notable climate anomalies and impacts namely a colder than usual central Europe (Cattiaux et al., 2010) and a wetter than usual Western Mediterranean region (Vicente-Serrano et al., 2011).

The complexity of the NAO behaviour and their impact has attracted the attention of a wide variety of scientific communities, with more than 3800 papers published on different NAO features during the last decade (SCOPUS database, visit: 15/11/2010). The association between the NAO and the Mediterranean climate variability has been well documented, showing that it is one of the main forcing factors in the region. Its impact on extreme events such as droughts, severe precipitations or heat and cold waves has been well established (e.g. Gallego et al., 2006; Garcia-Herrera et al., 2007; Della-Marta et al., 2007; López-Moreno and Vicente-Serrano, 2008). These changes have also an impact on the availability of water resources throughout the entire basin, affecting, not only river flows but also storage availability in lakes and reservoirs and snow cover (Trigo et al., 2004; Karabörk et al., 2005; López-Moreno et al., 2007; Küçük et al., 2009). The ecological dynamics of the region is also greatly affected, as has been shown through satellite imagery and tree rings (Gouveia et al., 2008; Vicente-Serrano and Heredia, 2004; Roig et al., 2009). This has an impact on the quality and quantity of crops and in the migration and welfare of animal populations (Gimeno et al., 2002; Rubolini et al., 2007). The marine environment is also affected by NAO through changes in sea level and fisheries dynamics (Woolf et al., 2003; Lloret et al., 2001; Maynou, 2008). Other reported impacts are landslides (Zêzere et al., 2008), air pollution and human health (Dayan and Lamb, 2008).

Current climate change scenarios obtained with state-of-the-art General Circulation Models (GCMs) predict large modifications in the NAO behaviour in the next decades, however a clear change towards more positive or negative NAO values has not been yet universally established (Solomon et al., 2007). Nevertheless,

the majority of the GCMs indicate as the most probable outcome an increase of the NAO values, with a decreased gradient of the Sea Level Pressures and more frequent positive NAO phases (Osborn et al., 1999; Osborn, 2004; Demuzere et al., 2009). However, the increase likelihood of positive NAO mode does not eliminate the high inter-annual variability, thus the possibility of occurrence of extreme negative values, which can trigger large impacts (Vicente-Serrano et al., 2011). In addition, GCMs predict a more robust and stable link between NAO and surface climate variables (Raible et al., 2006; Vicente-Serrano and López-Moreno, 2008b), suggesting that future climate variability, and their related impacts in the Mediterranean region might be controlled even more by the NAO pattern under a warmer world.

The NAO impacts on Northern Europe have been well reviewed and compiled (Hurrell et al., 2003), but no similar work has been attempted for the Mediterranean, which is the other key area impacted by NAO (Trigo et al., 2006). The aim of the book "Hydrological, socioeconomic and ecological impacts of the North Atlantic Oscillation in the Mediterranean region" is to serve as an updated reference text that covers the wide range of evidences on the NAO impacts in the Mediterranean regions and from a multidisciplinary perspective. We hope that this volume constitutes a unique document to present the state of the art of the numerous studies undertaken on the Hydrological, Socioeconomic and Ecological impact of the NAO, collecting the expertise of researchers from several complementary earth science fields (Geography, Hydrology, Remote-sensing, Climatology, Agriculture, Energy), but that have been lacking a common ground.

The first book chapter provides an updated overview on the internal spatial and temporal variability of the NAO and how external forcing factors may affect the future NAO pattern. Tim J. Osborn shows how multi-model ensemble of simulations under increasing anthropogenic forcing strengthens earlier findings of a shift in the mean state of the winter atmospheric circulation towards positive NAO conditions, but he indicates that if a shift towards positive NAO conditions is a realistic response to increasing anthropogenic forcing, then this signal has not yet emerged from the natural variability. He discuss that anthropogenic forcing could be altering the temporal or spatial character of the interannual NAO variability, though only relatively small changes in pattern are evident when considering a multi-model ensemble.

Some of the book chapters focus on the NAO impacts on climate extreme events, like droughts, which will also have subsequent environmental, hydrological and socioeconomic impacts. The paper by Sergio M. Vicente-Serrano and co-authors show the influence of the NAO on droughts in the entire Mediterranean region, focussing on the use of the Standardized Precipitation Evapotranspiration Index, a multi-scale drought indicator that allows to determine how the effects of the winter NAO are propagated for further months when long time scales of drought are analysed.

In terms of the hydrological impacts of the NAO, Ricardo M. Trigo presents that the strong control exerted by NAO on precipitation is very significant between October and March, being extensive to the river flow of major Iberian rivers, such as Douro/Duero, Tejo/Tagus and Guadiana. Correlation coefficient values computed between Iberian winter river flow and contemporaneous (and lagged) winter NAO

index show that the large inter-annual variability of the rivers flow is largely modulated by the NAO phenomena. In fact this mode controls inter-annual and decadal variability of both precipitation and river flow of large sectors of Iberia. In the East of the Mediterranean region the hydrological variability of river discharges and lake levels is also determined by the NAO. Ercan Kahya documents the impact of the NAO on the hydrology of eastern Mediterranean countries, such as Turkey, Iran, Kuwait, Oman, and Israel from a general perspective. Patterns of precipitation, streamflow, and lake levels in Turkey are discussed to show the NAO impacts. He devotes special attention to the NAO influences on the formation of streamflow homogeneous region and on the probability distribution functions of critical droughts. The results of his analyses clearly shows that the NAO signals are quite identifiable in various hydrologic variables in Turkey.

The NAO also affects the water stored in the form of snow with evident hydrological but also tourist implications, since the winter snow tourism represents an important income source for several mountain chains located around the Mediterranean basin. Juan I. López-Moreno and collaborators analyse the influence of NAO on the interannual evolution of winter temperature, precipitation and snowpack in these Mediterranean mountains. They show that the snow cover response to winter NAO may differ spatially as a consequence of the different influence of winter NAO on precipitation and temperature. In Switzerland, they show how the influence of NAO on snow is significant at the lowest elevation areas, where temperature is the main control on snowpack accumulation, but in the Pyrenees the highest correlation with snow is found at high elevations where precipitation controls mainly the accumulation of snow related to the NAO control of the interannual variability of precipitation.

Primary activities over land and sea are also being largely affected by NAO. On the one hand, several Mediterranean fisheries reveal inter-annual variations closely related to the NAO variability. Francesc Maynou shows that the NAO explains a large part of the variability in population abundance of red shrimp *Aristeus antennatus* in the western Mediterranean Sea, with a lag of 2–3 years. Moreover, the stocks of hake also respond to the NAO since positive NAO years enhance hake fishery production through increasing the individual size of recruits, as well as the individual weight and abundance of adult hake. The NAO signal is also observed in the crop productions as Simone Orlandini and collaborators illustrate. They stress that besides common meteorological information supplied by local stations, the use of the NAO may allow to forecast agricultural yields and production quality in regions of the Mediterranean basin.

The dynamic of ecosystems of the Mediterranean region is also largely determined by the NAO. In their chapter, Célia Gouveia and Ricardo Trigo analyse the relationship between NAO, vegetation dynamics and carbon absorption over Iberian Peninsula. They provides strong evidence that positive (negative) values of winter NAO induce low (high) vegetation activity in the following spring and summer seasons. These features are mainly associated with the impact of NAO on winter precipitation, together with the strong dependence of the spring and summer vegetation activity on water availability during the previous winter. Jesus J. Camarero

quantified the tree growth responses to NAO index across a climatic gradient in Northeastern Spain, considering ten tree species with contrasting habitats and plausibly different growth responses to climate. He shows that climatic variables and NAO indices explained on average 40.1 and 15.9% of the growth variance, respectively. He also shows how the growth responses to climate and NAO also changed through time, illustrating how long tree-ring chronologies of different species may serve as valuable monitors of the responses of forests to winter NAO. In terms of animal communities, Oscar Gordo and collaborators review more than 60 studies that have demonstrated the effects of the NAO on both terrestrial and aquatic Mediterranean ecosystems. They show how the NAO affects the condition and diet of mammals and disease-related mortality in amphibians and how the birds communities are affected by NAO impacts on the availability and extent of their habitat and by influencing dispersal decisions of individuals. Thus, they illustrate how the NAO plays an essential role in the migration of birds throughout the Mediterranean basin, and it is probably a reason for the observed advance of arrival dates during the spring in Europe.

Moreover, the environmental NAO impacts are also evident in terms of air quality and contaminants dispersion. Uri Dayan presents simulations of transport of anthropogenic CO for high and low phases of the NAO. He shows the different spatial and temporal influence played by the positive and negative phases of the NAO mode when controlling the dust tranport to the Mediterranean: the positive phase during summer over the western region and the negative one regulating dust transport over the Eastern Mediterranean in winter. He indicates that positive phases imply a reduced import of European trace gases, an enhancement of long range transport of air pollutants from North American sources and conditions in favor of mobilization and transport of North African dust mainly to the western part of this fragile basin.

Geomorphological processes in the Mediterranean region are also being largely affected by the NAO. The erosive capacity of rainfall is the factor that triggers different surface processes that finally determine soil erosion. Marta Angulo-Martínez and Santiago Beguería illustrate in a western Mediterranean area how the erosive power of rainfall is stronger during the negative phase of NAO and weaker during positive NAO conditions, a finding very useful in the implementation of soil conservation strategies. In addition, soil intensity and magnitude also triggers mass movements and landslides. José Luís Zêzere and Ricardo Trigo show that NAO have an impact on the landslide events that have occurred in the region located just north of Lisbon between 1956 and 2010. Thus, they show how many months with landslide activity are characterized by negative average values of the NAO index and high values of average precipitation (above 95 mm/month).

The volume ends showing the influence of the NAO on a key economic sector that at present is increasing its economic importance. David Pozo-Vazquez and collaborators explore the influence of the NAO on the solar and wind energy resources in the Mediterranean area, in particular, and over the whole North Atlantic area, in general. They show that interannual variability of the solar and wind energy resources in the Mediterranean area can reach values above 20% in winter and 10% in the annual case associated with changes in the NAO phase. They results are of interest

regarding the estimation of the expected interannual variability of the wind farms and solar plants production in the study region.

The set of chapters of this books provides the most complete overview about the impacts of the NAO in the Mediterranean region, highlighting the importance of this phenomenon not only in terms of the understanding of climate processes in the region, but also to know in depth the derived impacts that are affecting our environment, economy and natural resources.

Acknowledgements The scientific workshop that made possible this book was supported by the European Science Foundation, in the framework of the Mediterranean Climate Variability and Predictability (MedCLIVAR) program, the Spanish Ministry of Science and Innovation and the Aragón Government (Spain).

References

Cattiaux J, Vautard R, Cassou C, Yiou P, Masson-Delmotte V, Codron F (2010) Winter 2010 in Europe: a cold extreme in a warming climate. Geophys Res Lett 37. doi:10.1029/2010GL044613

Cook E, D'Arrigo RD, Mann ME (2002) A well-verified, multiproxy reconstruction of the winter North Atlantic Oscillation Index since A.D. 1400. J Clim 15:1754–1764

Dayan U, Lamb D (2008) Influence of atmospheric circulation on the variability of wet sulfate deposition. Int J Climatol 28:1315–1324

Della-Marta PM, Luterbacher J, von Weissenfluh H, Xoplaki E, Brunet M, Wanner H (2007) Summer heat waves over western Europe 1880–2003, their relationship to large-scale forcings and predictability. Clim Dyn 29:251–275

Demuzere M, Werner M, van Lipzig NPM, Roeckner E (2009) An analysis of present and future ECHAM5 pressure fields using a classification of circulation patterns. Int J Climatol 29: 1796–1810

Dünkeloh A, Jacobeit J (2003) Circulation dynamics of mediterranean precipitation variability 1948–98. Int J Climatol 23:1843–1866

Gallego MC, García JA, Vaquero JM, Mateos VL (2006) Changes in frequency and intensity of daily precipitation over the Iberian Peninsula. J Geophys Res D: Atmos 111:D24105. doi:10.1029/2006JD007280

García-Herrera R, Paredes D, Trigo RM, Trigo IF, Hernández H, Barriopedro D, Mendes MT (2007) The outstanding 2004–2005 drought in the Iberian Peninsula: the associated atmospheric circulation. J Hidrometeorol 9:483–498

García-Herrera R, Gallego D, Hernández E, Gimeno L, Ribera P (2001) Influence of the North Atlantic Oscillation on the Canary Island precipitation. J Clim 14:3889–3903

Gimeno L, Ribera P, Iglesias R, de la Torre L, Garcia R, Hernandez E (2002) Identification of empirical relationships between indices of ENSO and NAO and agricultural yields in Spain. Clim Res 21:165–172

Gouveia C, Trigo RM, DaCamara CC, Libonati R, Pereira JMC (2008) The North Atlantic Oscillation and European vegetation dynamics. Int J Climatol 28:1835–1847

Hurrell J (1995) Decadal trends in North Atlantic Oscillation and relationship to regional temperature and precipitation. Science 269:676–679

Hurrell JW, Kushnir Y, Ottersen G, Visbeck M (2003) The North Atlantic Oscillation: climate significance and environmental impact. Geophysical monograph series, vol 134. American Geophysical Union, Washington, DC

Hurrell J, Van Loon H (1997) Decadal variations in climate associated with the North Atlantic Oscillation. Clim Change 36:301–326

Jones PD, Jónsson T, Wheeler D (1997) Extension to the North Atlantic Oscillation using early instrumental pressure observations from Gibraltar and South-West Iceland. Int J Climatol 17:1433–1450

Karabörk MÇ, Kahya E, Karaca M (2005) The influences of the southern and North Atlantic Oscillations on climatic surface variables in Turkey. Hydrol Process 19:1185–1211

Küçük M, Kahya E, Cengiz TM, Karaca M (2009) North Atlantic Oscillation influences on Turkish lake levels. Hydrol Process 23:893–906

Lamb PJ, Peppler RA (1987) North Atlantic Oscillation: concept and application. Bull Am Meteorol Soc 68:1218–1225

Lloret J, Lleonart J, Solé I, Fromentin J-M (2001) Fluctuations of landings and environmental conditions in the north-western Mediterranean sea. Fisheries Oceanogr 10:33–50

Luterbacher J, Xoplaki E, Dietrich D, Jones PD, Davies TD, Portis D, González-Rouco JF, von Storch H, Gyalistras D, Casty C, Wanner H (2002) Extending North Atlantic Oscillation reconstructions back to 1500. Atmos Sci Lett 2:114–124

López-Moreno JI, Beguería S, Vicente-Serrano SM, García-Ruiz JM (2007) The influence of the NAO on water resources in Central Iberia: precipitation, streamflow anomalies and reservoir management strategies. Water Resour Res 43:W09411. doi:10.1029/2007WR005864

López-Moreno JI, Vicente-Serrano SM (2008) Extreme phases of the wintertime North Atlantic Oscillation and drought occurrence over Europe: a multi-temporal-scale approach. J Clim 21:1220–1243

Maynou F (2008) Influence of the North Atlantic Oscillation on Mediterranean deep-sea shrimp landings. Clim Res 36:253–257

Moses T, Kiladis N, Díaz H, Barry R (1987) Characteristics and frequency of reversals in mean sea level pressure in the north Atlantic sector and their relationship to long term temperature trend. J Climatol 7:13–30

Osborn TJ (2004) Simulating the winter North Atlantic Oscillation: the roles of internal variability and greenhouse gas forcing. Clim Dyn 22:605–623

Osborn TJ, Briffa KR, Tett SFB, Jones PD, Trigo RM (1999) Evaluation of the North Atlantic Oscillation as simulated by a coupled climate model. Clim Dyn 15:685–702

Pauling A, Luterbacher J, Casty C, Wanner H (2006) Five hundred years of gridded high-resolution precipitation reconstructions over Europe and the connection to large-scale circulation. Clim Dyn 26:387–405

Raible CC et al (2006) Climate variability-observations, reconstructions, and model simulations for the Atlantic-European and Alpine region from 1500–2100 AD. Clim Change 79:9–29

Roig FA, Barriopedro D, Herrera RG, Patón Domínguez D, Monge S (2009) North Atlantic Oscillation signatures in western Iberian tree-rings. Geogr Ann 91:141–157

Rubolini D, Ambrosini R, Caffi M, Brichetti P, Armiraglio S, Saino N (2007) Long-term trends in first arrival and first egg laying dates of some migrant and resident bird species in northern Italy. Int J Biometeorol 51:553–563

Solomon S, Qin D, Manning M, Chen Z, Marquis M, Averyt KB, Tignor M, Miller HL (2007) Climate change 2007: the physical science basis. Cambridge University Press, Cambridge, UK and New York, NY

Sousa P, Trigo RM, Aizpurua P, Nieto R, Gimeno L, Garcia-Herrera R (2011) Trends and extremes of drought indices throughout the 20th century in the Mediterranean. Nat Hazards Earth Syst Sci 11:33–51

Trigo IF (2006) Climatology and interannual variability of storm-tracks in the Euro-Atlantic sector: a comparison between ERA-40 and NCEP/NCAR reanalyses. Clim Dyn 26:127–143

Trigo RM et al (2006) Relations between variability in the Mediterranean region and Mid-Latitude variability. In: Lionello P, Malanotte-Rizzoli P, Boscolo R (eds) The Mediterranean climate: an overview of the main characteristics and issues. Elsevier, Amsterdam, pp. 179–226

Trigo RM, Osborn TJ, Corte-Real JM (2002) The North Atlantic Oscillation influence on Europe: climate impacts and associated physical mechanisms. Clim Res 20:9–17

Trigo RM, Pozo-Vazquez D, Osborn TJ, Castro-Diez Y, Gamiz-Fortis S, Esteban-Parra MJ (2004) North Atlantic Oscillation influence on precipitation, river flow and water resources in the Iberian Peninsula. Int J Climatol 24:925–944

Vicente-Serrano SM, Heredia A (2004) NAO influence on NDVI trends in the Iberian Peninsula (1982–2000). Int J Remote Sens 25:2871–2879

Vicente-Serrano SM, López-Moreno JI (2008a) The nonstationary influence of the North Atlantic Oscillation on European precipitation. J Geophys Res Atmos 113:D20120. doi:10.1029/2008JD010382

Vicente-Serrano SM, López-Moreno JI (2008b) Differences in the nonstationary influence of the North Atlantic Oscillation on European precipitation under different scenarios of greenhouse gases concentrations. Geophys Res Lett 35:L18710. doi:10.1029/2008GL034832

Vicente-Serrano SM, Trigo RM, López-Moreno JI, Liberato MLR, Lorenzo-Lacruz J, Beguería S, Morán-Tejeda E, El Kenawy A (2011) Extreme winter precipitation in the Iberian Peninsula in 2010: anomalies, driving mechanisms and future projections. Clim Res 46:51–65

Wanner H et al (2001) North Atlantic Oscillation—concepts and studies. Surv Geophys 22:321–382

Woolf DK, Shaw AGP, Tsimplis MN (2003) The influence of the North Atlantic Oscillation on sea-level variability in the North Atlantic region. Global Atmos Ocean Syst 9:145–167

Xoplaki E, González-Rouco JF, Luterbacher J, Wanner H (2003) Mediterranean summer air temperature variability and its connection to the large-scale atmospheric circulation and SSTs. Clim Dyn 20:723–739

Xoplaki E, González-Rouco JF, Luterbacher J, Wanner H (2004) Wet season Mediterranean precipitation variability: influence of large-scale dynamics and trends. Clim Dyn 23:63–78

Zêzere JL, Trigo RM, Fragoso M, Oliveira SC, Garcia RAC (2008) Rainfall-triggered landslides in the Lisbon region over 2006 and relationships with the North Atlantic Oscillation. Nat Hazards Earth Syst Sci 8:483–499

Zorita E, Kharin V, Von Storch H (1992) The atmospheric circulation and sea surface temperature in the North Atlantic area in winter: their interaction and relevance for Iberian precipitation. J Clim 5:1097–1108

Variability and Changes in the North Atlantic Oscillation Index

Tim J. Osborn

Abstract Most – or perhaps even all – of the observed variations in the winter North Atlantic Oscillation (NAO) index can be explained as internally-generated climate variability. The influence of external forcing factors on the observed NAO behaviour is still an open question. Two sea level pressure datasets yield different results for the strength of the winter NAO trend from the 1960s to the 1990s, though these trends from both datasets lie outside the 90% range of trends generated by the internal variability of climate models, and the latter is more precisely known now that over 8000 years of simulation under constant forcing is available for analysis. Similarly, a much expanded, multi-model ensemble of simulations under increasing anthropogenic forcing strengthens earlier findings of a shift in the mean state of the winter atmospheric circulation towards positive NAO conditions. There is considerable inter-model spread in both the magnitude of this response to increased forcing and in its regional structure, but of the 21 climate models analysed here, none showed an overall decrease in the mean level of the NAO index. If a shift towards positive NAO conditions is a realistic response to increasing anthropogenic forcing, then this signal has not yet emerged from the natural variability: observations since the 1990s show a return to lower values, and the 2009/2010 winter had the record negative NAO index in a record lasting almost two centuries. It is possible that anthropogenic forcing could be altering the temporal or spatial character of the interannual NAO variability, though only relatively small changes in pattern are evident when considering the multi-model ensemble as a whole and there is only weak evidence for an increase in the interannual variance.

Keywords Internal variability · External forcing · North Atlantic Oscillation index · Temperature

T.J. Osborn (✉)
Climatic Research Unit, School of Environmental Sciences, University of East Anglia, Norwich, UK
e-mail: t.osborn@uea.ac.uk

S.M. Vicente-Serrano, R.M. Trigo (eds.), *Hydrological, Socioeconomic and Ecological Impacts of the North Atlantic Oscillation in the Mediterranean Region*, Advances in Global Change Research 46, DOI 10.1007/978-94-007-1372-7_2,

1 Introduction

The interannual variability of the North Atlantic Oscillation (NAO) represents between 20 and 30% of the Northern Hemisphere winter atmospheric sea-level pressure (SLP) variance, and an even greater part of the variance in the Atlantic-European sector (e.g. Hurrell and Deser, 2009). For example, defining the NAO index as the leading principal component of winter (December to February or December to March, as here, are typically used) SLP in the domain 15–90°N and 110°W–70°E represents around 40% of the variance in this domain (Figs. 1a and 2a), updated from Osborn (2004). Taking instead a station-based index – simply the difference between standardised or raw SLP records near the Azores High and Iceland Low pressure systems can yield a longer record, but captures slightly less variance (because the principal component was optimised to maximise the variance captured). The Gibraltar minus Iceland record first developed by Jones et al. (1997) and updated here (Fig. 1b) now contains 187 complete winters (only those since 1887 are shown here).

This is all well established, as are the links between the NAO and some aspects of surface climate (Trigo et al., 2002; López-Moreno and Vicente-Serrano, 2008; and

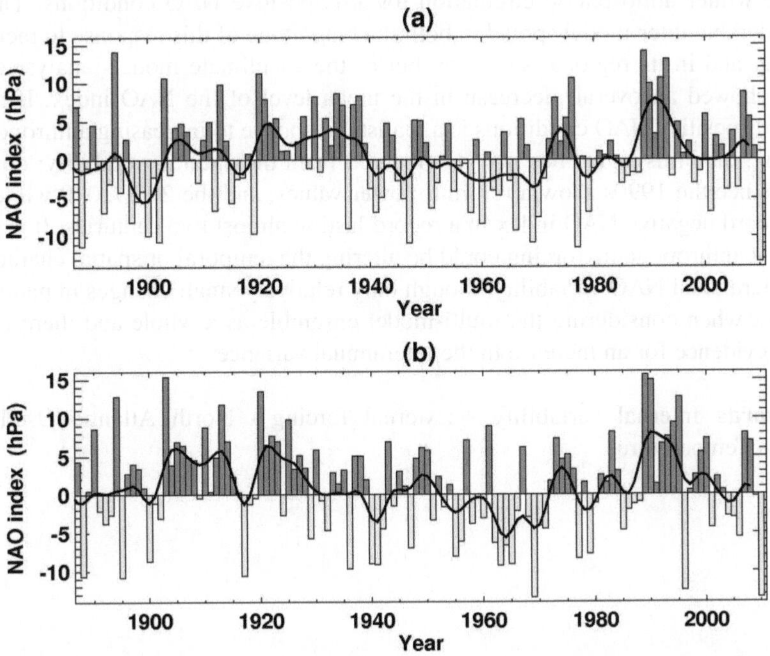

Fig. 1 Two alternative observed timeseries of the winter NAO index. (**a**) The leading principal component timeseries of Atlantic-European SLP, using the HadSLP2r dataset (Allan and Ansell, 2006). (**b**) The difference between Gibraltar and southwest Iceland SLP observations. The values represent the absolute pressure difference (hPa) between winter-mean values (**a**) from the maxima and minima of the EOF pattern, and (**b**) from Gibraltar and Iceland

many others). More recent research, coupled with longer observational records that enable empirical relationships to be better established, is increasing this knowledge and also extending it to new areas such as the consequences of NAO variability for various natural and societal systems. See the rest of this volume for many examples. What is considerably more uncertain is the link between internally-generated variability and externally-forced change, and whether the signature of the latter is evident in the observed record. The positive trend in the NAO index from the 1960s to the 1990s, evident in Fig. 1, has been a particular focus of such work (Osborn et al., 1999; Gillett et al., 2003; Osborn, 2004), though more general changes in the mean level of the NAO index have also been considered (Selten et al., 2004).

Others have investigated changes in the spatial structure of the NAO pattern (Fig. 2a). For example, a tendency for an eastward shift or extension of the NAO signature is apparent during part of the observed record (1978–1997 compared with 1958–1977; Jung et al., 2003) and also in the response to increased greenhouse gas concentrations simulated by some climate models (Ulbrich and Christoph, 1999). Multi-model analyses (Osborn, 2004; Kuzmina et al., 2005) suggest that such a result is model dependent, though the multi-model mean response in both cases did show a small north-eastward shift (Fig. 2b–d). A reconsideration of this is worthwhile using more recent climate models that typically have finer resolution and increasingly sophisticated representations of some physical and dynamical processes.

In addition to possible shifts in the mean level of the NAO index – i.e. corresponding to a change in the mean atmospheric circulation that has a structure that correlates positively with the NAO pattern (Fig. 2a) – or in the pattern of interannual NAO variability, it is feasible that external forcing could influence the *amplitude* of temporal variability. This is of topical interest because the winter 2009–2010 NAO index had a record low value in the 187-year record (Osborn, 2011). Though this could simply be a random event arising from internal variability, it is also important to assess the evidence for externally-forced changes in NAO variance which could alter the likelihood of such extreme events.

This chapter is structured as follows: Section 2 provides a brief description of the observed and climate model simulated datasets used in this study, while Section 3 compares the observed record with the range of internal variability simulated by the

Fig. 2 Spatial patterns of SLP variability associated with the winter NAO, estimated from the regression coefficients between local SLP and the NAO principal component index. (**a**) Observed pattern using HadSLP2r (Allan and Ansell, 2006). (**b**) Average simulated pattern from 7 older climate models used by Osborn (2004), using pre-industrial or twentieth-century control runs. (**c**) As (**b**), but using the 2050–2099 period under a scenario with increasing CO_2 concentrations (1% year^{-1}). (**d**) Difference (**c**) minus (**b**). (**e**) Average simulated pattern from 22 CMIP3 climate models, using the 1922–1999 period under historic forcing (anthropogenic only or anthropogenic and natural). (**f**) As (**e**), but using the 2050–2099 period under the SRES A1B scenario of increasing anthropogenic forcing. (**g**) Difference (**f**) minus (**e**). Note that the contour interval is 4 times smaller in *panel* (**d**) than in *panels* (**b**) and (**c**), and is 10 times smaller in *panel* (**g**) than in *panels* (**e**) and (**f**)

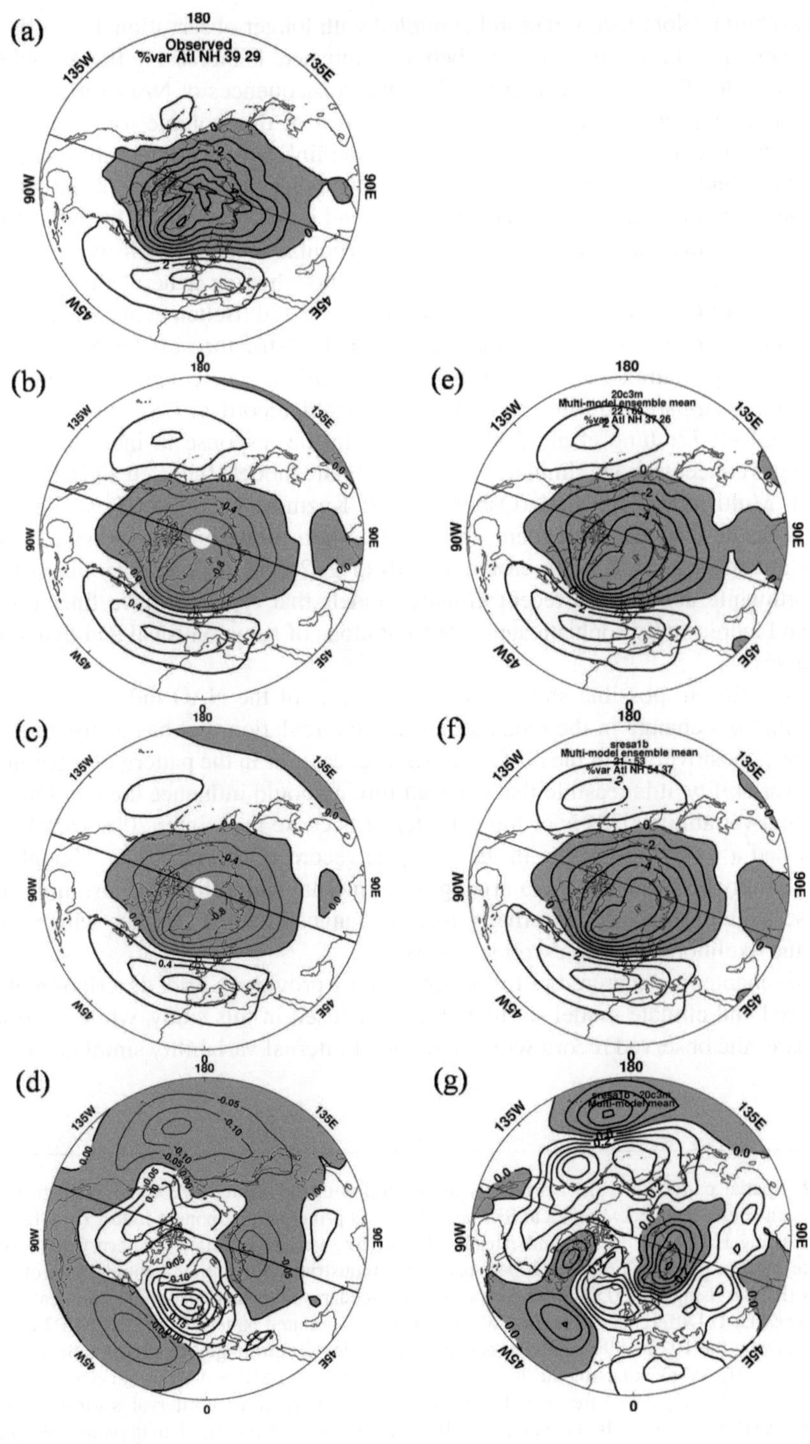

Fig. 2 (continued)

climate models. Then the simulated response to increasing greenhouse gases and other anthropogenic forcings is described, for the mean circulation (Section 4), the pattern of interannual variability (Section 5) and the amplitude of temporal variance (Section 6), before some concluding remarks in Section 7.

2 Observed and Simulated Data

2.1 Observed Data, NAO Pattern and NAO Index

Monthly-mean SLP fields from 1850 to 2010 from the HadSLP2r dataset (Allan and Ansell, 2006) were used. These consist of fields on a 5° latitude by 5° longitude grid based on ship and weather station observations till 2004, with near-real-time updates from 2005 to 2010 using NCEP/NCAR reanalysis data adjusted so that its climatological average matches the observed data. The EOF analysis used to define the NAO index was applied to the period from 1922 to 1999, because 1922 is the first year in the record when at least 20% of the Northern Hemisphere grid boxes contain at least 30 observations per month (Allan and Ansell, 2006). Note that the HadSLP2r dataset falls back below this threshold from 1941 to 1948; the dataset is spatially complete because it is based on interpolation from the available observations, but nevertheless the values of the NAO index prior to 1922 and between 1941 and 1948 should be considered with more caution.

The spatial pattern of the NAO (Fig. 2a, the leading EOF of the December-to-March seasonal mean SLP) is similar to the pattern found by Hurrell and Deser (2009) who used a longer period, a smaller spatial window focussed on the North Atlantic, and a different SLP dataset (Trenberth and Paolino, 1980, dataset with updates). One difference is in the relative strengths of the northern and southern parts of the dipole, which are more equal in Hurrell and Deser's pattern (bottom panel of their figure 8) compared with that shown in Fig. 2a here, perhaps because the EOF has not been calculated with area-weighting of the SLP data here. Hurrell and Deser (2009) also show that using the shorter December-to-February season (compare the top-left panel of their figure 6 with their figure 8) results in stronger loadings over the North Pacific – which is of interest because most climate models show an EOF pattern with that feature.

The observed NAO index obtained by projecting the HadSLP2r seasonal mean data fields onto the leading EOF is quite similar to the Gibraltar minus Iceland index (Fig. 1). This is in contrast to the earlier analysis of Osborn (2004), where the trend from the 1960s to the 1990s was more pronounced and the early 1990s values were distinctly higher than the preceding century. Osborn (2004) used SLP data based on the UK Met Office analyses, updated from Jones (1987). Hurrell and Deser (2009) use the leading EOF of the Trenberth and Paolino (1980) SLP data to obtain an NAO index that shows a more intermediate result. Although these NAO indices are highly correlated, they do have different characteristics (e.g. the strength of multi-decadal variations compared with shorter term variability) that can influence the statistical significance of trends. Further investigation is needed of the differences between

these SLP datasets (e.g. whether the reduced-space optimal interpolation algorithm used for HadSLP2r affects the amplitude of long-term variations).

2.2 *Simulated Data and NAO Patterns*

Osborn (2004) reported an analysis of the NAO as simulated by seven climate models whose primary references were published between 1997 and 2000 (table 1 of Osborn, 2004). Here, I report results of similar analyses applied to models that have been developed or improved more recently, using 22 models from the World Climate Research Programme's (WCRP's) Coupled Model Intercomparison Project phase 3 (CMIP3) multi-model dataset (Meehl et al., 2007). The focus is on the multi-model-mean results; detailed results from individual models will be presented elsewhere. The 22 models used here are listed in Table 1. The simulations used are control runs under constant pre-industrial forcings, those under historic transient forcing, beginning between 1850 and 1900 and extending to between 1999 and 2003, and those

Table 1 CMIP3 simulated data used in the analyses presented here

Model	Control run (year)	Historic forcing	Runs	Future forcing (SRES A1B)	Runs
bcc_cm1	–	1871–2003	4	–	–
bccr_bcm2_0	250	1850–1999	1	2000–2099	1
cccma_cgcm3_1	1001	1850–2000	5	2001–2200	5
cccma_cgcm3_1_t63	350	1850–2000	1	2001–2200	1
cnrm_cm3	390	1860–1999	1	2000–2200	1
csiro_mk3_0	380	1871–2000	3	2001–2200	1
gfdl_cm2_0	500	1861–2000	3	2001–2200	1
gfdl_cm2_1	500	1861–2000	3	2001–2200	1
giss_aom	502	1850–2000	2	2001–2100	2
giss_model_e_h	400	1880–1999	5	2000–2099	4
giss_model_e_r	500	1880–2003	9	2004–2200	5
iap_fgoals1_0_g	150	1850–1999	3	2000–2199	3
inmcm3_0	331	1871–2000	1	2001–2200	1
ipsl_cm4	–	1860–2000	1	2000–2100	1
miroc3_2_hires	100	1900–2000	1	2001–2100	1
miroc3_2_medres	500	1850–2000	3	2001–2200	1
				2001–2100	2
mpi_echam5	506	1860–2000	3	2001–2200	3
				2001–2100	1
mri_cgcm2_3_2a	350	1851–2000	5	2001–2200	1
				2001–2100	4
ncar_ccsm3_0	830	1870–1999	8	2000–2199	1
				2000–2099	3
ncar_pcm1	350	1890–1999	3	2000–2200	2
				2000–2099	2
ukmo_hadcm3	341	1860–1999	2	2000–2199	1
ukmo_hadgem1	–	1860–1999	2	2000–2099	1

under the SRES A1B future scenario of increasing greenhouse gas concentrations and other anthropogenic forcing to 2100. For most models, the SRES A1B simulations are extended beyond 2100 with fixed forcings, and model output up to 2200 is used here.

Multiple simulations under the same forcing, but different initial conditions and hence different realisations of internal variability, are available in some cases. These initial-condition ensembles are used by analysing the individual ensemble members separately, and then averaging the results to produce an ensemble mean. Table 1 indicates the availability of simulations, including the size of initial-condition ensembles, for each of the models used here.

Osborn (2004) explained the advantages of regridding all datasets to a common grid prior to analysis, so here all simulated data are regridded to the grid of the HadSLP2r observational data. The same methods described in Osborn (2004) for calculating and scaling the principal component NAO index time series from an EOF analysis of each model simulation are used here – refer to that paper for a detailed description.

The simulated spatial pattern of the NAO, when averaged across the 22 climate models with twentieth century simulations (Fig. 2e), bears a very close resemblance to the observed NAO pattern (Fig. 2a). As with the earlier models (Fig. 2b), the teleconnection between the NAO and the North Pacific is stronger than observed, resembling the Northern Annular Mode more closely (Miller et al., 2006). Note that the NAO is defined here from an EOF analysis of only the Atlantic half of the hemisphere, but the pattern shown in the maps is from the pattern of regression coefficients between the principal component NAO index and SLP from across the entire northern hemisphere. With this exception, the similarity with the observed pattern provides some justification for using these model experiments to investigate NAO behaviour.

3 Comparing the Observed Record with Internally-Generated Climate Variability

The detection of climate changes that might be caused by some external forcing requires the comparison of an observed record with an estimate of the variations that might typically be expected to arise from internal variability. In most cases, the amplitude of internal variability increases with decreasing timescale, so that even where a real climate change is present the likelihood of detecting it is rather small within a short (e.g. sub-decadal) record. But longer-term trends must be compared with an estimate of internally-generated variability on multi-decadal timescales, and a multi-century record is required to provide an adequate sample of such variations. The instrumental record is not long enough and is also "contaminated" by any responses to natural or anthropogenic forcings, rather than being a "pure" representation of internal variability. In most detection studies, therefore, the internal climate variability is estimated from climate model simulations, using

a multi-century control run (i.e., a simulation in which the external forcings are constant).

Detection results are thus completely dependent on whether the climate model(s) adequately simulate climate variability on the timescale of interest (e.g. multi-decadal fluctuations). These simulations can be evaluated by comparison with the observational record, but the focus is of course on the shorter timescales that are sampled within the limited length of the observations. Studies that go beyond detection and include an attempt to attribute some of the observed changes to particular external forcings also check for consistency between the observed changes and a combination of internal variability and attributed external effects. Although this can identify instances where the internal variability simulated by the climate model is inconsistent with the observed evidence, the power of the test is again limited for the crucial multi-decadal timescales.

Osborn (2004) compared the sub-decadal variance of simulated and observed NAO indices and found that three of the models had variances that were significantly higher than observed, while the other four were similar. Detection analyses were repeated, therefore, after scaling the simulated NAO indices to have the same sub-decadal variance as observed. This scaling step profoundly alters the meaning of the detection exercise, as noted by Osborn (2004), because the climate models are then only being used to estimate the *spectral shape* of the NAO indices, rather than the absolute magnitude of NAO variability.

The analysis of Osborn (2004) is repeated here using the CMIP3 pre-industrial control runs (Table 1), together with the NAO index from the HadSLP2r dataset. The 30-year trends evident in this observed NAO index, calculated using a sliding window, are compared with the distributions of 30-year trends simulated by these climate models with fixed, unchanging forcings (Fig. 3). These distributions are calculated from all the 30-year trends in the simulated NAO indices from the 19 climate models with pre-industrial control run data. The 5–95% ranges of these simulated trends are then determined either for each climate model separately (Fig. 3b) or after all the trends have been pooled together over the multi-model ensemble (Fig. 3a). This represents a considerable extension to the analysis of Osborn (2004), with 19 rather than 7 models, and total simulation time of over 8000 years rather than under 3000 years; in addition, many of the models are more sophisticated or have higher spatial and temporal resolution.

The results show that when the simulated NAO indices are scaled to have the same sub-decadal variance as the observed winter NAO index, the 30-year trends centred around 1980 (e.g. 1965–1994) lie outside the 90% range of internal variability simulated by these climate models. The result is the same whether the models are pooled together (Fig. 3a) or whether they are considered separately (Fig. 3b), even for the models with the strongest multi-decadal variability (relative to their sub-decadal variance). Note that Osborn (2004) showed the 95% range (from 2.5 to 97.5%) of the control run variability, whereas here I have used the 90% range (from 5 to 95%) because previous analysis of multi-model ensembles (e.g. Osborn, 2004; Kuzmina et al., 2005) now yields the a priori expectation of an increase in the NAO index, and thus 95% confidence in detecting an externally-forced trend can be

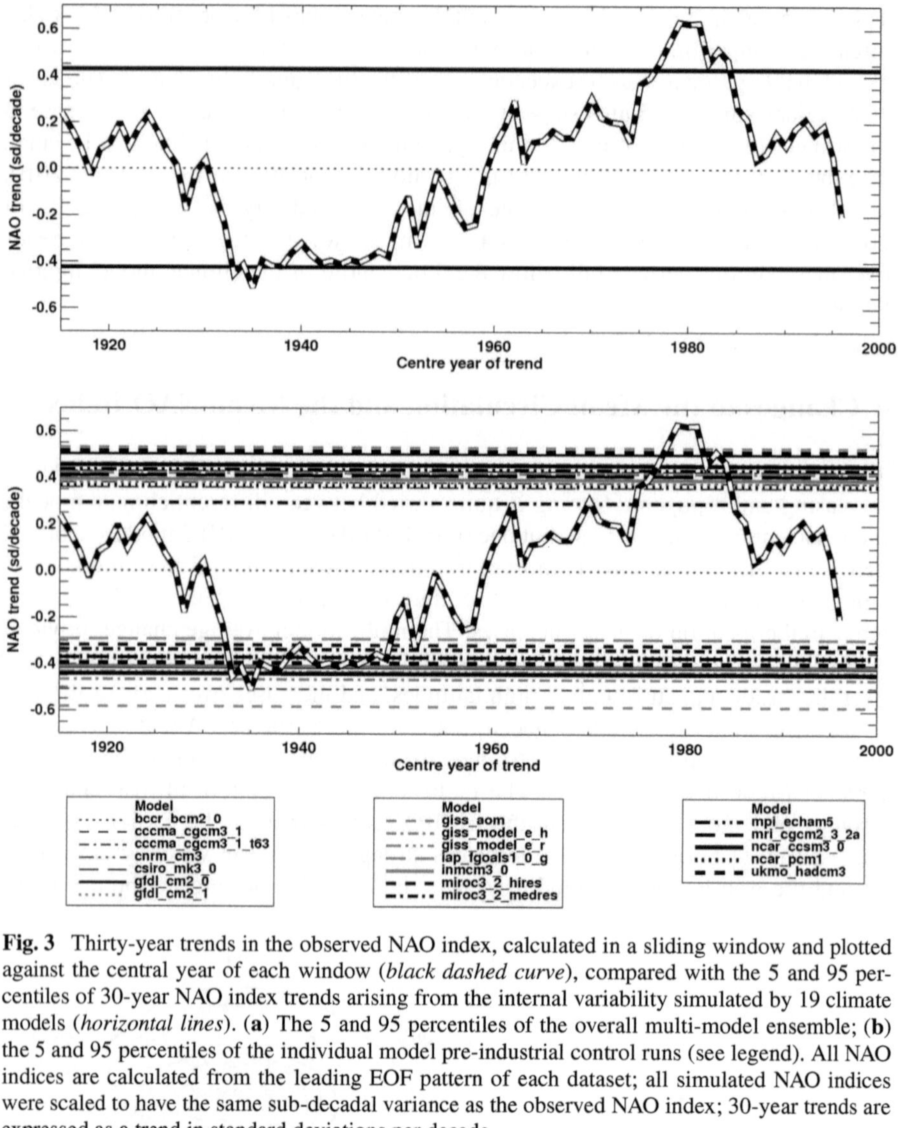

Fig. 3 Thirty-year trends in the observed NAO index, calculated in a sliding window and plotted against the central year of each window (*black dashed curve*), compared with the 5 and 95 percentiles of 30-year NAO index trends arising from the internal variability simulated by 19 climate models (*horizontal lines*). (**a**) The 5 and 95 percentiles of the overall multi-model ensemble; (**b**) the 5 and 95 percentiles of the individual model pre-industrial control runs (see legend). All NAO indices are calculated from the leading EOF pattern of each dataset; all simulated NAO indices were scaled to have the same sub-decadal variance as the observed NAO index; 30-year trends are expressed as a trend in standard deviations per decade

assessed using a one-tailed test. The observed trend also exceeds the 97.5 percentile (not shown) of the CMIP3 control run variability (for all individual models as well as for the multi-model ensemble), so this difference in choice of range is not critical.

The degree to which the observed trend exceeds the simulated internal variability is less in this study than found by Osborn (2004). This is mostly because the NAO index obtained from the HadSLP2r dataset has a considerably weaker trend than that found by Osborn (2004); the former peaks just above 0.6 standard deviations per decade, whereas the latter peaked just over 1.1 standard deviations per decade.

This was noted in Section 2.1. As a result, the behaviour of the observed NAO index might be compatible with the internal variability of the models, since we would expect the 95 percentile to be exceeded 5% of the time purely by chance. The lower NAO values since 1995 also suggest that any contribution to the earlier increase in the winter NAO index from continuing greenhouse gas forcing is likely to be small (see the discussion in Osborn, 2004). The most recent 30-year trend (1981–2010) is still comfortably within the range of internal variability, but analysing 15- or 20-year trends in the same way (not shown) shows that the most recent trends over these shorter periods fall below the 5 percentile of the climate model internal variability.

4 Changes in the Mean Circulation and the Mean NAO Index

The response of the Northern Hemisphere atmospheric circulation in winter to increasing anthropogenic forcing is quite varied between different climate models, though a feature common to all but one (GISS AOM) of the CMIP3 models analysed here is that SLP decreases over most of the Arctic. Most of these climate models also simulate an increase in SLP over parts of the subtropics, though there is less consistency in the location of these increases. The multi-model average change in winter SLP (Fig. 4) displays these consistent or semi-consistent features: decreases in SLP over the Arctic and surrounding high latitudes, and increases over parts of the subtropics, especially the Mediterranean, but also the Pacific and Atlantic. This is in agreement with Miller et al. (2006), who examined the behaviour of the Northern Annular Mode in these models. The multi-model mean pattern of SLP change is stronger and more consistent in the CMIP3 models than in the earlier ensemble analysed by Osborn (2004), and indicates strengthened mean westerly circulation around the mid-latitudes, including across the Atlantic and into northern Europe.

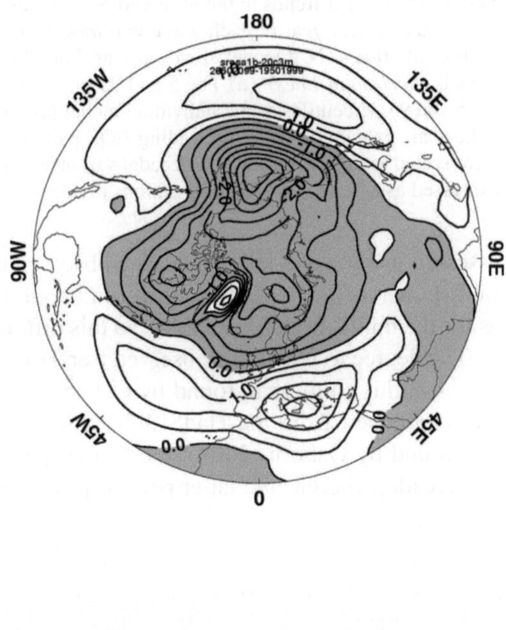

Fig. 4 Differences in long-term mean December-to-March SLP (hPa) simulated during the 1950–1999 period of the historic simulations and the 2050–2099 of the simulations under the SRES A1B scenario, averaged across all available model simulations

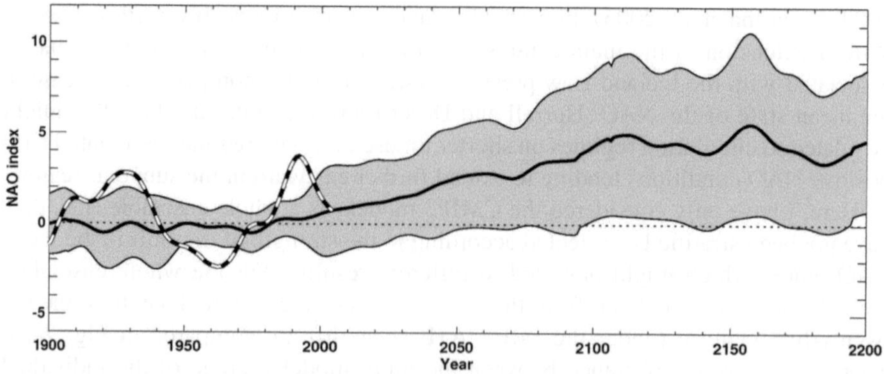

Fig. 5 Timeseries of observed (*black dashed line*) and simulated December-to-March NAO index, smoothed with a 30-year low-pass filter. Only the mean and estimated 5–95% range of the ensemble of individual climate model results are shown, indicated by the *thick black line* and the *grey shading*. The simulations are driven by historic forcing estimates up to around 2000, then by the increasing anthropogenic forcing of the SRES A1B scenario to 2100, followed by constant forcings to 2200. The NAO indices are calculated from the leading EOF pattern of each dataset, and then scaled to represent the change in absolute pressure difference (hPa) between the maximum and minimum of the EOF pattern

There are also strong changes in the western North Pacific, though they are not relevant to the current paper.

Given this multi-model pattern of SLP change, and some consistency between models in the large-scale structure (though not in the regional details for individual models – not shown here), it is not surprising that the simulated NAO indices show a long-term shift towards positive values (Fig. 5). The individual model results (not shown) provide a large range of future projections, from almost no change up to values in excess of a 5 hPa increase in the pressure difference between the locations of maximum and minimum of the EOF patterns (note that these do not correspond exactly with the maximum and minimum in the present-day SLP in this region). The GISS AOM model simulates the lowest NAO index during the twenty-first century, but even for this model there is no statistically significant decrease in the NAO index. The inter-model spread is used to estimate a 5–95% range of NAO projections from the CMIP3 ensemble, and this is shown in Fig. 5. The observed NAO index is also shown, though it is not strictly comparable because some of the model results are based on the mean of an initial-conditions ensemble (Table 1), which reduces the spread arising from internal variability.

5 Changes in the Pattern of Interannual NAO Variability

Previous work (summarised in Section 1) identified the possibility that the spatial pattern of interannual variability represented by the NAO might change when the mean climate changes (e.g. Jung et al., 2003; Ulbrich and Christoph, 1999; Osborn,

2004; Kuzmina et al., 2005). In particular, an eastward (or possibly north-eastward) shift or extension of the main centres of action, especially the northern anomalies associated with the Iceland Low pressure system, might accompany an increase in the mean state of the NAO. Hurrell and Deser (2009) also discuss how this might be related to circulation regimes on shorter timescales, with regimes associated with positive NAO conditions tending to extend further eastward in the subpolar region.

Here, I have only considered the CMIP3 models as a whole ensemble (i.e., they have not been stratified or selected according to the strength of the shift in the mean NAO index, which might have led to different results). For the whole ensemble, the EOF patterns calculated from the last 50 years of the twentieth century and the twenty-first century (under the SRES A1B scenario) are compared in Fig. 2e, f. There is very little difference between the multi-model average of the individual EOFs during the two periods. The difference between the EOF patterns (Fig. 2g) shows some minor differences (note that the contour interval in Fig. 2g is ten times smaller than in 2e and 2f). The small negative changes (shaded) in the loadings over Scandinavia and north-west Russia, coupled with the small positive changes (unshaded) south of Iceland, represent a minor eastward shift in the northern centre of action. The change in the multi-model mean EOF pattern is, however, considerably smaller than obtained by Osborn (2004) from the smaller ensemble of older climate models (Fig. 2b–d).

6 Changes in the Amplitude of Interannual NAO Variability

Few changes in the amplitude of NAO variability simulated by individual CMIP3 climate models are statistically significant. This is not surprising, however, because higher-order statistics, such as the temporal standard deviation, are associated with greater sampling variability and changes in those statistics are, therefore, more "difficult" to detect than changes in the mean climate (i.e. even when a real change in variability is present, the likelihood of detecting it is smaller than for an equivalent change in the mean). Treating them as a multi-model ensemble can help to identify relatively small changes, providing the changes are consistent across models. In this case, however, there is little consistency between models. Comparing the intradecadal standard deviations of the simulated NAO indices during the last 50 years of the twentieth century with the last 50 years of the twenty-first century (under the SRES A1B scenario), seven models show an increase of at least 10%, three show a decrease of at least 10%, and 11 show changes of less than 10%. Averaging across all 21 models shows a minor increase of 4% in the standard deviation of the NAO index under the SRES A1B scenario.

Another measure of interest is the fraction of overall SLP variability that is associated with the NAO. Within the Atlantic half of the hemisphere, and over the same periods used above, there is also little change under increased anthropogenic forcing. Averaged across all 21 models, the SLP variance associated with the leading EOF increases from 39 to 41%. However, when assessed for individual models, there are nearly as many that show a decrease in the variance explained as show an increase.

7 Conclusions

Fluctuations in atmospheric circulation related to the North Atlantic Oscillation are an important driver of climate variability over Europe and the Mediterranean region. Yet it is rather uncertain whether anthropogenic forcings might cause a shift in the mean atmospheric circulation that resembles the NAO, or in some way alter the characteristics of the internal variability of the NAO (e.g. the spatial pattern or the temporal variance). This cannot be confidently determined solely from an empirical analysis of the observed record, due to difficulties in separating the response to individual external forcings from internal variability within the available observational record. Other approaches to learning more about NAO behaviour include examining processes within the atmosphere (e.g. stratosphere-troposphere interactions, Perlwitz and Harnik, 2004) that might influence the NAO.

An alternative approach, taken in this study, is to analyse ensembles of multiple climate models (based on general circulation models of the atmosphere and oceans). Here the CMIP3 ensemble of around 20 climate models shows that increasing anthropogenic forcings (principally greenhouse gases) might drive a shift of the mean circulation towards positive NAO conditions, with enhanced westerly flow across the Atlantic and into northern Europe. The magnitude of the simulated shift is very model dependent, and whether such an externally-forced response is needed to explain the observed NAO record is uncertain. The observed increase in the winter NAO index from the 1960s to the 1990s is large compared with the models' internally-generated variability, but the dependence of the magnitude of this trend on the sea level pressure dataset used, plus the return to lower values of the NAO index since the early 1990s, reduce the likelihood that a strong influence of greenhouse gas forcing is already evident in the observed NAO record. This study significantly extends earlier multi-model analyses of NAO behaviour (Osborn, 2004; Kuzmina et al., 2005) by using many more climate models, including some with finer resolution and more sophisticated representations of physical and dynamical processes. It also complements multi-model studies of related atmospheric circulation features such as the annular modes (Miller et al., 2006).

Acknowledgements This work was supported by a Research Councils UK Academic Fellowship. The modelling groups, the Program for Climate Model Diagnosis and Intercomparison (PCMDI) and the WCRP's Working Group on Coupled Modelling (WGCM) are acknowledged for their roles in making available the WCRP CMIP3 multi-model dataset. Support for CMIP3 dataset is provided by the Office of Science, U.S. Department of Energy.

References

Allan RJ, Ansell TJ (2006) A new globally complete monthly historical mean sea level pressure data set (HadSLP2): 1850–2004. J Clim 19:5816–5842

Gillett NP, Graf HF, Osborn TJ (2003) Climate change and the North Atlantic Oscillation. In Hurrell JW, Kushnir Y, Ottersen G, Visbeck M (eds) North Atlantic Oscillation: climatic significance and environmental impact. Geophysical Monograph, vol 134. American Geophyscial Union, Washington, DC, pp 193–209

Hurrell JW, Deser C (2009) North Atlantic climate variability: the role of the North Atlantic Oscillation. J Mar Syst 78:28–41

Jones PD (1987) The early twentieth century Arctic high—fact or fiction? Clim Dyn 1:63–75. doi:10.1007/BF01054476

Jones PD, Jonsson T, Wheeler D (1997) Extension to the North Atlantic Oscillation using early instrumental pressure observations from Gibraltar and South-West Iceland. Int J Climatol 17:1433–1450

Jung T, Hilmer M, Ruprecht E, Kleppek S, Gulev SK, Zolina O (2003) Characteristics of the recent eastward shift of interannual NAO variability. J Clim 16:3371–3382

Kuzmina SI, Bengtsson L, Johannessen OM, Drange H, Bobylev LP, Miles MW (2005) The North Atlantic Oscillation and greenhouse-gas forcing. Geophys Res Lett 32:L04703. doi:10.1029/2004GL021064

López-Moreno JI, Vicente-Serrano SM (2008) Positive and negative phases of the wintertime North Atlantic Oscillation and drought occurrence over Europe: a multitemporal-scale approach. J Clim 21:1220–1243

Meehl GA, Covey C, Delworth T, Latif M, McAvaney B, Mitchell JFB, Stouffer RJ, Taylor KE (2007) The WCRP CMIP3 multi-model dataset: a new era in climate change research. Bull Am Meteorol Soc 88:1383–1394

Miller RL, Schmidt GA, Shindell DT (2006) Forced annular variations in the 20th century Intergovernmental Panel on Climate Change Fourth Assessment Report models. J Geophys Res 111:D18101. doi:10.1029/2005JD006323

Osborn TJ (2004) Simulating the winter North Atlantic Oscillation: the roles of internal variability and greenhouse gas forcing. Clim Dyn 22:605–623

Osborn TJ (2011) Winter 2009/2010 temperatures and a record-breaking North Atlantic Oscillation index. Weather 66:19–21

Osborn TJ, Briffa KR, Tett SFB, Jones PD, Trigo RM (1999) Evaluation of the North Atlantic Oscillation as simulated by a coupled climate model. Clim Dyn 15:685–702

Perlwitz J, Harnik N (2004) Downward coupling between the stratosphere and troposphere: the relative roles of wave and zonal mean processes. J Clim 17:4902–4909

Selten FM, Branstator GW, Dijkstra HA, Kliphuis M (2004) Tropical origins for recent and future Northern Hemisphere climate change. Geophys Res Lett 31:L21205. doi:10.1029/2004GL020739

Trenberth KE, Paolino DA (1980) The Northern Hemisphere sea level pressure data set, trends, errors and discontinuities. Mon Wea Rev 108:855–872

Trigo RM, Osborn TJ, Corte-Real J (2002) The North Atlantic Oscillation influence on Europe: climate impacts and associated physical mechanisms. Clim Res 20:9–17

Ulbrich U, Christoph M (1999) A shift of the NAO and increasing storm track activity over Europe due to anthropogenic greenhouse gas forcing. Clim Dyn 15:551–559

The NAO Impact on Droughts in the Mediterranean Region

Sergio M. Vicente-Serrano, Juan I. López-Moreno, Jorge Lorenzo-Lacruz, Ahmed El Kenawy, Cesar Azorin-Molina, Enrique Morán-Tejeda, Edmond Pasho, Javier Zabalza, Santiago Beguería, and Marta Angulo-Martínez

Abstract This chapter shows the influence of the North Atlantic Oscillation (NAO) on droughts in the entire Mediterranean region between 1901 and 2006. The analysis has been based on identification of positive and negative NAO winters and also detection of the anomalies of drought severity by means of the Standardized Precipitation Evapotranspiration Index (SPEI). The analysis is focussed on the winter NAO. Nevertheless, given that the SPEI drought indicator can be obtained at different time-scales, the study shows how the effects of the winter NAO on droughts are propagated for the following months when long time scales are considered. In general, during the positive phases, the negative SPEI averages are recorded in Southern Europe (the Iberian Peninsula, Italy and the Balkans), areas of Turkey and northwest Africa. On the contrary, the SPEI averages are found positive in northeast Africa. The opposite configuration, but with some differences in the spatial patterns and the magnitude of the SPEI averages, is found during the negative NAO years. The findings of this study should be of great applicability in terms of developing early warning systems. The established relationships between NAO phases and drought indices seem appropriate for drought prediction over large areas of the Mediterranean basin.

Keywords Standardized Precipitation Evapotranspiration Index (SPEI) · Drought index · Drought trends · Time-scales · Potential evapotranspiration · North Atlantic Oscillation (NAO)

1 Introduction

Droughts are first-order hazards of high frequency and intensity in the Mediterranean region. They are one of the main natural causes of agricultural, economic and environmental damage (Burton et al., 1978; Wilhite, 1993). In contrast to other extreme events such as floods, which are typically restricted to small

S.M. Vicente-Serrano (✉)
Instituto Pirenaico de Ecología, Spanish National Research Council (CSIC), Zaragoza, Spain
e-mail: svicen@ipe.csic.es

S.M. Vicente-Serrano, R.M. Trigo (eds.), *Hydrological, Socioeconomic and Ecological Impacts of the North Atlantic Oscillation in the Mediterranean Region*, Advances in Global Change Research 46, DOI 10.1007/978-94-007-1372-7_3,

regions and well-defined temporal intervals, droughts are difficult to pinpoint in time and space. In particular, droughts can affect wide areas over long periods of time. It is also very difficult to determine the beginning of a drought event until human activities, or the environment, are affected. Moreover, the influences of a drought can persist over longer periods after it has ended (Changnon and Easterling, 1989). Although droughts are complex phenomenon related to a multitude of factors (Wilhite and Glantz, 1985), they always have a climatic origin. This has promoted scientific research to investigate the behaviour and characteristics of climatic droughts in the Mediterranean region, including their driving mechanisms (Van del Schrier et al., 2006).

In the Mediterranean region, droughts are a climatic risk that have severe consequences for agriculture and natural vegetation (Austin et al., 1998; Lázaro et al., 2001; Reichstein et al., 2002; Iglesias et al., 2003), increasing the frequency of fires (Colombaroli et al., 2007; Pausas, 2004) and significantly reducing water availability for urban and tourist consumption (Morales et al., 2000). Droughts are frequent, but not spatially uniform in the Mediterranean region (Briffa et al., 1994; López-Moreno and Vicente-Serrano, 2008), with significant spatial differences, even at regional scale (Vicente-Serrano et al., 2004; Vicente-Serrano, 2006).

The North Atlantic Oscillation (NAO) is the primary atmospheric circulation mode that determines the climate of the Mediterranean region (Hurrell et al., 2003), with large and well known impacts on temperature, precipitation, cloudiness, etc. (e.g., Hurrell et al., 2003; Hurrell and Van Loon, 1997; Trigo et al., 2002). Different studies have focused on analysing the role of the NAO on droughts in South of Europe and the Mediterranean region (Briffa et al., 1994; Lloyd-Hughes and Saunders, 2002; Van del Schrier et al., 2006). These studies made significant contributions to literature in terms of analyzing the spatial and temporal patterns of drought using one of the most widely used drought indices: the Palmer Drought Severity Index (PDSI; Palmer, 1965).

In addition, drought is widely accepted as a multi-scalar phenomenon. McKee et al. (1993) clearly illustrated this essential characteristic of droughts through consideration of usable water resources including: soil moisture, ground water, snowpack, river discharges, and reservoir storages. The time period from the arrival of water inputs to availability of a given usable resource differs considerably. Thus, the time scale over which water deficits accumulate becomes extremely important, and functionally separates hydrological, environmental, agricultural and other droughts. For example, the response of hydrological systems to precipitation can markedly vary as a function of time (Changnon and Easterling, 1989; Elfatih et al., 1999; Pandey and Ramasastri, 2001). This is determined by the different frequencies of hydrologic/climatic variables (Skøien et al., 2003). For this reason, drought indices must be associated with a specific timescale to be useful for monitoring and management of different usable water resources. Thus, the 9–12-month fixed time scale considered by the PDSI (Guttman, 1998) limits the ability of the user to analyze certain characteristics of droughts. This is mainly because the response of different water sources to drought is very sensitive to the chosen time scale. López-Moreno and Vicente-Serrano (2008) used the Standardised Precipitation

Index (SPI; McKee et al., 1993) to determine the impact of the NAO on drought conditions, quantified at different time-scales, in Europe. They found that the response of droughts to the NAO varies spatially, and differs largely according to the month of the year and the time scale of the analysis.

At present there is no study that analyses the role of the NAO on droughts at different time scales across the entire Mediterranean region. In this regard, the study conducted by López-Moreno and Vicente-Serrano (2008) only focussed on the Mediterranean Europe employing the SPI, which neglects the role of other variables different to precipitation, to determine drought severity. Different empirical studies have shown that temperature rise markedly affects the severity of droughts and their derived impacts (e.g., Abramopoulos et al., 1988; Breshears et al., 2005; Ciais et al., 2005; Cai and Cowan, 2008; Gerten et al., 2008). Owing to its general increase in the Mediterranean region (0.5–2°C) during the last past 150 years (Jones and Moberg, 2003), it is reasonably expected to have adverse consequences of temperature on drought conditions, with an increase in water demand due to evapotranspiration (Sheffield and Wood, 2008). For this purpose, and to analyse the NAO impact on drought at different time-scales, we have used a new drought index (the Standardized Precipitation Evapotranspiration Index; SPEI). This index is advantageous in that it is sensitive to changes in evaporation demand and can also be obtained at different time-scales.

The objective of this chapter is to show the influence of NAO phases on drought conditions in the Mediterranean region both spatially and temporally, and also to include the influence on different time-scales of drought.

2 Methods

To analyse the NAO impact on droughts throughout the Mediterranean basin, we used the SPEIbase, which is based on the Standardised Precipitation Evapotranspiration Index (SPEI). The data set covers the period 1901–2006 considering time-scales of drought from 1 to 24 months at a spatial resolution of 0.5°, which represent the variability of a variety of water usable sources (Szalai et al., 2000; Sims et al., 2002; Vicente-Serrano and López-Moreno, 2005; Patel et al., 2007; Khan et al., 2008; Lorenzo-Lacruz et al., 2010) and may have very different impacts on natural ecosystems and crops (Ji and Peters, 2003; Vicente-Serrano, 2007; Quiring and Ganesh, 2010). More details about the SPEI and the corresponding data set can be found in Vicente-Serrano et al. (2010), Beguería et al. (2010) and Vicente-Serrano et al. (2010b). From the SPEIbase we selected the area comprised between 25.5–48.5°N and 10°W–45.5°E.

The analysis of the NAO impact on droughts was based on a composite procedure by means of the selection of the positive and negative NAO winters. In the present study, we used the NAO index calculated from the difference between the standardized pressure anomalies measured at Gibraltar (south of Spain) and Reykjavík (Iceland). The advantage of using these stations is discussed in depth by Jones et al. (1997), who noted that Gibraltar appears to better represent the southern

part of the NAO dipole than other commonly used stations, such as Lisbon or Ponta Delgada in the Azores. The NAO index was compiled by the Climate Research Unit (CRU), University of East Anglia (http://www.cru.uea.ac.uk/cru/data/nao.htm). We calculated the average wintertime NAO index for months from December to March (DJFM) because the NAO has higher activity at this time of the year (Hurrell et al., 2003). Interdecadal variability of the NAO appears to be the most coherent when the December–March season is used, rather than using other months (Osborn et al., 1999). Since the NAO index is calculated by the CRU using data from the beginning of the nineteenth century and using the period 1950–1980 as reference, the average of the winter NAO index for the period 1901–2006 is not 0. Therefore, in order to have the same number of positive and negative winters and a magnitude comparable, the resulting winter NAO series obtained from the average of the monthly values was standardised to have an average of 0 and a standard deviation equal to 1. Based on the obtained series, we identified the positive and negative NAO winters as those with values that differed from the average by at least $\pm\sigma$. Consequently, 16 years were identified as having positive phases and 16 as having negative phases. The positive phases are found in 1903, 1905, 1910, 1913, 1920, 1925, 1957, 1961, 1967, 1983, 1989, 1990, 1992, 1994, 1995, and 2000; while the negative phases occur in 1917, 1936, 1940, 1941, 1947, 1955, 1956, 1963, 1964, 1965, 1969, 1971, 1977, 1979, 1996 and 2006.

For both positive and negative phases of the NAO, we calculated the average values of the SPEI at different time scales for each 0.5° series. We then used the Wilcoxon–Mann–Whitney rank-sum test (Siegel and Castelan, 1988; Wilks, 2006) to determine whether contrasted NAO phases (positive/negative) lead to statistically significant differences in SPEI averages (dry/humid conditions) at different time scales. The Wilcoxon–Mann–Whitney rank-sum test is based on ranks that do not require normally distributed samples; it is slightly less powerful than parametric tests such as the t test (Helsel and Hirsch, 1992). Although the SPEI is a normalized variable for the entire series, the normality of the sample cannot be guaranteed, and the distribution may be biased when a sample of years is extracted (i.e., the SPEI values during positive and negative years). For this reason, the use of a non-parametric test is more convenient. The sample that comprised SPEI values for each of the months of the positive/negative years was compared with the values of SPEI for months of normal years (those years within the rank $+\sigma >$ NAO $<-\sigma$) and months of years with the opposite sign. The significance level was established at $\alpha < 0.05$.

Given that SPEI anomalies for the positive and negative NAO years were obtained for a range of time scales and months for each 0.5° series, we summarised the information by means of Principal Component Analysis (PCA) applied to the matrix of month/time-scale SPEI anomalies for each series. PCA has been widely used to determine the general temporal and spatial patterns of climatic variables. It allows common features to be identified and specific local characteristics to be determined (Jolliffe, 1990). In this case, the purpose was to identify the general patterns of the month/time-scale SPEI anomaly matrixes for the positive and negative NAO phases. The areas represented by each mode (component) were identified

by mapping the factorial loadings. The number of components was selected in accordance with the criteria of an eigenvalue larger than 1.

3 Results

Figure 1 shows, as a representative example, the evolution of the SPEI at the time scales of 3-, 6-, 9-, 12-, 18- and 24-month in a region of central France (46.5°N, 8°E). Negative SPEI index values indicate higher severity of droughts. On the shorter time scales (3 or 6 months), the dry and humid periods are short and occur with high-frequency. At a time scale of 12 months, droughts were less frequent, but they lasted longer. At longer time scales (18 or 24 months), droughts lasted longer, but were less frequent, with very few dry or humid periods recorded. The index allows to identify the main drought periods affected this region in the decades of 1940, 1990 and 2000.

Figure 2 shows the SPEI anomalies corresponding to the entire Mediterranean region from the average Mediterranean SPEI series in the positive and negative NAO phases. The pattern is very different among both phases, with general negative (positive) SPEI anomalies across time scales and months during the positive (negative) NAO phases. The most negative anomalies recorded for the positive

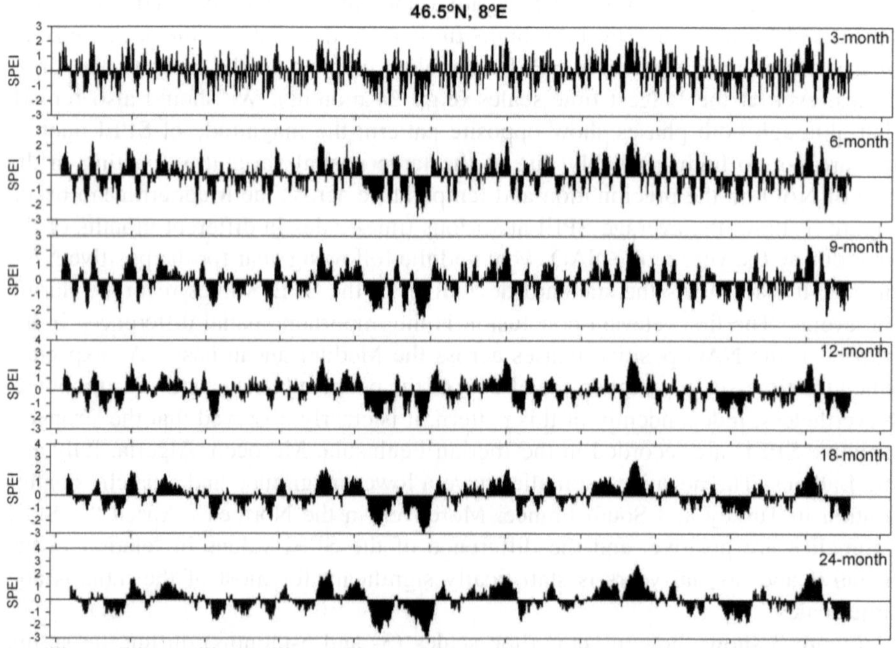

Fig. 1 Evolution of the 3-, 6-, 9-, 12-, 18- and 24-month SPEI in central France (46.5°N, 8°E)

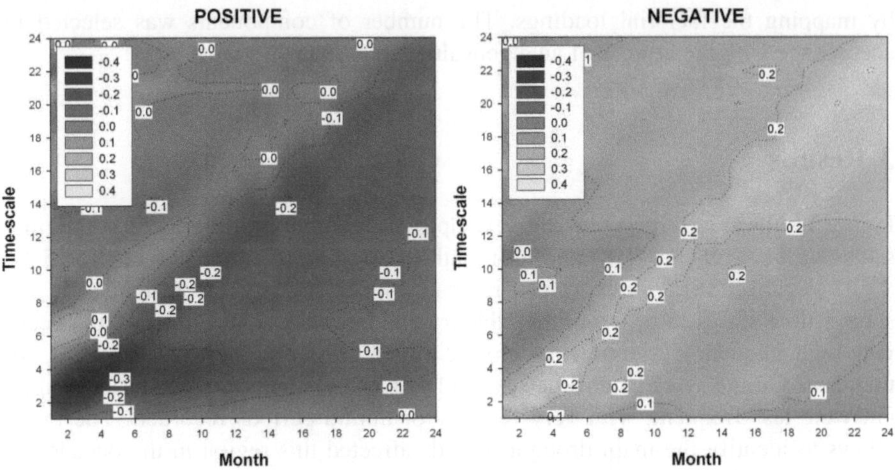

Fig. 2 Average SPEI anomalies for the entire Mediterranean region corresponding to the positive and negative years of the NAO. The month 1 corresponds to December of the NAO phase and the month 24 is November of two months after

NAO phases is observed at the time scales of 2- to 5-month considering the period between February and June. Nevertheless, although not so intense as observed in late winter and early spring, the negative anomalies are propagated throughout longer SPEI time-scales several months further. On the contrary, the negative NAO phases show general positive SPEI anomalies in most of the months and time-scales, even at the longest time-scales (e.g., 24 months). We should also remark that although both phases show opposite pattern, the magnitude of SPEI anomalies are not so high, probably due to the large spatial variability existing in the role of NAO on the precipitation and temperature across the Mediterranean basin. Figure 3 shows the average SPEI at various time-scales in different months of the year during the year of the NAO event and the following year for the positive NAO phases. It also shows the standard deviations of the SPEI values recorded during the events. The first relevant conclusion is the important spatial differences in the impact of the NAO positive phases across the Mediterranean basin. As expected, when the time-scale is higher, the effects of the positive NAO events are smoothed. Nevertheless, independently of this pattern, it is clearly observed that the strongest negative SPEIs are recorded in the Iberian Peninsula, Morocco, Algeria, Italy and the Balkans. The negative anomalies have a lower magnitude and a patchy configuration in Turkey and South France. Moreover, in the Northeast Africa the SPEI anomalies are positive, and the difference of the SPEI values in relation to the negative and normal years is statistically significant for most of the months and time-scales.

Figure 3 shows how at short time scales (3- and 5-months) during the spring immediately after the winter NAO, the SPEI averages are very strong: negative in the north and southwest and positive in the southeast of the basin. In most of the

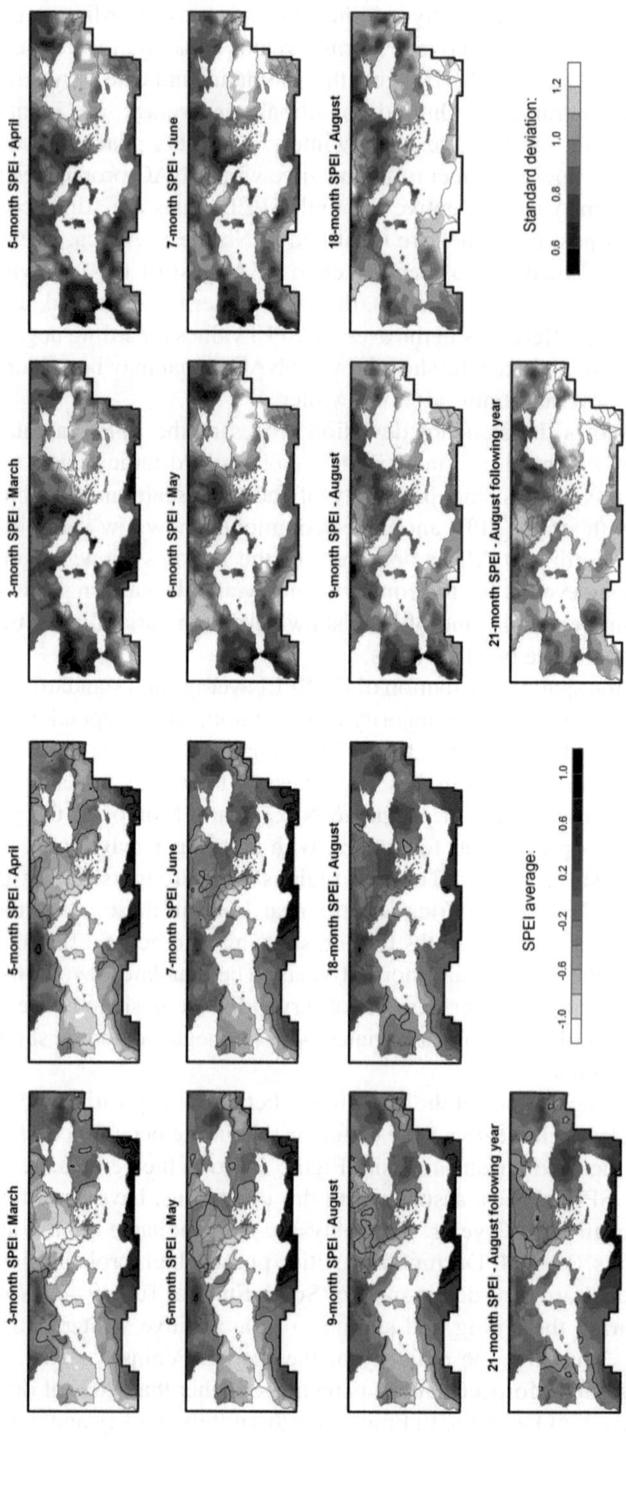

Fig. 3 *Left*: Spatial distribution of the average SPEI at different time-scales for some months of the positive NAO years and in August of the following year. *Black lines* isolate areas in which the average SPEI for the positive NAO years is significantly different to the rest of the years. *Right*: Standard deviation of the SPEI values during the positive NAO winters

basin these averages show statistically significant differences regarding to the SPEI averages recorded during the negative or normal years. Thus, in most of the Iberian Peninsula and large regions of the Balkans the anomalies indicate very strong dry conditions. On the contrary, very humid conditions, on average, are identified in northeast Africa during the positive NAO winters. Droughts quantified at longer time-scales also show how the effect of the positive winter NAO propagates several months after. Thus, in August negative 9-month SPEI values are still recorded in large areas, whereas positive values are identified in Northeast Africa. Even during the summer of the following year (see the chart of August of the following year considering the time-scale of 21 months), the same pattern is identified, and large areas show significant differences in the average SPEI values regarding negative and normal years. Therefore, the results show how the NAO signal may be evident in the drought conditions several months after the winter.

Figure 3 also shows the standard deviation values of the SPEI calculated for the 16 positive NAO years. Low standard deviations would indicate low variability among the positive years. Independently of the time scale and the month of the year, regions with strong SPEI anomalies commonly show low variability. The Iberian Peninsula, Northwest Africa and areas of the Balkans show, in general, a low variability among events. On the contrary, Turkey and the eastern sectors of the Balkans, the Iberian Peninsula, and of Italy show very high variability of the SPEI values corresponding to the positive years.

Figure 4 shows the spatial distribution of the SPEI average and standard deviation for the negative NAO years. In the majority of the region, and independently of the time-scale and the month of the year, the SPEI anomalies tend to be positive. Thus, the magnitude of the average anomalies, regardless of their sign, tends to be much higher for the positive than for the negative NAO years. Moreover, the effect of the negative phases is propagated for longer, with a stronger magnitude, than the observed for the positive phases. The two regions with droughts during positive NAO phases are the Northeast Africa and the near East. In these areas, the NAO effects are recorded for several months further, showing significant differences with the SPEI values of the positive and normal years. The standard deviations of the SPEI values are, in general, lower than the observed for the positive phases, which indicates that the humid conditions associated with the negative phases show a low variability among events.

The different average values of the SPEI found between the positive and the negative NAO years clearly indicate a very different risk in the occurrence of drought conditions across the Mediterranean basin. Figure 5 shows the percentage of years in which negative SPEI values, associated to dry conditions, have been found for the positive and negative NAO years. The analysis covers the same months and time scales shown in Figs. 3 and 4. During the positive phases, high probabilities of dry conditions occurrence are recorded in most of South Europe, Turkey and Northwest Africa, mainly during the spring and summer of the positive winters. In March, at a time scale of 3 months, the majority of the Iberian Peninsula, Italy, western Turkey, the Balkans and Morocco show a percentage higher than 90% of dry conditions during positive NAO winters. In France, northern Italy, Turkey and Algeria the

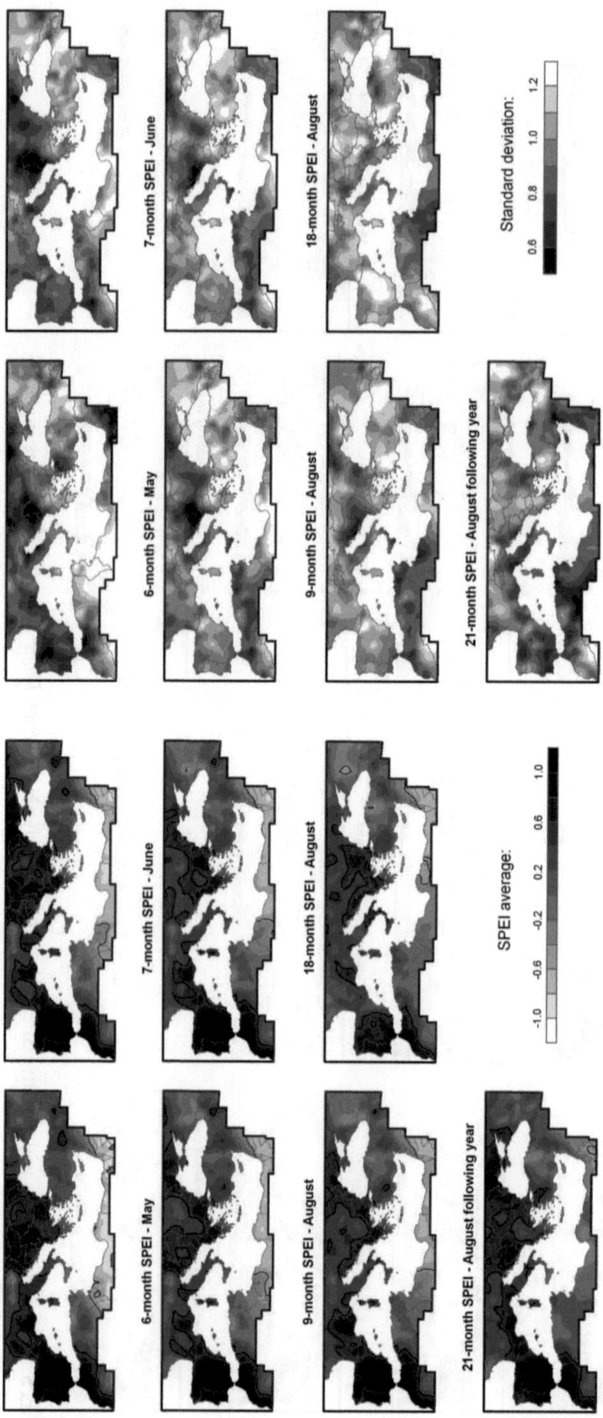

Fig. 4 *Left*: Spatial distribution of the average SPEI at different time-scales for some months of the negative NAO years and in August of the following year. *Black lines* isolate areas in which the average SPEI for the negative NAO years is significantly different to the rest of the years. *Right*: Standard deviation of the SPEI values during the negative NAO winters

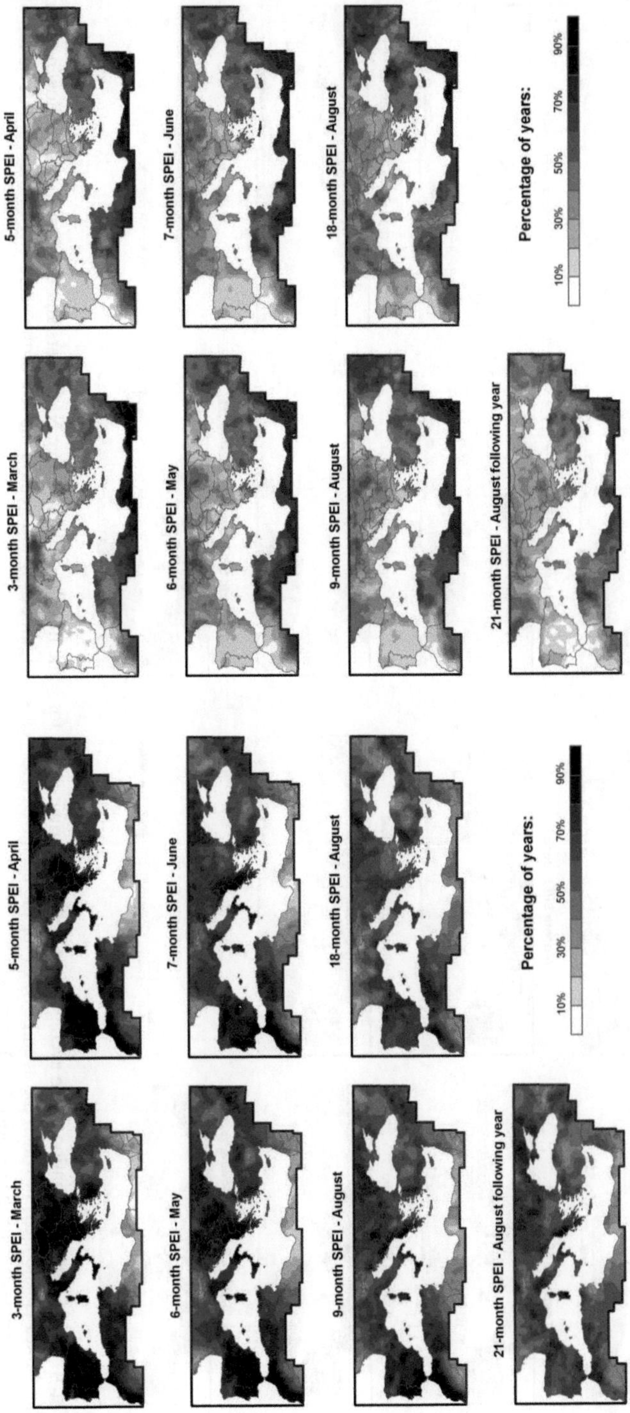

Fig. 5 Percentage of years with negative SPEI values. *Left*: Positive NAO years. *Right*: Negative NAO years

probability is lower but still relatively high (> 65%). On the contrary, in Northeast Africa the probability of having dry conditions at this time scale is close to 0. In April, and at the time-scale of 5-months, the spatial pattern of probability is similar to March. Thus, until June the probability of having dry conditions is higher than 80% in the Iberian Peninsula and North of Morocco considering a time-scale of 7 months. During the negative NAO years (Fig. 5 left) the probability of having dry conditions is very low in the majority of the basin; thus in some areas of southwestern Iberian Peninsula the value is close to 0 during the spring and summer. Only in the Northeast Africa and the Near East the probability is high (>80%) for most of the months of the year and time-scales.

We have summarised the information of the entire Mediterranean region by means of a principal component analysis (see methods) for the positive and the negative NAO years. The purpose was to determine the most general time-scale/month matrices of the SPEI anomalies corresponding to both phases, in addition to the regions represented by each pattern. The results of the matrix are shown in standardised units (PC-scores), comparable to the SPEI magnitudes.

Figure 6 shows the results for the positive phases. Three PCs were extracted, which account for 69.3% of the total variance. The component 1 shows some resemblance with the average pattern indicated in Fig. 2: negative SPEI anomalies at the time scales of 2–6 months between January and August of the positive year. Moreover, the negative values, indicative of dry conditions, propagate for further months and are dominant until, at least, 24 months after the positive NAO winter. According to the loadings of the first PC, this pattern of time-scale/month anomalies mainly represents the entire Iberian Peninsula, the southern coastland of France, most of Italy, the majority of the Balkans, the Anatolian Peninsula and the northwest Africa. Therefore, in these regions the positive NAO years are associated with a propagation of the dry conditions for most of the months and SPEI time scales. In northeast Africa the PC-loadings are dominantly negative and close to –1, which indicates that these areas are represented by the inverse of the obtained matrix during the positive NAO years. The second and third components represent the northernmost areas of France, Italy, the Balkans; and the northern basin of the Black Sea, and the near East and parts of North Africa, respectively. They contribute to a lower portion of the total variance (15.5 and 9.7%, respectively). In the northernmost areas of the basin (Component 2) the positive NAO phases are also related to the generalised negative SPEI values during the winter and spring, but there is no clear propagation throughout time-scales and months.

For the negative NAO years, the first PC explains 49.2% of the total variance, and represents most of the Iberian Peninsula, south of France, Italy, Morocco, south of the Balkans and western Turkey (Fig. 7). The pattern of the time-scale/month SPEI anomalies reveals the first pattern of the positive NAO phases, but with opposite signs to SPEI values. Thus, the negative NAO phases tend to cause humid conditions, which are propagated to the SPEI values several months after the winter. This pattern shows negative loadings in northeast Africa and the near East, suggesting the opposite behaviour in these areas: a propagation of the dry conditions. The

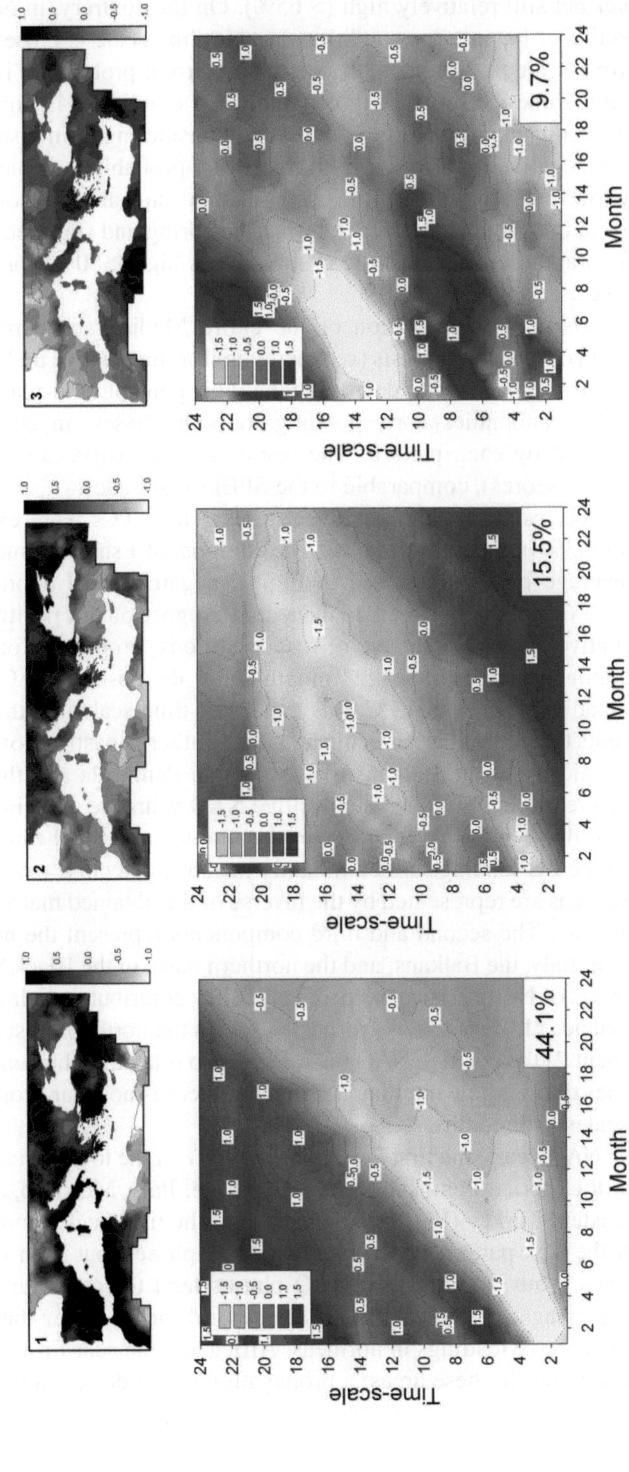

Fig. 6 Results of the principal component analysis for the positive NAO years. The *upper figures* represent the PC loadings of each component. The *lower figures* represent the principal components that summarise the matrix of NAO-SPEI correlations at different time-scales and months

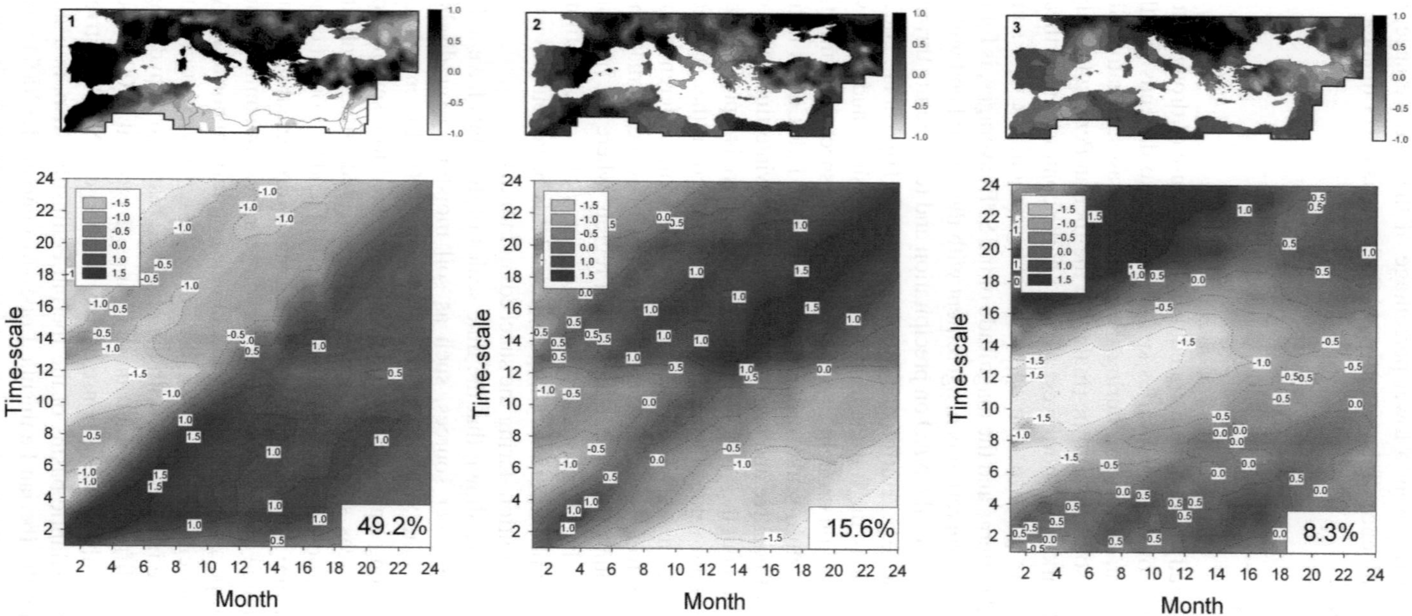

Fig. 7 Results of the principal component analysis for the negative NAO years. The *upper figures* represent the PC loadings of each component. The *lower figures* represent the principal components that summarise the matrix of NAO-SPEI correlations at different time-scales and months

components 2 and 3 represent a lower percentage of the total variance, including areas in which the influence of the negative NAO phases is lower than the observed for the component 1.

4 Conclusions

The results of this study have demonstrated that the response of droughts to positive and negative phases of the NAO varies spatially and also depends greatly on the time scale of drought. In general, during the positive phases, negative SPEI averages are recorded in southern Europe (mainly in the Iberian Peninsula, Italy and the Balkans), parts of Turkey and northwest Africa. On the contrary, in northeast Africa the SPEI averages are positive. The opposite configuration, but with some differences in the spatial patterns and the magnitude of the SPEI averages, is found for the negative years. This observation is in agreement with the general and widely known behaviour of the effects of the NAO on precipitation and temperature throughout the Mediterranean region (e.g., Hurrell and Van Loon, 1997; Trigo et al., 2002).

This study highlights the outstanding influence of positive and negative phases of the wintertime NAO on drought conditions during the succeeding months, quantified by means of a multi-scalar drought indicator. This index gives indication of the important role of positive and negative phases of the wintertime NAO on usable water sources throughout the year. We have found that there is a prevalence of either maximum or minimum values of the SPEI along the diagonal that summarised the principal component analysis in Figs. 6 and 7, with some differences depending on the regions. This fact is related to the importance of winter climate in terms of the total annual amounts. In the Mediterranean region, where the bulk of the precipitation comes mostly during winter, differences in winter precipitation between positive and negative phases are very important, and this behaviour has a noticeable impact on drought condition during the succeeding months.

Previous studies have shown that the time scales of droughts have a direct relationship with usable water sources, such as soil moisture, river discharge, and reservoir storage (Vicente-Serrano and López-Moreno, 2005; Lorenzo-Lacruz et al., 2010). In general, long time scales of the SPI are related to reservoir storage and river discharge, whereas short time scales are connected with variability in soil moisture. Therefore, positive phases of the NAO could lead to drought conditions in terms of soil moisture during winter and spring and hydrological droughts (reduction in river discharge and the levels of reservoir storages) in summer and autumn over southern Europe and northwest Africa. On the contrary, in northeast Africa and the near East the deficits will be mainly driven by the negative phases of the NAO. The above results are in agreement with the observed response of hydrological systems to the NAO in different Mediterranean regions. Trigo et al. (2004) documented a lag time of 2–3 months in the response of river discharge to NAO positive and negative phases in the Atlantic basins of the Iberian Peninsula. Also, in a study of the Tagus basin within the Iberian Peninsula, López-Moreno et al. (2007) reported that the impact of positive and negative phases of the NAO on river discharge and levels

of storage in reservoirs is neither instantaneous nor restricted to the winter months. Thus, the decrease in the amount of water stored in the reservoirs during negative phases of the NAO is especially severe during the next summer and autumn, whereas in winter and spring the amount of stored water is similar to that of normal years.

Nevertheless, the expected impact is not only restricted to the direct usable water resources since different natural systems and crops show a response to longer drought time scales. Vicente-Serrano (2007) found a high correlation between the vegetation activity in semi-arid regions of northeast Spain and drought time-scales between 6 and 12 months. Within dry-lands of this region, crop production of early summer responded considerably at time-scales between 6 and 12 months (Vicente-Serrano et al., 2006). Therefore, in these regions the NAO variability will have a great impact on the vegetation activity, directly driven by the soil moisture availability (Vicente-Serrano and Heredia-Laclaustra, 2004; Gouveia et al., 2008). In this way, in the coastland areas of Egypt and the near East, where large agricultural areas depend on winter rainfall, crop productions will be driven by drought variability, which is highly conditioned by the NAO positive phases.

Finally, we must highlight that although teleconnection indices such as the NAO have been widely used to predict drought conditions with a high degree of reliability (e.g., Rodwell, 2003), they may be useful only in those regions in which a general stability of the SPEI values for the positive or negative SPEI phases is found (Figs. 3 and 4). In other areas (mainly in Turkey and South France), the large variability among events observed during the positive NAO phases does not make the prediction and management of droughts an easy task, particularly those based solely on the wintertime NAO index. Given that the NAO configurations and the NAO-climate connectivity have had large changes in the last decades (Vicente-Serrano and López-Moreno, 2008), the regions less affected by the direct influences of the NAO dipole show a large variability among events (Massei et al., 2007).

The findings reported here should be of great use in the development of early warning systems and in mitigating the effects of drought in an area that is especially prone to such phenomena. The development of measures for drought planning and preparedness is a priority to reduce drought hazards (Wilhite, 1996; Wilhite et al., 2007; Prabhakar and Shaw, 2008). A critical component in drought planning is the provision of timely and reliable climate information on which management decisions are made (Svoboda et al., 2004). Hence, drought monitoring is crucial for the implementation of drought plans and the use of the results found here, in terms of the lagged impacts of NAO events throughout different time-scales and months, offers a unique mechanism for drought prediction. The established relationships between NAO phases and drought indices therefore seem appropriate for drought prediction over large areas of the Mediterranean basin.

Acknowledgements This work has been supported by the research projects CGL2008-01189/BTE and CGL2006-11619/HID financed by the Spanish Commission of Science and Technology and FEDER, EUROGEOSS (FP7-ENV-2008-1-226487) and ACQWA (FP7-ENV-2007-1- 212250) financed by the VII Framework Programme of the European Commission, and "La nieve en el Pirineo aragonés: Distribución espacial y su respuesta a las condiciones climáticas" Financed by "Obra Social La Caixa" and the Aragón Government.

References

Abramopoulos F, Rosenzweig C, Choudhury B (1988) Improved ground hydrology calculations for global climate models (GCMs): soil water movement and evapotranspiration. J Clim 1: 921–941

Austin RB, Cantero-Martínez C, Arrúe JL, Playán E, Cano-Marcellán P (1998) Yield-rainfall relationships in cereal cropping systems in the Ebro river valley of Spain. Eur J Agron 8:239–248

Beguería S, Vicente-Serrano SM, Angulo M (2010) A multi-scalar global drought data set: the SPEIbase: a new gridded product for the analysis of drought variability and impacts. Bull Am Meteorol Soc 91:1351–1354

Breshears DD, Cobb NS, Rich PM et al (2005) Regional vegetation die-off in response to global-change-type drought. PNAS 102:15144–15148

Briffa KR, Jones PD, Hulme M (1994) Summer moisture variability across Europe, 1892–1991: an analysis based on the Palmer drought severity index. Int J Climatol 14:475–506

Burton I, Kates RW, White GF (1978) The environment as hazard. Oxford University Press, New York, 240 pp

Cai W, Cowan T (2008) Evidence of impacts from rising temperature on inflows to the Murray-darling basin. Geophys Res Lett 35:L07701. doi:10.1029/2008GL033390

Changnon SA, Easterling WE (1989) Measuring drought impacts: the Illinois case. Water Resour Bull 25:27–42

Ciais Ph, Reichstein M, Viovy N et al (2005) Europe-wide reduction in primary productivity caused by the heat and drought in 2003. Nature 437:529–533

Colombaroli D, Marchetto A, Tinner W (2007) Long-term interactions between Mediterranean climate, vegetation and fire regime at Lago di Massaciuccoli (Tuscany, Italy). J Ecol 95:755–770

Elfatih A, Eltahir B, Yeh PJF (1999) On the asymmetric response of aquifer water level to floods and droughts in Illinois. Water Resour Res 35:1199–1217

Gerten D, Rost S, von Bloh W, Lucht W (2008) Causes of change in 20th century global river discharge. Geophys Res Lett 35:L20405. doi:10.1029/2008GL035258

Gouveia C, Trigo RM, DaCamara CC, Libonati R, Pereira JMC (2008) The North Atlantic Oscillation and European vegetation dynamics. Int J Climatol 28:1835–1847

Guttman NB (1998) Comparing the Palmer drought index and the Standardized Precipitation Index. J Am Water Resour Assoc 34:113–121

Helsel DR, Hirsch RM (1992) Statistical methods in water resources. Elsevier, New York, 522 pp

Hurrell JW, Kushnir Y, Ottersen G, Visbeck M (2003) An overview of the North Atlantic Oscillation. In Hurrell JW, Kushnir Y, Ottersen G, Visbeck M (eds) The North Atlantic Oscillation: climate significance and environmental impact. Geophysical Monograph Series, vol 134, pp 1–36

Hurrell J, Van Loon H (1997) Decadal variations in climate associated with the North Atlantic Oscillation. Clim Change 36:301–326

Iglesias E, Garrido A, Gomez-Ramos A (2003) Evaluation of drought management in irrigated areas. Agric Econ 29:211–229

Ji L, Peters AJ (2003) Assessing vegetation response to drought in the northern Great Plains using vegetation and drought indices. Rem Sens Environ 87:85–98

Jolliffe IT (1990) Principal component analysis: a beginner's guide. Part I: Introduction and application. Weather 45:375–382

Jones PD, Jónsson T, Wheeler D (1997) Extensions to the North Atlantic Oscillation using early instrumental pressure observations from Gibraltar and southwest Iceland. Int J Climatol 17:1433–1450

Jones PD, Moberg A (2003) Hemispheric and large-scale surface air temperature variations: an extensive revision and an update to 2001. J Clim 16:206–223

Khan S, Gabriel HF, Rana T (2008) Standard precipitation index to track drought and assess impact of rainfall on watertables in irrigation areas. Irrigat Drain Syst 22:159–177

Lloyd-Hughes B, Saunders MA (2002) A drought climatology for Europe. Int J Climatol 22:1571–1592

Lorenzo-Lacruz J, Vicente-Serrano SM, López-Moreno JI, Beguería S, García-Ruiz JM, Cuadrat JM (2010) The impact of droughts and water management on various hydrological systems in the headwaters of the Tagus River (central Spain). J Hydrol 386:13–26

Lázaro R, Rodrigo FS, Gutiérrez L, Domingo F, Puigdefábregas J (2001) Analysis of a 30-year rainfall record (1967–1997) in semi-arid SE spain for implications on vegetation. J Arid Environ 48:373–395

López-Moreno JI, Beguería S, Vicente-Serrano SM, García-Ruiz JM (2007) The influence of the NAO on water resources in central Iberia: precipitation, streamflow anomalies and reservoir management strategies. Water Resour Res 43:W09411. doi: 10.1029/2007WR005864

López-Moreno JI, Vicente-Serrano SM (2008) Extreme phases of the wintertime North Atlantic Oscillation and drought occurrence over Europe: a multi-temporal-scale approach. J Clim 21:1220–1243

Massei N, Durand A, Deloffre J, Dupont JP, Valdes D, Laignel B (2007) Investigating possible links between the North Atlantic Oscillation and rainfall variability in Northwestern France over the past 35 years. J Geophys Res 112:D09121. doi:10.1029/2005JD007000

McKee TBN, Doesken J, Kleist J, (1993) The relationship of drought frequency and duration to time scales. 8th conference on applied climatology, Anaheim, CA. Amer Meteor Soc, pp 179–184

Morales A, Olcina J, Rico AM (2000) Diferentes percepciones de la sequía en España: adaptación, catastrofismo e intentos de corrección. Investigaciones Geográficas 23:5–46

Osborn TJ, Briffa KR, Tell SFB, Jones PD, Trigo RM (1999) Evaluation of the North Atlantic Oscillation as simulated by a climate model. Clim Dyn 15:685–702

Palmer WC (1965) Meteorological droughts. U.S. Department of Commerce Weather Bureau Research Paper 45, 58 pp

Pandey RP, Ramasastri KS (2001) Relationship between the common climatic parameters and average drought frequency. Hydrol Process 15:1019–1032

Patel NR, Chopra P, Dadhwal VK (2007) Analyzing spatial patterns of meteorological drought using standardized precipitation index. Meteorol Appl 14:329–336

Pausas JG (2004) Changes in fire and climate in the eastern Iberian Peninsula (Mediterranean basin). Clim Change 63:337–350

Prabhakar SVRK, Shaw R (2008) Climate change adaptation implications for drought risk mitigation: a perspective for India. Clim Change 88:113–130

Quiring SM, Ganesh S (2010) Evaluating the utility of the Vegetation Condition Index (VCI) for monitoring meteorological drought in Texas. Agr Forest Meteoro 150:330–339

Reichstein M, Tenhunen JD, Roupsard O, Ourcival J-M, Rambal S, Miglietta F, Peressotti A, Valentini R (2002) Severe drought effects on ecosystem CO_2 and H_2O fluxes at three Mediterranean evergreen sites: revision of current hypotheses? Global Change Biol 8:999–1017

Rodwell MJ (2003) On the predictability of the North Atlantic climate. In: Hurrell JW, Kushnir Y, Ottersen G, Visbeck M (eds) The North Atlantic Oscillation: climate significance and environmental impact. Geophysical Monograph Series, American Geophysical Union (AGU), Washington DC, vol 134, pp 173–192

Sheffield J, Wood EF (2008) Projected changes in drought occurrence under future global warming from multi-model, multi-scenario, IPCC AR4 simulations. Clim Dyn 31:79–105

Siegel S, Castelan NJ (1988) Nonparametric statistics for the behavioral sciences. McGraw-Hill, New York

Sims AP, Nigoyi DS, Raman S (2002) Adopting indices for estimating soil moisture: a North Carolina case study. Geophys Res Lett 29:1183. doi:10.1029/2001GL013343

Skøien JO, Blöschl G, Western AW (2003) Characteristic space scales and timescales in hydrology. Water Resour Res 39:1304. doi:10.1029/2002WR001736

Svoboda MD, Hayes MJ, Wilhite DA, Tadesse T (2004) Recent advances in drought monitoring. 14th conference on applied climatology, Seattle, WA, Jan 12–16. American Meteorological Society, pp 5237–5240

Szalai S, Szinell CS, Zoboki J (2000) Drought monitoring in Hungary. In: Early warning systems for drought preparedness and drought management. World Meteorological Organization, Lisbon, pp 182–199
Trigo RM, Osborn TJ, Corte-Real JM (2002) The North Atlantic Oscillation influence on Europe: climate impacts and associated physical mechanisms. Clim Res 20:9–17
Trigo RM, Pozo-Vazquez D, Osborn TJ, Castro-Diez Y, Gamiz-Fortis S, Esteban-Parra MJ (2004) North Atlantic Oscillation influence on precipitation, river flow and water resources in the Iberian Peninsula. Int J Climatol 24:925–944
Van del Schrier G, Briffa KR, Jones PD, Osborn TJ (2006) Summer moisture variability across Europe. J Clim 19:2818–2834
Vicente-Serrano SM (2006) Spatial and temporal analysis of droughts in the Iberian Peninsula (1910–2000). Hydrol Sci J 51:83–97
Vicente-Serrano SM (2007) Evaluating the impact of drought using remote sensing in a mediterranean, semi-arid region. Nat Hazards 40:173–208
Vicente-Serrano SM, Beguería S, López-Moreno JI (2010) A multi-scalar drought index sensitive to global warming: the standardized precipitation evapotranspiration index – SPEI. J Clim 23:1696–1718
Vicente-Serrano SM, Beguería S, López-Moreno JI, Angulo M, El Kenawy A (2010b) A new global 0.5° gridded dataset (1901–2006) of a multiscalar drought index: comparison with current drought index datasets based on the Palmer Drought Severity Index. J Hydrometeorol 11:1033–1043
Vicente-Serrano SM, Cuadrat JM, Romo A (2006) Early prediction of crop productions using drought indices at different time scales and remote sensing data: application in the Ebro valley (North-East Spain). Int J Remote Sens 27:511–518
Vicente-Serrano SM, González-Hidalgo JC, de Luis M, Raventós J (2004) Spatial and temporal patterns of droughts in the Mediterranean area: the Valencia region (East–Spain). Clim Res 26:5–15
Vicente-Serrano SM, Heredia-Laclaustra A (2004) NAO influence on NDVI trends in the Iberian Peninsula (1982–2000). Int J Remote Sens 25:2871–2879
Vicente-Serrano SM, López-Moreno JI (2005) Hydrological response to different time scales of climatological drought: an evaluation of the standardized precipitation index in a mountainous Mediterranean basin. Hydrol Earth Syst Sci 9:523–533
Vicente-Serrano SM, López-Moreno JI (2008) The nonstationary influence of the North Atlantic Oscillation on European precipitation. J Geophys Res Atmos 113:D20120. doi:10.1029/2008JD010382
Wilhite DA (1993) Drought assessment, management and planning: theory and case studies. Kluwer, Boston, MA
Wilhite DA (1996) A methodology for drought preparedness. Nat Hazards 13:229–252
Wilhite DA, Glantz MH (1985) Understanding the drought phenomenon: the role of definitions. Water Int 10:111–120
Wilhite DA, Svoboda MD, Hayes MJ (2007) Understanding the complex impacts of drought: a key to enhancing drought mitigation and preparedness. Water Resour Manag 21:763–774
Wilks DS (2006) Statistical methods in the atmospheric sciences. Academic Press, San Diego, CA, 648 pp

The Impacts of the NAO on Hydrological Resources of the Western Mediterranean

Ricardo M. Trigo

Abstract The present study analyses the impact of the North Atlantic Oscillation (NAO) on the precipitation and river flow regime of the western Mediterranean region. The spatial pattern of the NAO impact over western Europe and the Mediterranean is evaluated on a monthly basis showing that the NAO impact is well apparent and significant between December and March, and not so well defined between October and November. I have focused the attention on the impact of the NAO pattern on the river flow regimes for the three largest transboundary Iberian river basins, namely the Douro (north), the Tejo (centre) and the Guadiana (south). Results show that the large inter-annual variability of these three rivers flow is largely modulated by the NAO phenomena. Moreover, the magnitude of the relationship between NAO and river discharges increases substantially for the last period considered (1973–1998), with values being statistically significant at the 1%. Major changes in the precipitation regime of Iberia have occurred in the last 5 decades, with an outstanding decrease in precipitation being noticed in March since the 1960s, which is significantly associated with increasing probabilities of positive values of the NAO index. It is also shown that this decline of precipitation is inducing a significant decrease of river flow for the three Iberian river basins considered. Finally, it is well known that the precipitation and river flow regimes in Iberia present large values of inter-annual variability, being characterised by large disparities between wet and dry years. These characteristics will probably exacerbate in the coming decades as, according to modelling results, the entire Mediterranean basin is bound to become hotter and drier under an increased frequency of positive NAO winters. Therefore, if we consider the rising use of water in both Iberian countries (Portugal and Spain) for agricultural, touristic, hydroelectricity production and urban purposes, one can foresee an increasing stressed situation for the water resources management in the region directly driven by the NAO conditions.

Keywords NAO · Precipitation · Water resources · River flow · Iberian Peninsula · Trends

R.M. Trigo (✉)
Instituto Dom Luiz, Universidade de Lisboa, Campo Grande, Faculdade de Ciencias, Edificio C8, Piso 3, 1749-016 Lisbon, Portugal
e-mail: rmtrigo@fc.ul.pt

S.M. Vicente-Serrano, R.M. Trigo (eds.), *Hydrological, Socioeconomic and Ecological Impacts of the North Atlantic Oscillation in the Mediterranean Region*, Advances in Global Change Research 46, DOI 10.1007/978-94-007-1372-7_4,

1 Introduction

The western Mediterranean region is roughly positioned at the southern edge of the North Atlantic storm tracks. Thus this sector is particularly susceptible to interannual shifts in the trajectories of mid-latitude cyclones and associated fronts that can lead to remarkable anomalies of precipitation and, to a lesser extent, of temperature (Trigo et al., 2006). Given the seasonal characteristics of the Atlantic storm-tracks, this is particularly true in winter when the influence of mid-latitude variability is at its greatest. Storm-track variability impacts primarily the western Mediterranean, but it has also a signature clearly present in the eastern Mediterranean (e.g. Türkeş and Erlat, 2003; Xoplaki et al., 2004). The Iberian Peninsula orography has a strong influence on the way that most low pressure systems affect the climate at a more local scale, as mountain ranges shield regions from oceanic moisture advection (Gimeno et al., 2010). Thus the precipitation regime in the northern and western sectors of the Iberian Peninsula is mostly controlled by the mean annual cycle of the location of the Atlantic storm track, while the interior and the eastern sectors of Iberia are partially affected by large-scale synoptic systems with Atlantic origin, but also by convective precipitation with Mediterranean influence (Trigo et al., 1999, 2000). The amount and distribution of precipitation in the Iberian Peninsula is highly irregular in both the spatial and temporal dimensions (Esteban-Parra et al., 1998; Serrano et al., 1999; Trigo and DaCamara, 2000).

The variability of precipitation over large sectors of Iberia are known to be closely connected to the large-scale atmospheric circulation modes, including the North Atlantic Oscillation (von Storch et al., 1993; Serrano et al., 1999; Trigo and Palutikof, 2001; López-Moreno et al., 2007). The North Atlantic Oscillation (NAO) has been described since the pioneering works of Walker (1924) in the early twentieth century and corresponds to one of the major patterns of atmospheric variability in the Northern Hemisphere (Wallace and Gutzler, 1981; Barnston and Livezey, 1987). This important circulation mode was poorly studied in the following decades until relatively recently when it has been characterised in more detail (e.g., Wallace and Gutzler, 1981; Barnston and Livezey, 1987; Hurrell, 1995; Hurrell et al., 2003). In simple terms, the NAO corresponds to a large-scale meridional oscillation of atmospheric mass between the subtropical anticyclone near the Azores and the subpolar low pressure system near Iceland (Hurrell, 1995; Trigo et al., 2002). Recently, a number of different studies have shown the relevance of the NAO to the winter surface climate of the Northern Hemisphere in general and over the Atlantic/European sector in particular (Hurrell, 1995; Hurrell and van Loon, 1997; Trigo et al., 2002). Furthermore several studies have established links between different NAO modes and changes in the associated activity of North-Atlantic storm tracks (Osborn et al., 1999; Trigo et al., 2006). The existence of two contemporaneous and possibly related winter-time trends over the last 3 decades: a trend towards the positive phase of the NAO and a trend towards warmer Northern Eurasian land temperatures has now been established (Hurrell and van Loon, 1997). Nevertheless it should be noted that the influence of the NAO pattern in the Mediterranean is considerably higher for precipitation

than for temperature, where it plays a relatively minor role (Castro-Díez et al., 2002).

The strong control exerted by the NAO on precipitation over the Mediterranean basin is bound to be reflected in the seasonal flow of rivers across the region. It should be noted that the river basin system acts as an effective spatial and temporal integrator of precipitation (rain and snow), temperature, and related evapotranspiration over a specific region. Therefore, seasonal to interannual streamflow variability in many river-basins can be controlled by corresponding changes in large scale atmospheric circulation patterns. The relationship between these climatic regimes and its hydrological response, through its streamflow, presents different grades of complexity according to the physical characteristics of the basin (e.g. topography, land cover, soil types, lithology, etc.). Thus, the spatial variability in the response of river discharges to the NAO can be very high, with strong differences between neighbour rivers as a consequence of the basin characteristics (Morán-Tejeda et al., 2011).

During the last decade several works have been published focusing on the impact of the NAO pattern on river flow regimes of many Mediterranean river basins (e.g. Struglia et al., 2004; Trigo et al., 2004; López-Moreno et al., 2007) but also for adjacent Middle Eastern rivers, including the Tigris, Euphrates and Jordan (Cullen and deMenocal, 2000; Cullen et al., 2002), and the large central European river Danube (Rîmbu et al., 2002). While the statistical significance of results varies among river basins, these studies show that river flow tends to be lower (higher) when the NAO index is in its positive (negative) phase.

The main objective of this chapter is to summarize different impacts of the NAO investigated by the author in the last decade, namely on the European/Mediterranean climate (Trigo and Palutikof, 2001, 2002, 2004, Trigo et al., 2006), with a particular emphasis on the Iberian water resources (Trigo et al., 2004; Trigo, 2008), but also on changes of precipitation and river flow regime (Paredes et al., 2006; Trigo et al., 2008).

2 Results

2.1 *Impact in Large-Scale Circulation and Precipitation*

There are a number of NAO indices available in the literature and provided by researchers that have implemented different methodological approaches (e.g. Barnston and Livezey, 1987; Hurrell, 1995; Jones et al., 1997). Results presented here were based on the NAO index developed at the Climatic Research Unit (University of East Anglia, UK) and defined, on a monthly basis, as the difference between the normalized surface pressure at Gibraltar (southern tip of Iberian Peninsula) and Stykkisholmur in Iceland (Jones et al., 1997). Taking into account that the NAO index for winter months presents a positive trend during the last 3 decades of the twentieth century its distribution is dominated by positive values,

with monthly averages above zero (Jones et al., 1997). Therefore it is advisable to normalize the entire winter NAO index (average of DJFM values) so it has zero mean (for the normal period 1961–2000) and standard deviation one. Finally, we defined the seasonal high NAO composite to be a combination of all winters with NAO index greater than 0.5, while the low NAO composite corresponds to all winters characterised with an NAO index value smaller than –0.5.

Following the approach adopted in Trigo et al. (2004), we show the large-scale impact of winter NAO on atmospheric circulation and precipitation for the North Atlantic and European sector (Fig. 1). This is performed using monthly average values of sea level pressure (SLP) and precipitation rate fields obtained from the NCEP/NCAR (National Centers for Environmental Prediction/National Center for Atmospheric Research) reanalyses dataset (Kalnay et al., 1996) for the period spanning between 1961 and 2000. Anomaly fields of SLP and precipitation rate for winter (DJFM) months characterized by high (NAO index > 0.5) and low (NAO index < –0.5) values are shown in Fig. 1a, b, respectively. The spatial signature of the NAO is given by the difference in SLP between composites (Fig. 1c). The corresponding precipitation rate anomalies are also represented (grey scale), wherever those anomalies are significantly different (at the 5% level) from climatology (Fig. 1a, b) or between composites (Fig. 1c). Several aspects in this figure are worth to be mention, namely:

(a) The typical NAO pattern can be recognised in the SLP anomalies obtained for both the positive (Fig. 1a) and negative (Fig. 1b) composites that reveal the expected dipole between the Iceland and the Azores regions so well depicted when the difference between composites is computed (Fig. 1c). However, it should be noted that the southern centre of action is not located over the strait of Gibraltar (where the station used for computing the NAO index is placed), because the variance of SLP is lower there than over the Azores, and thus the magnitude of the composite anomaly is greater over the Azores (Trigo et al., 2002);
(b) The impact of the NAO on the North Atlantic winter precipitation field, for both phases of the NAO index, is not restricted to the European continent but extends considerably over large sectors of the, mostly uninhabited, North Atlantic. In fact Fig. 1c shows quasi-zonal bands of opposite anomaly signs, with large positive differences concentrated in the northern latitudes, extending from eastern Greenland to Scandinavia. At lower latitudes, a significant band of negative differences extends from south of Newfoundland crosses the Azores archipelago and extends throughout the northern Mediterranean basin until Turkey, with larger differences located west of Iberia.

Interestingly, the vast majority of works that have been published evaluating the impact of NAO in Europe/Mediterranean have been performed on a seasonal basis, usually confined to the winter months (e.g. DJF of DJFM). Following the approach adopted in Trigo et al. (2008) the monthly spatial correlation pattern between the European precipitation and the corresponding monthly NAO pattern is presented

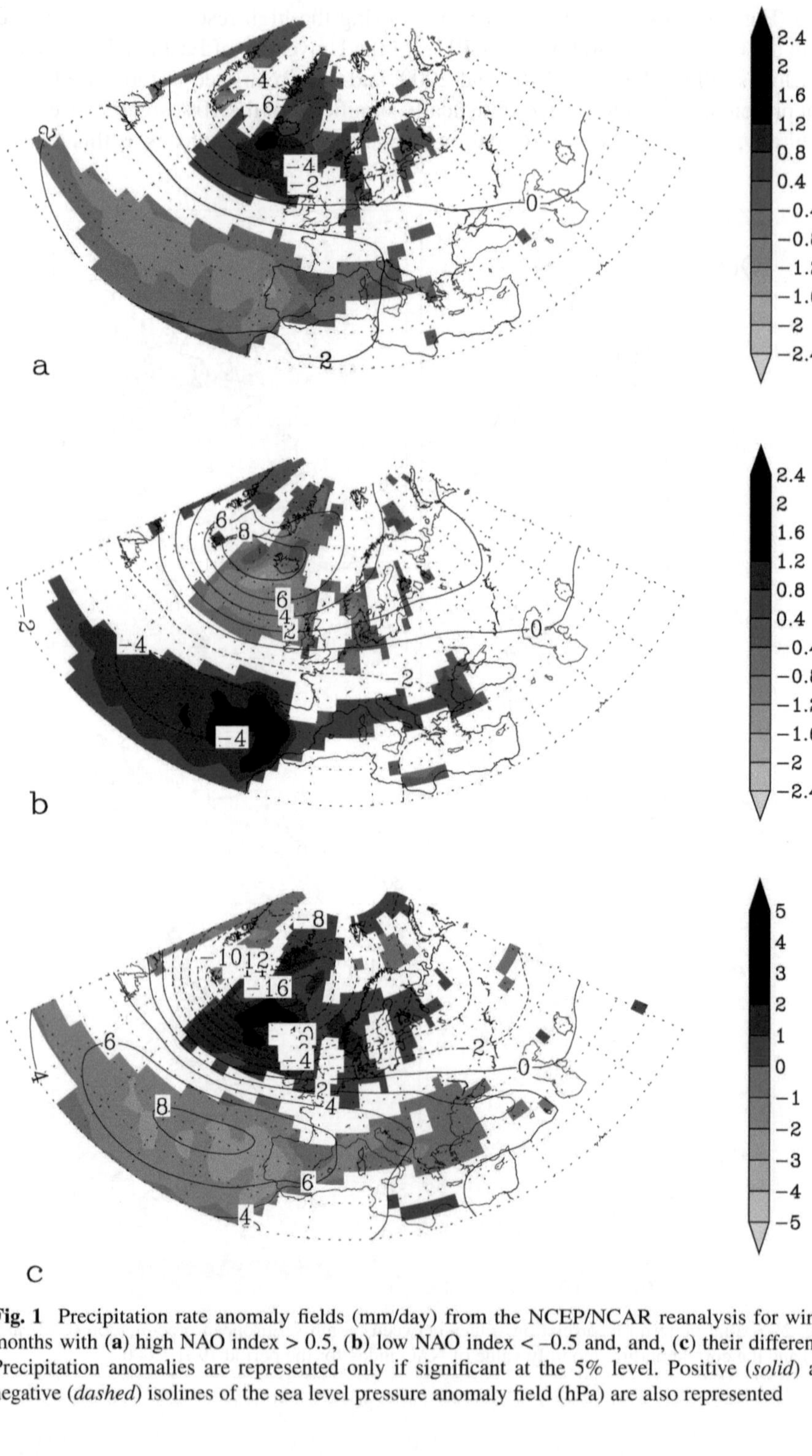

Fig. 1 Precipitation rate anomaly fields (mm/day) from the NCEP/NCAR reanalysis for winter months with (**a**) high NAO index > 0.5, (**b**) low NAO index < –0.5 and, and, (**c**) their difference. Precipitation anomalies are represented only if significant at the 5% level. Positive (*solid*) and negative (*dashed*) isolines of the sea level pressure anomaly field (hPa) are also represented

here (Fig. 2). This assessment is obtained using the high resolution (0.5°) dataset from the Climate Research Unit – CRU – of the University of East Anglia (Mitchell and Jones, 2005), between 1961 and 2000, and only restricted to those grid boxes that present a correlation value statistically significant at the 10% significance level. Naturally, no precipitation information is available over the oceans as this data set

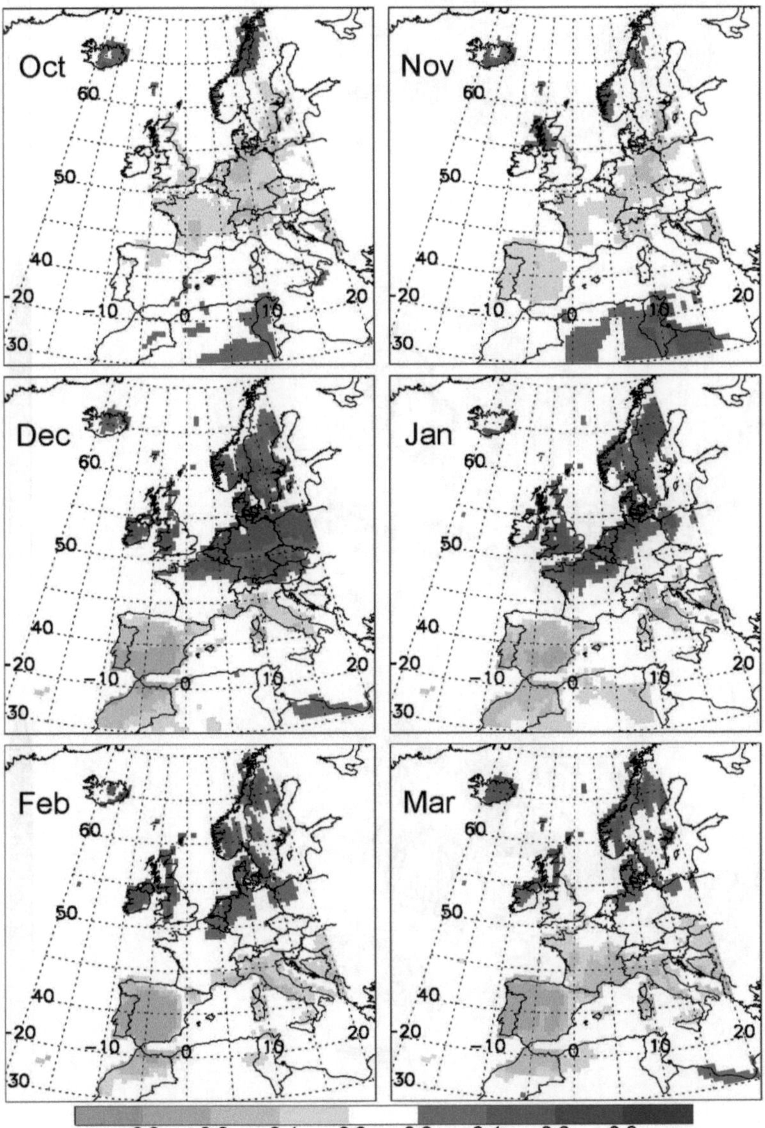

Fig. 2 Spatial pattern of correlation values between monthly European precipitation (CRU data set) and the corresponding monthly value of NAO index for the period 1961–2000

is constructed with precipitation values observed in rain gauges. The typical winter NAO-precipitation pattern shown in Fig. 2 confirms that the most prominent correlations can be observed between the core winter months that span December and March, with large negative values over Iberia ($r < -0.7$), northern Africa and central Mediterranean, while positive values dominate over northern Europe, albeit with a smaller magnitude ($r < 0.6$). A slightly different pattern emerges between October and November with weak, but still statistically significant, negative correlation values covering most of Europe between Iberia and southern Scandinavia, while positive values remained confined to small-scattered patches of UK and Scandinavia and most of northern Africa.

2.2 Impact in Western Mediterranean Water Resources

As stated in the introduction, it is expectable that the strong impact of the NAO on the precipitation regime over the western and central sectors of the Mediterranean should be also observed on the stream flow of rivers located within these sectors. This section focuses on three large transboundary Iberian rivers (Douro, Tejo and Guadiana) that correspond almost to 50% of the entire Iberian surface (Fig. 3) and flow westwards, i.e. from Spain to Portugal. It should be stressed that water resources in this south-western European region are characterised by present and future challenges, namely:

- As a consequence of high interannual precipitation variability most Iberian rivers show relatively high coefficients of interannual stream flow variation, that decrease from rivers located in the north (~50% for Douro) to those in the southern sector (~100% for Guadiana).
- There has been a steady decline of Iberian precipitation in late winter (February) and early spring (March) since the 1960s (González-Hidalgo et al., 2010). This decrease is driven by latitudinal shifts of storm-tracks over the Atlantic that are ultimately associated with an increasing trend of the NAO index (Paredes et al., 2006; Trigo et al., 2008). This trend in precipitation is noticeable in river flow in March and April and diminishes the soil water moisture available in the following spring and summer seasons.
- Both Iberian countries present an increasing demand of water supply in recent decades. Besides hydroelectricity production, the touristic (e.g. golf courses) and agricultural sectors correspond to the two main socio-economic activities responsible for this increment. The strong socio-economic impacts of the recent extreme drought observed during the 2004/2005 hydrological year further emphasises the necessity to develop long-term planning tools (García-Herrera et al., 2007).
- The latest IPCC report (Solomon et al., 2007) confirms what many studies based on global and regional climate models have been suggesting in the last decade; i.e. that the Mediterranean area will register a general trend towards less precipitation during the twenty-first century (e.g. Giorgi and Lionello, 2008;

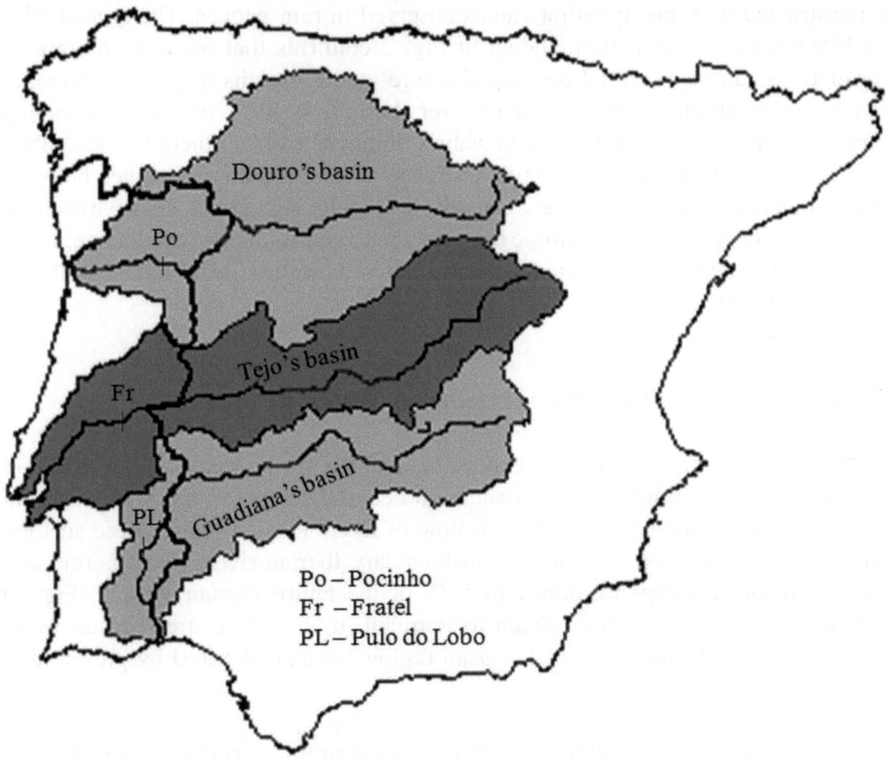

Fig. 3 Location of the three Iberian river basins considered. Small vertical lines show the location of river flow gauges used in each basin

López-Moreno et al., 2009). The combined effects of this future precipitation decrease and the widely accepted future increment in the surface temperature on the Mediterranean will bear important changes in the region's water cycle (Mariotti et al., 2008).

Following the work developed in Trigo et al. (2004) and updated in Trigo (2008) we used monthly river flow data for the three rivers depicted in Fig. 3, namely the Guadiana, the Tejo (Tagus or Tajo) and the Douro (or Duero). The datasets were kindly provided by the Portuguese Institute of Water (INAG). The location of the river gauges, corresponds to large dams located within the Portuguese section of the rivers as shown in Fig. 3. Monthly river flow anomalies during and following winters with high (low) NAO index values are represented by black (grey) bars in Figs. 4a, 5a and 6a for the Douro, Tejo and Guadiana rivers, respectively. Taking into account the very dry months that characterize Iberia between July and September we show results using the hydrological year, i.e. between October and September of the following year. It is clear that all three cases are characterized by a clear separation of river flow between the high and low winter NAO composites

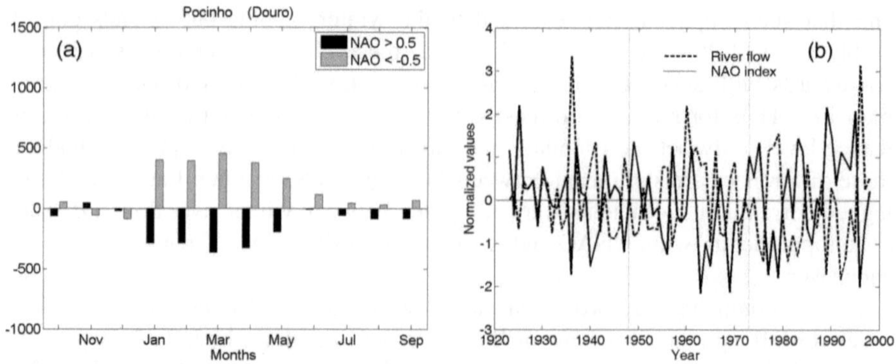

Fig. 4 (**a**) Monthly river flow anomalies (hm^3/month) of river Douro at Pocinho during and following winters with high NAO index (*black bars*) and winters with low NAO index (*grey bars*), (**b**) Interannual variability of the mean winter (JFM) river flow (*solid curve*), for river Douro at Pocinho, and the lagged winter (DJF) NAO index. Both *curves* have been normalised

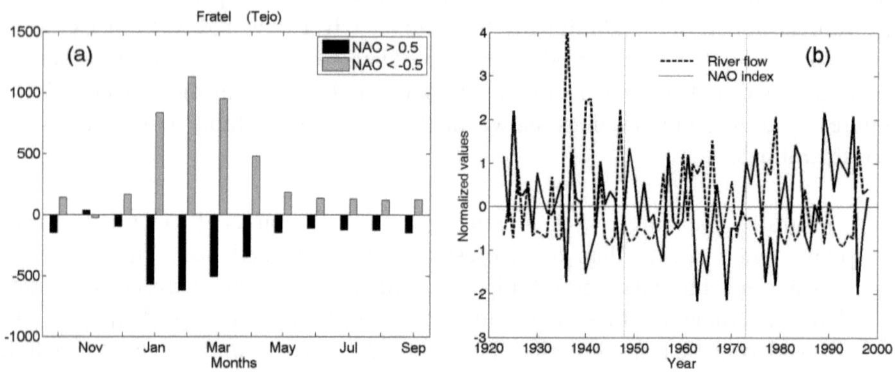

Fig. 5 The same as in Fig. 4 but for river Tejo at Fratel

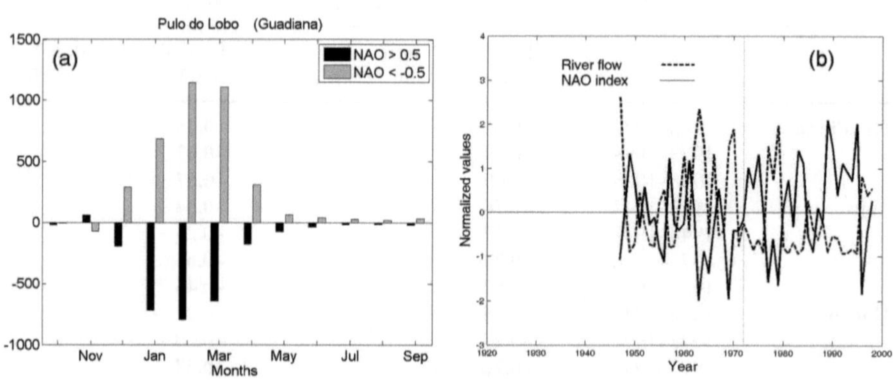

Fig. 6 The same as in Fig. 4 but for river Guadiana at Pulo do Lobo

and that such impact is not restricted to the winter months but extends considerably to the following seasons. Moreover, these changes increase as we move southwards with successive larger ranges of positive (black) and negative (grey) bars size. Thus for river Douro these differences are significant (at the 5% significance level) only between January and April, while for rivers Tejo and Guadiana these differences are significant between January and September. Lagged and non-lagged correlation coefficients were computed between winter river flow (DJF) and contemporaneous winter NAO index, using the full period of data available for each river.

The 1-month lagged correlation values were computed considering the NAO index for DJF and the river flow for JFM. In order to evaluate how stationary results are correlation coefficients were computed for three consecutive sub-periods; 1923–1947, 1948–1972 and 1973–1998 for rivers Tejo and Douro rivers while, due to shorter time series availability, results for Guadiana have to be restricted for the two most recent sub-periods (Table 1). The inter-annual variability of winter (JFM) river flow against the winter (DJF) NAO index is represented in Figs. 4b, 5b and 6b for the Douro, Tejo and Guadiana rivers, respectively. Sub-periods are delimited by vertical dotted lines, additionally both curves (river flow and NAO index) were normalized (for the total period of available data for each river) and represented between 1920 and 2000 to facilitate comparisons between all three rivers. These figures (and correlation values shown in the companying Table 1) provide some considerations:

(1) Generally speaking the 1 month-lagged correlation coefficient values are consistently higher than the corresponding non-lagged ones. In fact, when considering the entire period of data available for all three rivers, the correlation coefficient values increase when the NAO index is lagged by 1 month. Considering the last sub-period (i.e. 1973–1998) the 1-month lagged correlation

Table 1 Seasonal 1-month lagged (lag 1) and simultaneous (lag 0) correlation coefficients between winter NAO index and river flow in Pulo do Lobo (Guadiana), Fratel (Tejo) and Pocinho (Douro). Statistically significant results at the 5% level are highlighted in bold

		River flow Lag 1 NAO (DJF), flow (JFM)	River flow Lag 0 NAO (DJF), flow (DJF)
Guadiana	1948–1972	**–0.60**	**–0.45**
	1973–1998	**–0.79**	**–0.67**
	1948–1998	**–0.69**	**–0.57**
Tejo	1923–1947	**–0.56**	**–0.54**
	1948–1972	–0.37	–0.25
	1973–1998	**–0.77**	**–0.61**
	1923–1998	**–0.52**	**–0.45**
Douro	1923–1947	**–0.48**	–0.27
	1948–1972	–0.29	–0.10
	1973–1998	**–0.76**	**–0.58**
	1923–1998	**–0.55**	–0.35

coefficients achieve very significant values, namely; –0.76 for Douro, –0.77 for Tejo and –0.79 for Guadiana. These results have opened promising prospects for the use of these relationships for constructing monthly to seasonal forecasting models.

(2) The longer time series available for the Douro and Tejo rivers allow further insight on the long term behavior of these link between NAO and river flow. Interestingly, both rivers reveal a certain decrease between the first and the second sub-periods that is followed by a major increment between the second and third sub-periods. The long term evolution of this relationship can be also appreciated by the change in coherence between both time series (Figs. 4b, 5b and 6b).

The increase in the strength of the impact of NAO in the river flow since the 1970s reflects most probably two different aspects that contribute to such an outcome. On the one hand the link between precipitation and the NAO patterns is not stationary, but changes at decadal scale, having increased significantly between the mid-twentieth century to the last few decades (e.g. Trigo et al., 2004; Vicente-Serrano and López-Moreno, 2008). Additionally, it has been suggested that part of this increment in strength is related to the increase in water storage volume associated with the construction of major dams in the 1950s and 1960s (Trigo et al., 2004), given that there are evidences that variability in the water stored by reservoirs of the Tagus basin is more determined by NAO than river discharges, both in the duration and magnitude of the anomalies (López-Moreno et al., 2007). In fact, results obtained in Table 1 show that the correlation increase is larger at 1-month lag than it is for simultaneous correlations.

2.3 Precipitation and River Flow Trends Associated with the NAO

A number of studies has shown that winter (DJFM) precipitation has been decreasing steadily since the 1960s over western Mediterranean, particularly for the month of March (e.g. Trigo and DaCamara, 2000; Paredes et al., 2006; González-Hidalgo et al., 2010), with latter analysis showing that such decline is becoming to be extensive to the month of February (Trigo et al., 2008). These authors show that March monthly-accumulated precipitation for the central and western regions of the Iberian Peninsula presents a very significant decline of circa 50% during the 1960–2000 period. Naturally these changes are not confined to Iberia but correspond to the most visible aspect of a larger phenomenon affecting the entire North-Atlantic/European sector. In fact, while Iberia and the majority of the western sector of the Mediterranean experiences this decreasing trend in March, the northern regions of Scandinavia and British Isles present a remarkable increase (Trigo et al., 2008). It should be notice that the significance of these trends varies according to the period of data considered.

As expected such trends are bound to be connected with changes of the atmospheric circulation over the North Atlantic and European sectors. In this regard a

number of different approaches were used to quantify the links with large-scale and smaller scale features (Paredes et al., 2006; Trigo et al., 2008). Based on the Iberian Weather Types (WTs) classification developed by Trigo and DaCamara (2000) it was shown that, at the synoptic scale, the major contributors towards the March decline are associated with a corresponding decrease in the frequency of a few specific WTs. These "wet" WTs are characterized by higher than average values of daily precipitation and include the cyclonic (C), the western (W) and south-western (SW). On the contrary, March is also characterized by a notorious increment in the frequency of "dry" anticyclonic (A) type. In contrast the corresponding "wet" WTs for the British Isles display increasing frequencies justifying the increment of Precipitation observed in England and Scotland (Paredes et al., 2006). Within a larger context, a Lagrangean approach, based on the analysis of storm-tracks over the entire North Atlantic region, reveals dramatic changes in the location of cyclones since the early 1960s that coincides with the corresponding precipitation trends in Europe (Trigo et al., 2008). It is of no surprise that the NAO is somehow related with both phenomena, i.e. changes in the preferred path of most low pressure perturbations and contemporaneous changes of monthly precipitation totals for March. In fact, for Iberia, the region displaying the maximum correlation values between NAO and March precipitation is almost exactly coincident with the region presenting the largest decreases in precipitation. Using data for the 1960–2000 period the spatially averaged series of precipitation shows a correlation value of –0.60, that decreases to –0.48 when computed with the detrended series (Paredes et al., 2006).

It should be emphasised that the region with significant NAO induced decline of precipitation in March encloses the three major international Iberian river basins referred in the previous section, i.e. rivers Douro (North), Tejo (Center) and Guadiana (South). Thus, it is highly expectable that such a significant change in precipitation must have an imprint on the long-term evolution of monthly river flow for these rivers. That is precisely what can be seen for rivers Douro and Guadiana (Fig. 7), with both cases exhibiting very strong declines (significant at 1%) of river flow in March, computed between 1956 and 2000, similar results were obtained for river Tejo (not shown). Moreover, this decline is not restricted to the month of

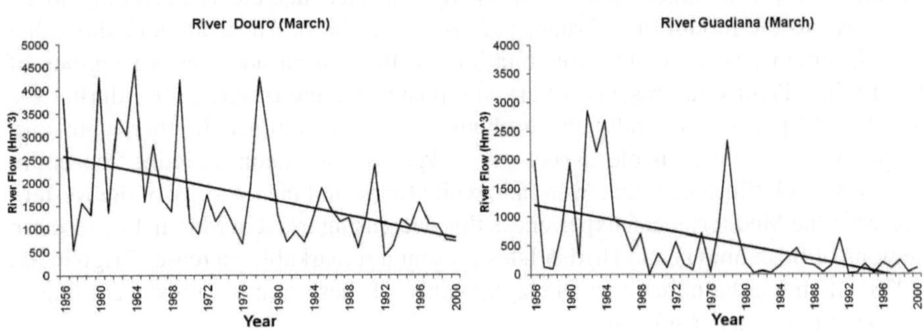

Fig. 7 Inter-annual variability of March river flow for Douro and Guadiana rivers. The corresponding linear regression trend is also represented

March but can be observed also in the following months of April and May (albeit less pronounced) due to the relatively high values of persistence that characterize river flow time series.

Finally it must be stressed that these changes can have significant socio-economic impacts for both Portugal and Spain economies. Large amounts of water from these three rivers are used for various economic activities as well as for water supply to urban areas. As an example, these three rivers together account for a very significant amount of hydroelectric production in both Iberian countries, with roughly 55% (70%) of Spanish (Portuguese) annual average (Trigo et al., 2004).

3 Conclusions

Western Mediterranean precipitation regime is strongly influenced by the NAO pattern with roughly 40% of the winter precipitation being related to the NAO index. Because this large-scale atmospheric circulation phenomenon is characterized by a marked interannual variability, it implies that the Iberian Peninsula is particularly prone to large inter-annual variability of precipitation. Therefore NAO is a key component of the relatively high frequency of wet and drought periods observed in Iberia (Vicente-Serrano, 2006; Sousa et al., 2011). Two recent extreme years exemplify perfectly this dichotomy, the extreme dry hydrological year of 2004/2005 and the wet winter of 2009/2010. The year of 2004–2005 was considered the driest for many stations in Iberia (including Lisbon and Madrid) since the beginning of the twentieth century (García-Herrera et al., 2007). This extreme event was induced by several favoring atmospheric circulation anomalies including positive values of the NAO index between November 2004 and February 2005. On the contrary the winter of 2009/2010 was characterized by record breaking negative values of the NAO index at the monthly and seasonal scales (Cattiaux et al., 2010; Wang et al., 2010). This anomalous circulation triggered particularly cold temperatures over central and northern Europe (Cattiaux et al., 2010) and simultaneously, large parts of the Iberian Peninsula (including Gibraltar and Lisbon) have registered the highest amounts of winter precipitation since the beginning of regular measurements in the second half of the nineteenth century (Vicente-Serrano et al., 2011).

As shown here the influence of NAO on the winter Iberian precipitation is extensive to the river flow of major Iberian rivers. Moreover it is possible to state that the magnitude of the NAO impact is considerably larger for Iberian rivers than what has been obtained by other authors for rivers Tigris and Euphrates (Cullen and deMenocal, 2000) and the Danube (Rîmbu et al., 2002). In fact, according to the results obtained by Struglia et al. (2004) for many river basins in the Mediterranean there are very few European rivers presenting a seasonal correlation coefficient value with – either with NAO or other large-scale circulation mode – similar to the values obtained here (~0.75). Recent results obtained by Spanish colleagues have extended the results presented here to many more gauges of Douro and Tejo, including several subsidiaries (López-Moreno et al., 2007; Morán-Tejeda et al., 2011). These authors showed that the impact of positive and negative phases of the NAO on river

discharge and levels of storage in reservoirs is neither instantaneous nor restricted to the winter months, extending well into the rest of the year. Thus, the decrease in the amount of water stored in the reservoirs during positive phases of the NAO is especially severe during the following summer and autumn, whereas in winter and spring the amount of stored water is similar to that of normal years. In this regard it is now clear that the NAO during winter months holds a significant capacity to constrain the observed river flow and water capacity for most of the year in most Iberian rivers, thus representing an important factor to be accounted for all water resources managers in general, and by agricultural and hydroelectric power production companies in particular.

Acknowledgements This work was supported by the Portuguese Science Foundation (FCT) through project ENAC (Evolution of North Atlantic Climate; the role of Blocking and Storm-tracks in the Past, Present and Future climate of Southern Europe) PTDC/AAC-CLI/103567/2008. Many thanks to Pedro Sousa for helping with the editing of this chapter.

References

Barnston AG, Livezey RE (1987) Classification, seasonality and persistence of low-frequency atmospheric circulation patterns. Mon Wea Rev 115:1083–1127

Castro-Díez Y, Pozo-Vázquez D, Rodrigo FS, Esteban-Parra MJ (2002) NAO and winter temperature variability in southern Europe. Geophys Res Lett 29. doi:10.1029/2001GL014042

Cattiaux J, Vautard R, Cassou C, Yiou P, Masson-Delmotte V, Codron F (2010) Winter 2010 in Europe: a cold extreme in a warming climate. Geophys Res Lett 37:L20704. doi:10.1029/2010GL044613

Cullen HM, Kaplan A, Arkin PA, DeMenocal PB (2002) Impact of the North Atlantic Oscillation on the Middle Eastern climate and streamflow. Clim Change 55:315–338

Cullen HM, deMenocal PB (2000) North Atlantic influence on Tigris-Euphrates streamflow. Int J Climatol 20:853–863

Esteban-Parra MJ, Rodrigo FS, Castro-Díez Y (1998) Spatial and temporal patterns of precipitation in Spain for the period 1880–1992. Int J Climatol 18:1557–1574

García-Herrera R, Paredes D, Trigo RM, Trigo IF, Hernández E, Barriopedro D, Mendes MA (2007) The outstanding 2004/05 drought in the Iberian Peninsula: associated atmospheric circulation. J Hydrometeorol 8:483–498

Gimeno L, Drumond A, Nieto R, Trigo RM, Sthol A (2010) On the origin of continental precipitation. Geophys Res Lett 37:L13804. doi:10.1029/2010GL04371

Giorgi F, Lionello P (2008) Climate change projections for the Mediterranean region. Glob Planet Change 63:90–104

Gonzalez-Hidalgo JC, Brunetti M, de Luis M (2010) Precipitation trends in Spanish hydrological divisions, 1946–2005. Clim Res 43:215–228

Hurrell JW (1995) Decadal trends in the north Atlantic oscillation: regional temperatures and precipitation. Science 269:676–679

Hurrell JW, Kushnir Y, Ottersen G, Visbeck M (2003) The North Atlantic Oscillation: climate significance and environmental impact. Geophysical Monograph Series, vol 134. American Geophysical Union, Washington, DC

Hurrell JW, van Loon H (1997) Decadal variations in climate associated with the north Atlantic Oscillation. Clim Change 36:301–326

Jones PD, Jónsson T, Wheeler D (1997) Extension to the North Atlantic Oscillation using early instrumental pressure observations from Gibraltar and South-West Iceland. Int J Climatol 17:1433–1450

Kalnay E, Kanamitsu M, Kistler R, Colins W, Deaven D, Gandin L, Iredell M, Saha S, White G, Wollen J, Zhu Y, Leetmaa A, Reynolds R, Chelliah M, Ebisuzaki W, Higgins W, Janowiak J, Mo KC, Ropelewski C, Wang J, Jenne R, Joseph D (1996) The NCEP/NCAR 40-years reanalyses project. Bull Am Meteorol Soc 77:437–471

López-Moreno JI, Beguería S, Vicente-Serrano SM, García-Ruiz JM (2007) The influence of the NAO on water resources in Central Iberia: precipitation, streamflow anomalies and reservoir management strategies. Water Resour Res 43:W09411. doi: 10.1029/2007WR005864

López-Moreno JI, Vicente-Serrano SM, Gimeno L, Nieto R (2009) The stability of precipitation regimes in the Mediterranean region: observations since 1950 and projections for the twenty-first century. Geophys Res Lett 36:L10703

Mariotti A, Zeng N, Yoon JH, Artale V, Navarra A, Alpert P, Li L (2008) Mediterranean water cycle changes: transition to drier 21st century conditions in observations and CMIP3 simulations. Environ Res Lett 3(4):044001. doi:10.1088/1748-9326/3/4/044001

Mitchell TD, Jones P (2005) An improved method of constructing a database of monthly climate observations and associated high-resolution grids. Int J Climatol 25:693–712

Morán-Tejeda E, López-Moreno JI, Ceballos-Barbancho A, Vicente-Serrano SM (2011) Evaluating Duero's basin (Spain) response to the NAO phases: spatial and seasonal variability. Hydrol Process 25:1313–1326

Osborn TJ, Briffa KR, Tett SFB, Jones PD, Trigo RM (1999) Evaluation of the North Atlantic Oscillation as simulated by a climate model. Clim Dyn 15:685–702

Paredes D, Trigo RM, Garcia-Herrera R, Trigo IF (2006) Understanding precipitation changes in Iberia in early Spring: weather typing and storm-tracking approaches. J Hydrometeorol 7:101–113

Rîmbu N, Boroneanţ C, Carmen B, Mihai D (2002) Decadal variability of the Danube river flow in the lower basin and its relation with the North Atlantic Oscillation. Int J Climatol 22:1169–1179

Serrano A, García JA, Mateos VL, Cancillo ML, Garrido J (1999) Monthly modes of variation of precipitation over the Iberian Peninsula. J Clim 12:2894–2919

Solomon S, Qin D, Manning M, Alley RB, Berntsen T, Bindoff NL, Chen Z, Chidthaisong A, Gregory JM, Hegerl GC, Heimann M, Hewitson B, Hoskins BJ, Joos F, Jouzel J, Kattsov V, Lohmann U, Matsuno T, Molina M, Nicholls N, Overpeck J, Raga G, Ramaswamy V, Ren J, Rusticucci M, Somerville R, Stocker TF, Whetton P, Wood RA, Wratt D (2007) Technical summary. In: Solomon S, Qin D, Manning M, Chen Z, Marquis M, Averyt KB, Tignor M, Miller HL (eds) Climate change 2007: the physical science basis. Contribution of Working Group I to the 4th assessment report of the Intergovernmental Panel on Climate Change. Cambridge University Press, Cambridge, UK and New York, NY, USA

Sousa P, Trigo RM, Aizpurua P, Nieto R, Gimeno L, Garcia-Herrera R (2011) Trends and extremes of drought indices throughout the 20th century in the Mediterranean. Nat Hazards Earth Syst Sci 11:33–51. doi:10.5194/nhess-11-33-2011

Struglia MV, Mariotti A, Filograsso A (2004) River discharge into the Mediterranean Sea: climatology and aspects of the observed variability. J Clim 17:4740–4751

Trigo IF, Davies TD, Bigg GR (1999) Objective climatology of cyclones in the Mediterranean region. J Clim 12:1685–1696

Trigo IF, Davies TD, Bigg GR (2000) Decline in Mediterranean rainfall caused by weakening of Mediterranean cyclones. Geophys Res Lett 27:2913–2916

Trigo RM et al (2006) Relations between variability in the Mediterranean region and Mid-latitude variability. In: Lionello P, Malanotte-Rizzoli P, Boscolo R (eds) The Mediterranean climate: an overview of the main characteristics and issues. Elsevier, Amsterdam, pp 179–226

Trigo RM (2008) Quantifying the impact of the North Atlantic Oscillation on western Iberia. In: Soares A, Pereira MJ, Dimitrakopoulos R (eds) geoENV VI – geostatistics for environmental applications. Springer, Berlin, pp 235–246

Trigo RM, DaCamara C (2000) Circulation weather types and their impact on the precipitation regime in Portugal. Int J Climatol 20:1559–1581

Trigo RM, Osborn TJ, Corte-Real JM (2002) The North Atlantic Oscillation influence on Europe: climate impacts and associated physical mechanisms. Clim Res 20:9–17

Trigo RM, Palutikof JP (2001) Precipitation scenarios over Iberia: a comparison between Direct GCM output and different downscaling techniques. J Clim 14:4422–4446

Trigo RM, Pozo-Vazquez D, Osborn TJ, Castro-Diez Y, Gámis-Fortis S, Esteban-Parra MJ (2004) North Atlantic Oscillation influence on precipitation, river flow and water resources in the Iberian Peninsula. Int J Climatol 24:925–944

Trigo RM, Valente MA, Trigo IF, Miranda PM, Ramos AM, Paredes D, García-Herrera R (2008) North Atlantic wind and cyclone trends and their impact in the European precipitation and Atlantic significant wave height. Ann NY Acad Sci 1146:212–234. doi:10.1196/annals.1446.014

Türkeş M, Erlat E (2003) Precipitation changes and variability in Turkey linked to the North Atlantic Oscillation during the period 1930–2000. Int J Climatol 23:1771–1796

Vicente-Serrano SM (2006) Spatial and temporal analysis of droughts in the Iberian Peninsula (1910–2000). Hydrol Sci J 51:83–97

Vicente-Serrano SM, López-Moreno JI (2008) The nonstationary influence of the North Atlantic Oscillation on European precipitation. J Geophys Res Atmos 113:D20120. doi:10.1029/2008JD010382

Vicente-Serrano SM, Trigo RM, Liberato MLR, López-Moreno JI, Lorenzo-Lacruz J, Beguería S, Morán-Tejeda H, El Kenawy A (2011) Extreme winter precipitation in the Iberian Peninsula, 2010: anomalies, driving mechanisms and future projections. Clim Res 46:51–65. doi:10.3354/cr00977

von Storch H, Zorita E, Cubasch U (1993) Downscaling of global climate change estimates to regional scales: an application to Iberian rainfall in wintertime. J Clim 6:1161–1171

Walker GT (1924) Correlations in seasonal variations of weather, IX. Mem Indian Meteorol Dept 24:275–332

Wallace JM, Gutzler DS (1981) Teleconnections in the geopotential height field during the Northern Hemisphere winter. Mon Wea Rev 109:784–812

Wang C, Liu H, Lee S (2010) The record-breaking cold temperatures during the winter of 2009/2010 in the Northern Hemisphere. Atmos Sci Lett 11:161–168

Xoplaki E, González-Rouco JF, Luterbacher J, Wanner H (2004) Wet season Mediterranean precipitation variability: influence of large-scale dynamics. Clim Dyn 23:63–78

The Impacts of NAO on the Hydrology of the Eastern Mediterranean

Ercan Kahya

Abstract The impacts of the NAO on the hydrology of eastern Mediterranean countries, such as Turkey, Iran, Kuwait, Oman, and Israel, are documented here from a general perspective. Results for the eastern Mediterranean countries differ from one location to another in terms of consistent NAO signal. Patterns of precipitation, streamflow, and lake levels in Turkey are discussed to show the NAO impacts. A special attention is devoted to the NAO influences on the formation of streamflow homogeneous region and on the probability distribution functions of critical droughts. The results of all these analyses clearly showed that the NAO signals are quite identifiable in various hydrologic variables in Turkey. For example, the NAO during winter was found to influence precipitation and streamflow patterns. In contrast temperature patterns appeared to be less sensitive to the NAO. The results of wavelet analysis showed that the Tuz, Sapanca, and Uluabat lakes reflect strong NAO influences. In southwest Iran, the October–December season is influenced mostly by the NAO in both dry and wet spells. Positive significant correlation values were found between the NAO and the total rainfall in the centre and southern Israel and with some of the rainfall categories. High correlations between the winter mode of the NAO and temperature and sea level pressure in Israel were also noted. In conclusion, current evidences have shown that the NAO has detectable influences on the hydrology of eastern Mediterranean countries with different magnitudes.

Keywords NAO · Eastern Mediterranean · Middle East · Hydrology · Climate

1 Introduction

The intensification hypothesis of the hydrologic cycle at the global scale was reported by climate modellers; however, knowledge of these changes will provide little guidance for developing long-term strategies in water resources management from a water resources perspective. Most water resources issues are regional in nature, and decisions are often made at river basin to catchment scales. Therefore,

E. Kahya (✉)
Istanbul Technical University, Istanbul, Turkey
e-mail: kahyae@itu.edu.tr

S.M. Vicente-Serrano, R.M. Trigo (eds.), *Hydrological, Socioeconomic and Ecological Impacts of the North Atlantic Oscillation in the Mediterranean Region*, Advances in Global Change Research 46, DOI 10.1007/978-94-007-1372-7_5,

it is import to note that future climate impacts are appropriately observed and studied at regional scales to make relevant and accurate information available for decision-making purposes (Sorooshian et al., 2003).

Studies in recent decades have shown that the North Atlantic Oscillation (NAO), a dominant mode of Atlantic sector climate variability, has the far-field influence on interannual to decadal variations of main climate and hydrological variables (i.e., temperature, precipitation and streamflow) in the eastern Mediterranean region. The Middle East (ME) region is commonly neglected in North Atlantic studies because of its proximity to the monsoonal region and the relatively scarceness of hydrological data. However the form of the NAO signature, with enhanced precipitation in the western Europe during its high index phases, and in the eastern Mediterranean during its low index phases, has been noted. A relationship between the eastern Mediterranean and Atlantic sector is expected since the NAO controls Atlantic heat and moisture fluxes into the Mediterranean region. In this context, it was shown that the first principal component of December through March streamflow variability reflects changes in the NAO using a sea surface temperature (SST) based index of the NAO and available streamflow data from some Middle Eastern rivers (Cullen et al., 2002).

A summary of climate characteristics of the ME where a complicated relation between land and climate exists is worthwhile to mention here. The ME region is known to experience a Mediterranean climate type (Csa in the Köeppen classification) with wet winters and dry summers in most parts; however, spatial gradients in climate are sharper than in the broad prototype Csa region to the west (Evans et al., 2004). That is, northward transition from desert through steppe to cool highland climate all take places within 400 km along the 40°N meridian. Presence of many coastlines and mountain ranges function in the modification of local climates in the ME elsewhere. Specifically speaking, upslope seasonal precipitation occurs in mountain ranges of southern the Black and Caspian Seas, and eastern coast of the Mediterranean Sea. Similarly orographic precipitation in the Taurus and Zagros Mountains feeds the Euphrates and Tigris Rivers, a source integrated watershed system for the Mesopotamia region. As prevailing sources of water vapour, the Red Sea and Persian Gulf activate little precipitation locally owing to descending air mass in the Hadley cell. Interior steppe and deserts extending Syria, Iraq, Jordan and Saudi Arabia, cause drier climate conditions by means of surrounding mountain ranges. Dry summers render only winter rainfall (which is highly changeable) possible for source of renewable water in the eastern Mediterranean region. Developing economies in the region heavily depend upon water, so they are extremely vulnerable to droughts. Krichak et al. (2002) proclaimed that the NAO and the East Atlantic-West Russian oscillation (EA-WR) have a combined effect, so that when they are both in their positive phase, a rainy season occurs in the southeastern region, and vice versa.

For regional atmospheric patterns governing rainfall in Israel, Ziv et al. (2006) explained that the synoptic-scale system affecting the eastern Mediterranean that is responsible for most of the annual rainfall is an extratropical cyclone, known as the Cyprus Low. Rainfall is formed within cold air masses originated from Europe, entering the region from the northwest. Air masses gain moisture from warm water

and become conditionally unstable. The strong thermal affects significantly cyclone dynamics over this region, in turn, resulting in intensive rainfall over Israel.

A relationship between the eastern Mediterranean and Atlantic sector is to be expected since the NAO controls Atlantic heat and moisture fluxes into the Mediterranean region (Türkes, 1996). Since these winter cyclones are the dominant source of Middle Eastern rainfall and streamflow, changes in Atlantic westerly heat/moisture transport and Atlantic/Mediterranean SSTs in relation to the NAO are expected to influence the ME climate (Cullen et al., 2002). Thus both fundamental climatic processes and their critical socio-economical consequences are manifested in the eastern Mediterranean region, and offer the motivation for this chapter. Presentation of typical pictures of the NAO impacts on the country and regional scales in the eastern Mediterranean region is the purpose of this chapter.

2 Results

In the last 2 decades, climate modelling studies have been undertaken to simulate the climate of the Middle East/eastern Mediterranean region. For example, Evans et al. (2004) had the same aim and showed that processes controlling the seasonality of precipitation differ in different subregions and are often different from the processes controlling interannual variability in the Middle East. More recently, in a regional climate modelling study of the winter climate of Central–Southwest Asia (CSWA), focusing on the mean model climatology of temperature and precipitation, Syed et al. (2010) analysed the simulated storm characteristics and the effects of ENSO and NAO on storm activity and precipitation. Their detailed storm analysis demonstrated that they are mostly associated with an intensification of western disturbances originating in the eastern Mediterranean and Middle East regions and moving eastward across a mid-troposphere (500 hPa) trough situated over the CSWA during the positive NAO and warm ENSO phases. The NAO and ENSO signals are reinforced by enhanced moisture sources from the Mediterranean, Caspian and Arabian Seas.

To overview the NAO signals on two essential hydrologic variables (precipitation and streamflow) in the eastern Mediterranean region, it would be reasonable to address the issue at both regional and country scale under the lights of some major studies. For the first scale, we prefer to rely on our review on two diagnostic studies by Cullen and her colleagues in this section. For the second scale, the order of naming country in the region is selected according to clockwise direction starting from the north.

2.1 *Regional Scale Review: NAO Impacts on the Hydrology of Middle East Region*

2.1.1 Precipitation Variability Associated with the NAO

Cullen et al. (2002) described the influence of the NAO on interannual to decadal precipitation variability in the Middle East by computing composite

precipitation anomalies using the 21-year CMAP monthly data set over the region (90°W–60°E; 20°N–70°N) extending from the east coast of the U.S. to the Middle East for months with positive and negative values of the NAO index for the complete period 1979–1998. Their composite analysis clearly showed evident appearance of the expected signature of the NAO in the form of enhanced precipitation in north Europe during high index periods and in the Mediterranean region during low index periods (Fig. 1). Another important clue visualised in this figure is said to be an axis of enhanced monthly wintertime precipitation that is clearly observable during low index periods extending from southeastern U.S. across the Atlantic to the Iberian Peninsula having negative anomalies more or less parallel to the north and south. In contrast a reversed pattern with departures from strict linearity is virtually evident during high index periods. More specifically this pattern comprises of the enhanced precipitation axis during low index months extending across the northern Mediterranean nearly to the Caspian Sea with an enhanced secondary maximum along the west coast of Spain and Portugal and over Turkey.

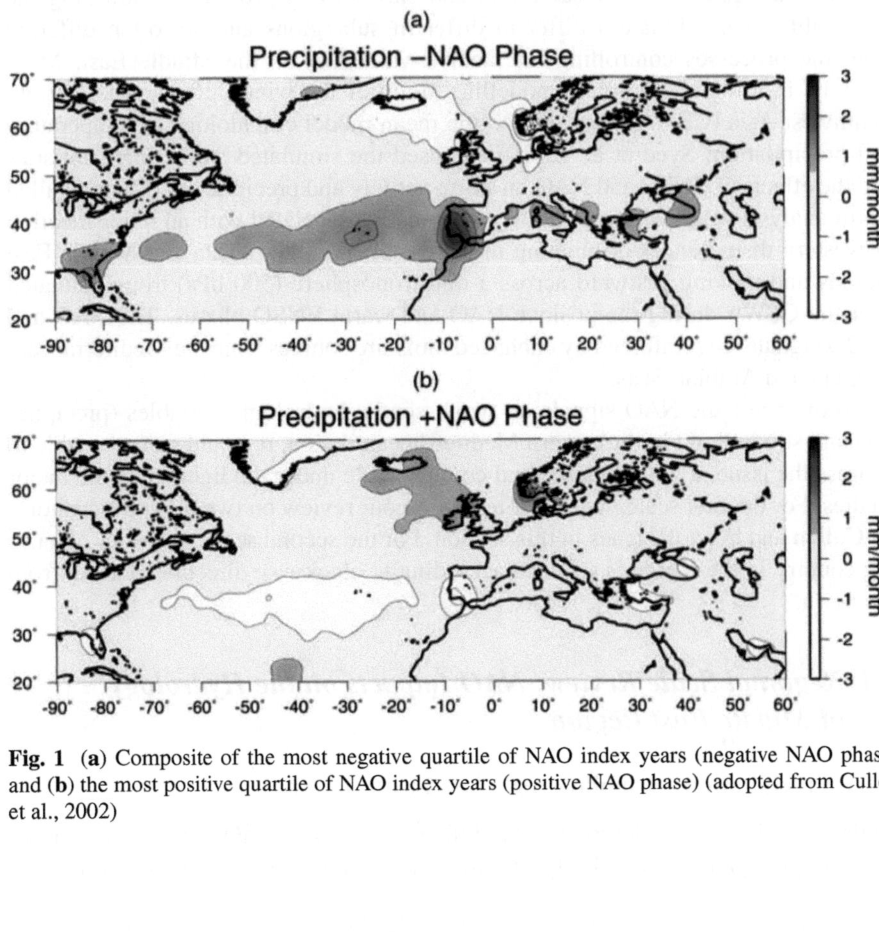

Fig. 1 (**a**) Composite of the most negative quartile of NAO index years (negative NAO phase) and (**b**) the most positive quartile of NAO index years (positive NAO phase) (adopted from Cullen et al., 2002)

2.1.2 Streamflow Variability Associated with the NAO

Cullen et al. (2002) examined streamflow data in relation to the NAO from five Middle Eastern rivers, namely Ceyhan (Turkey), Euphrates (Turkey), Karun (Iran), Tigris (Iraq), and Yarmouk (Jordan). They demonstrated that the first principal component of December through March streamflow variability reflects changes in the NAO. The hydrology of these rivers revealed two primary flooding periods: the first due to rainfall-driven runoff from December through March, regulated on inter-annual to decadal timescales by the NAO, and the second period due to spring snowmelt from April through June, contributing in excess of 50% of annual runoff. They also noted that the latter witness no significant NAO teleconnection and a less direct relationship with local climatic factors.

2.2 NAO Impacts on the Hydrology of Turkey

Climate response in Turkey to variability of the NAO, in particular, has been a major concern in last decade. The results of Türkeş and Erlat (2003) indicated that negative correlation coefficients between Turkish precipitation and the NAO indices were stronger in winter and autumn, and weaker in spring and almost non-existent in summer. They stated that annual and seasonal composite precipitation means were mostly characterized by wetter than longterm average during the weak NAO phase, whereas the strong NAO responses mostly exhibited drier than long-term average annually and in all seasons except summer. Spatially coherent and statistically significant changes in precipitation amounts during the extreme NAO phases were more evident in western and middle Turkey.

Henceforth this section will emphasize on the findings of the author and his colleagues' earlier and current works in regard to the NAO impacts on hydrological surface variables in Turkey. As a first start of presenting our results based upon basic statistical analyses, we examined Turkish climatic variables (precipitation, streamflow and maximum and minimum temperatures) in association with the NAO using correlation analysis with various lag combinations (Karabörk et al., 2005). In this study, we used monthly precipitation totals (compiled from 94 climate stations with a period 1951–1993), monthly streamflow (compiled from 76 streamflow gauging stations with a period 1964–1994) and monthly maximum and minimum temperature (compiled from 54 climate stations with a period 1951–1994) observations. In general, simultaneous and lag cross-correlation analyses showed that precipitation, streamflow, maximum temperature and minimum temperature patterns over Turkey have significant negative correlations with the NAO index. The results of some lag correlations also indicated a considerable prediction potential for precipitation and streamflow variables. A portion of this potential for streamflow could be readily explained by the carry-on effects of catchments, nevertheless there is a need for further investigation to elucidate the NAO and precipitation relationship that has been identified to be valid for the time lag of several months in some cases. Correlation analysis revealed less sensitive characteristics of Turkish temperature

patterns against the NAO. The consequences of the present study verify far reaching influences of the NAO in eastern Mediterranean basin and the Middle East region. Moreover, we stated that the SOI series was negatively correlated with both Turkish precipitation and streamflow patterns; however, these correlations were not as significant when compared with those between the NAO and precipitation and streamflow separately.

Of particular interest in Hurrell (1996) is said to be the contribution of the NAO variability to the recent trend in global hemispheric mean wintertime temperature. It was shown that both the NAO and SO extremes are linearly related to the climate trend, with the NAO dominating a larger land area than the SO. When he used only the NAO and the SO indices in his regression models, more confidence was observed in the estimated coefficients and 44% of the variance of the extra-tropical Northern Hemisphere temperatures was explained.

This diagnosis also appeared, to some extent, in the results of our next investigation (Kalayci and Kahya, 2006), involving the systematic modes of spatial and temporal variation in monthly streamflow records in Turkey using the principal components (PCs) analysis. The PC1 reproduces 38.2% of the total variance of the entire streamflow data set. In other words, 142 of 372 observed anomaly map patterns during the study period 1964–1994 have a similar map pattern to that of PC1 loadings (Fig. 2), illustrating only positive sign across the study domain. It represents a basic streamflow anomaly pattern having above-normal streamflow anomalies everywhere when the PC1 scores are positive or below-normal streamflow anomalies everywhere when the scores are negative values. The homogeneous appearance of PC1 is a common feature of the PC analysis. The contoured values in Fig. 2 are the PC1 loadings, corresponding to the values of correlation coefficients between streamflow and the PC values. It is important to note that strongest loadings indicate greatest departures from the mean streamflow (Lins, 1985). We then targeted seasonal correlations between the PC scores and the indices of large-scale atmospheric circulation (i.e., the NAO and SO) to elucidate the physical meaning of the components. For this purpose, we carried out various correlation analyses between the scores of each PC and the series of the NAO index at the annual and seasonal bases. The relation between PC1 scores and NAO index was witnessed as –0.491 (significant at 99% level) for annual average; –0.424 (significant at 95% level) for DJF seasonal average; –0.387 (significant at 95% level) for MAM seasonal average; and –0.506 (significant at 99% level) for JFM seasonal average. There were no similar significant correlations for other PCs. For the JFM season, Karabörk et al. (2005) also found a highly significant relation in conjunction with the NAO in the western Turkey. These indications clearly imply the sensitivity of the Turkish streamflow climatology to the NAO variations.

More recently we considered the NAO impacts on the hydrology of Turkey using different variables through new but uncommon perspectives and arrived to following interesting results. Let's consider first drought variables, representing complex events which may impair social, economic, agricultural, and other activities in a society. We were concerned in Şarlak et al. (2009) with a hydroclimatological study, which shed light on the NAO impacts on regional drought characteristics in Turkey,

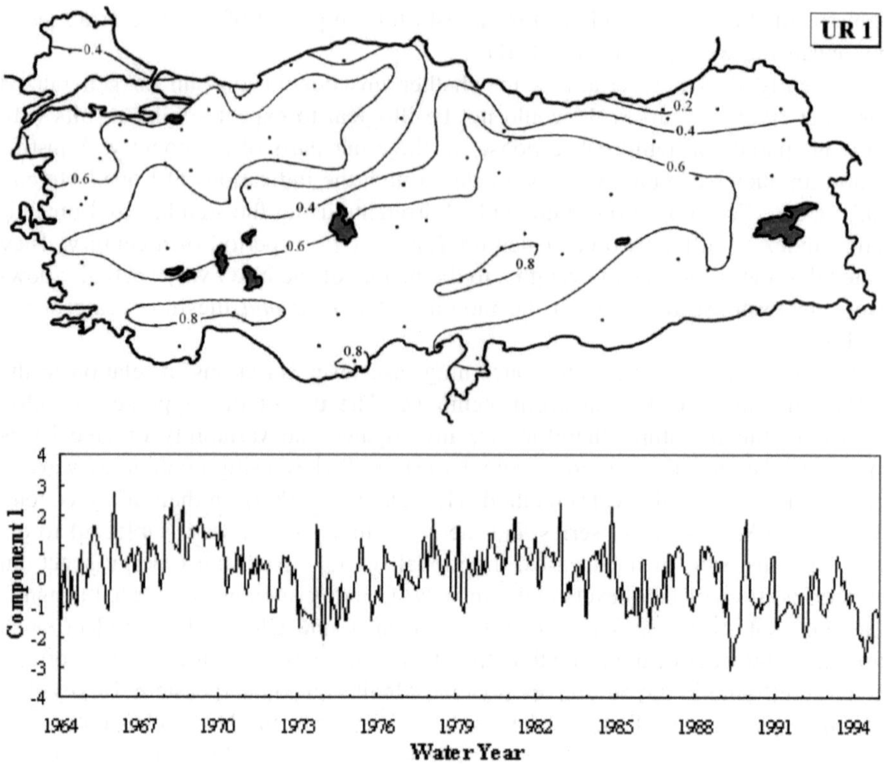

Fig. 2 The map pattern of the first component loadings and time series of the first component scores (adopted from Kalayci and Kahya, 2006)

consisting of three stages: first a Markov chain model based on annual flows at Göksu river (located in the mid-southern Turkey); second a critical drought analysis; and third the influences of the NAO on the probability distribution functions (PDFs) of critical droughts. Drought duration was taken as a key parameter in our Markov chain model. It is noteworthy to recall that the critical drought duration is the possible maximum duration likely to occur over the economic life of any water resources system. The expectations of the critical drought duration are provided for different transitional probabilities. We examined the influences of NAO on the PDFs of critical droughts for two opposite phases; namely, the period dominated by negative NAO index values (1936–1971) and the period controlled by positive NAO index values (1972–2000). The analysis of results clearly indicated that the NAO mode has quantifiable influences on the transition probabilities and expectation of the critical drought duration for annual flows at Göksu River. For example, the probability *P(d/d)* corresponding to the characteristic of negative anomaly followed by another negative anomaly, takes on a remarkably high value of 0.81 for the period of positive NAO index phase causing dry conditions in Turkey. It is striking that this

value resided at a magnitude as low as 0.40 for the period of negative NAO index phase causing wet conditions in Turkey.

Currently it is not certain as to whether this conclusion can be generalized for other rivers in Turkey. It would not be illogical to expect similar results with varying spatial and temporal response for the other parts of the country. A justification for such expectation can be made through the indications of López-Moreno and Vicente-Serrano (2008) who studied droughts using the standardized precipitation index at different time scales for Europe over a period of a century. They stated that the responses of droughts to the phases of the NAO vary spatially; however, the response depends on the month of the year *and* the time scale of the analysis.

In our other study we were concerning lake-level variations in relation to the NAO using the wavelet transform technique. The use of this approach is rather limited in the literature, therefore we investigated the variability of lake levels in seven lakes mostly scattered around western Turkey using continuous wavelet transforms and global spectra methods (Küçük et al., 2009). In their analysis, long winter (DJFM) lake-level series and the NAO index series were subjected to the wavelet transform. In most cases, the global wavelet spectrum (energy spectrum of periodicities) of lake levels and winter NAO index anomalies revealed a significant correlation. It was shown that Tuz, Sapanca, and Uluabat lakes reflect much stronger influences of the NAO than the other four lakes. In contrast, weak correlations were found in the coastal areas of the Mediterranean and eastern Turkey. The periodic structures of Turkish lake levels in relation to the NAO revealed a spectrum between the 1-year and 10-year scale level. They stated that the periodicities of more than 10-year scale levels were detected; however, explanation of such significant relations between the NAO and these long-term periodicities still remains a challenging task.

Since we were curious about different aspects of the NAO impacts on different hydrological variables in Turkey and have been releasing the relevant results from our previous studies, we later wondered whether or not it is possible to discover a noticeable difference due to the NAO forcing on hydrological regionalization in Turkey. This might be thought as an aggregate hydrologic response to the NAO; consequently we have started to conduct a research based on cluster analysis of streamflow data in order to depict maps related to positive and negative phase of the NAO. Following a similar logic used by Şarlak et al. (2009) and after rigorous inspection of available continuous streamflow series, we were able to form two different data sets; one with a period 1962–1971 and other with a period 1970–1997. Needless to say, the former period represents the negative phase period of the NAO whereas the latter stands for the opposite phase. The number of stations included for the analysis of each case is not the same; that is, 40 stations come out for the negative phase period as 83 stations were selected for the positive phase period. The results of analysis for monthly streamflow means are illustrated in Fig. 3. Differences between the two maps could be highlighted as follows: (i) elongated regions in the middle and east Turkey for the negative phase of the NAO and (ii) additional small region appearance (located in western Black Sea coast shown by pink colour) and

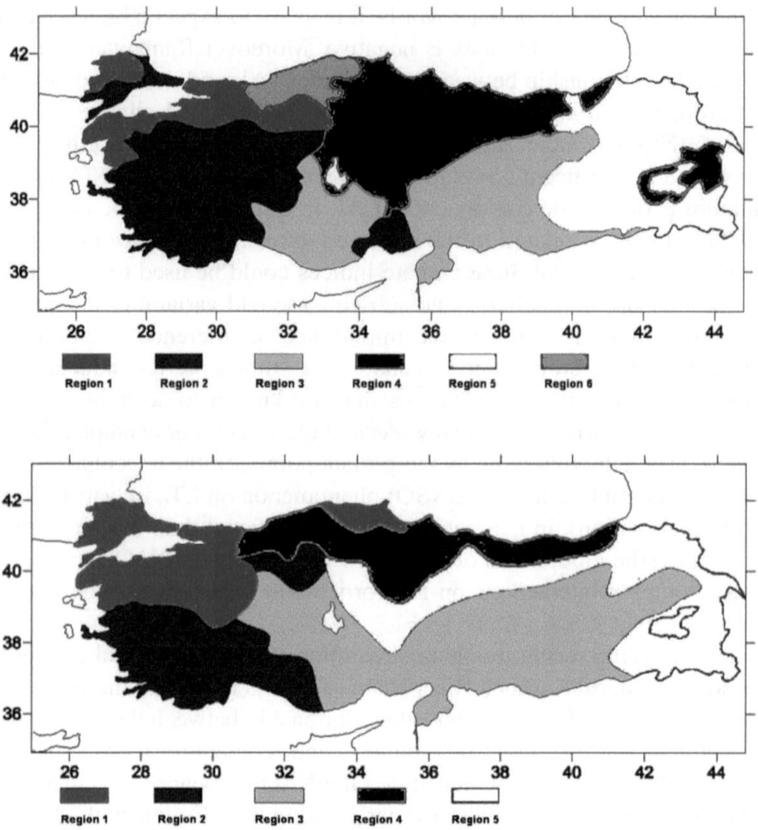

Fig. 3 The map patterns of hydrological regionalization based on the positive phase of NAO (*top panel*) and the negative phase of NAO (*lower panel*)

patchy form of Marmara region for the positive phase of the NAO. As a result this comparison shows a potential of the NAO in controlling mechanisms for hydrologic regionalization in Turkey.

2.3 NAO Impacts on the Hydrology of Iran

The influence of the NAO on historically observed rainfed crop yields of both barley and wheat in the northeast of Iran was investigated by Bannayan et al. (2010a) for the period 1983–2005. In central part (Mashhad), the NAO provided a firm basis for a measurable teleconnection between crop yields and climate factors. They noted significant correlation between the NAO and wheat only during October, but no such significant relationship for barley was detected for any month in Mashhad. During the period of positive NAO index, yields increased whereas the reverse happened

during the opposite phase. In other words, it is likely to expect a higher wheat yield in Mashhad when the NAO index is negative. Moreover Bannayan et al. (2010b) studied that the relationship between aridity index (AI) and detrended crop yield of selected crops (wheat and barley) and the influence of three climate indices (AO, NAO and NINO-3.4) were assessed for Khorasan province in northeast of Iran. They observed a significant correlation between the NAO and NINO-3.4 with AI. Precipitation is one of the components of AI, in turn; the AI response to the NAO and NINO-3.4 can be related to the observed association between this index and precipitation. It seems that these climate indices could be used to construct useful tools to monitor drought patterns and subsequent yield variability in some regions of Khorasan province. Owing to the importance of reference evapotranspiration (ET_O) in determining crop water demand, some studies were conducted to assess the impacts of different climate factors that are known to account for important fraction of climate variability in many areas of the world. For example, Sabziparvar et al. (2010) stated the following two important points: (i) the teleconnection impact of El Nino/Southern Oscillation (ENSO) phenomenon on ET_O in warm arid regions (southern parts of Iran), in general, was more significant than that in warm humid regions, and (ii) the modulation of the ENSO impacts by the NAO variability might also cause complex interactions on ET_O process in some sites in the north coasts of Iran.

Besides analyzing recent trends in streamflow, precipitation and temperature in mountainous, semiarid Karkheh River basin (which is located in the western part of Iran), Masih et al. (2010) studied also the relationship between the NAO index with the local climate variable (e.g., precipitation and temperature). Some earlier studies for the Middle East have demonstrated the NAO influences on temperature and precipitation regime during winter and early spring (e.g. Cullen et al., 2002; Mann, 2002; Evans et al., 2004). Following these works Masih et al. (2010) examined the magnitude of correlation values between monthly NAO index and the monthly precipitation and temperature during the winter months from December to March. Their findings regarding precipitation appeared to be different than those of Cullen et al. (2002) who found strong impact of changes in the NAO on the streamflow, precipitation and temperature during December–March for the Euphrates-Tigris River system. However, the results of Masih et al. (2010) confirmed those of Evans et al. (2004) who indicated that the NAO index alone could not be a predictor of the local climate in the Zagros Mountains, Iran.

In order to make more reliable decisions in water resources management, information on regional characteristics of drought duration and intensity is thought to provide critical insights in arid and semiarid regions. In this regard, determination of appropriate predictors for seasonal rainfall is important in regional analyses of meteorological droughts, since rainfall is the sole variable to quantify the characteristics of such drought. Dezfuli et al. (2010) concerned the regional rainfall relationships with the SOI and NAO at the seasonal time scale to find the appropriate regional drought predictors over southwest Iran. They found that the NAO is negatively correlated with autumn rainfall such that it is least likely for an extreme autumn drought to occur when June–August NAO is negative. A spring drought

is usually preceded by an October–December NAO greater than 0.5. In contrast winter droughts were not lag-correlated with either NAO or SOI. In addition to the findings for droughts, these indices considerably influenced wet seasons. A wet autumn tends to occur when either May–July SOI is less than −0.5 or June–August NAO is less than about −0.3. It is also noted that the extreme wet springs were absent during positive October–December NAO. This season is influenced most by NAO in both dry and wet spells. Finally, Defluzi et al. concluded that, similar to droughts, the wet winter seasons were not in association with either SOI or NAO.

Ghasemi and Khalili (2006), who studied the impact of the Arctic Oscillation (AO) on the winter surface air temperature (SAT) over Iran, found a significant relationship between the NAO index and the winter SAT for about 53% of 51 stations in Iran. Significant correlation coefficients between the NAO index and the winter SAT vary from –0.34 in Tabriz to –0.58 in Abadan. Using seasonal average of the NAO, there were some attempts to develop a forecasting model for climate variables in Iran. A model developed by Setoodeh et al. (2004) was able to predict rainfall and temperature one season ahead with reasonable error based on neural network-based models in Shiraz.

2.4 NAO Impacts on the Hydrology of Kuwait, Oman and Israel

Because of small geographical extent and limitation in data coverage and length country wise, there is no ample amount of studies that could reflect different aspects of NAO signals in Oman and Kuwait. However recent results introduced by Marcella and Eltahir (2008) may shed light on possible teleconnection between the NAO and Kuwait's hydrology. Kuwait, surrounded by the Syrian and Arabian Deserts to the west and the Persian Gulf to the east, is located in a semiarid climate zone. Its location lies in a transitional area between the tropics and midlatitudes; hence, the contributions of both climate zones to Kuwait's rainfall intrigues for answer to how sensitive the country and the surrounding region's rainfall is to large-scale circulation indices. Among other issues in their study, Marcella and Eltahir (2008) attempted to explain changes in the strength of the NAO in affecting Kuwait's annual rainfall. Nevertheless their investigation revealed no significant correlation on a monthly or yearly time scale for the relationship between Kuwait rainfall and NAO. These results are similar to those of Evans et al. (2004), who did not find a significant correlation between the NAO and rainfall in the Middle East. As far as it can be seen from their paper, the authors have used normalized annual rainfall anomalies in Kuwait from CRU (Climate Research Unit: 1976–2001) and GPCP (Global Precipitation Climatology Project: 1979–2001). The use of gridded datasets (that have used with few original stations in their assemblage) casts some doubts on the significance of these results and might explain why these are fundamentally different from the relatively well established impacts of NAO over Turkey, Iran and Oman.

A recent work by Charabi and Al-Hatrushi (2010) provides some evidence for Oman, having studied the relation between winter rainfall in Oman and the large-scale circulation as well as synoptic activity on a monthly basis. Their investigation focused on wet and dry spells that occurred during the period 1984–2007 based on composite analysis and the role of the main Eurasian oscillations; the EA/WR and the NAO in controlling the Oman winter rainfall. Charabi and Al-Hatrushi obtained the best results for the NAO in such a way that the wet spells are negatively correlated with NAO. The negative NAO phase corresponds to enhanced westerly flow (850-hPa up to 200-hPa), bringing moisture from the Atlantic ocean across the Mediterranean area towards Oman. However, the correlation detected between dry spells and positive phase of the NAO was relatively weak. The teleconnection showed that the extreme wet spells in northern Oman during winter correspond to the extreme negative index of NAO. Such prominent examples were noted in December 1989 (regional rainfall index = 1.45 vs. NAO index = −1.15) and in December 1995 (regional rainfall index = 3.32 vs. NAO index = −1.67). It is also worth to note that the influence of a weak NAO in winter precipitation may tend to cause a widespread wet event rather than an expected widespread drought condition in northern Oman during some winter months (e.g., December 1991, December 1992, January 1998, January 2001, February 1993, and March 1987). The bulk of rainfall in northern Oman occurs in the winter *seif* season from November/December to April and accounts for 58–83% of the average annual rainfall. These are caused by midlatitude westerly depression by polar front jet stream (Andy et al., 2009). An essential conclusion drew by Charabi and Al-Hatrushi (2010) was that the winter rainfall in Oman showed, in general, some linkages with the negative phase of NAO.

Israel is located on a climatic border, between the temperate and the arid regions. There have been extensive studies to put forward detailed aspects of the NAO impacts on hydrology and climatology of Israel and some major works in the context of this section are reviewed here. Ziv et al. (2006) aimed to find a linkage between the interannual variations of the rainfall measured in twelve stations in the northern half of Israel (the southern Levant) and regional to global scale atmospheric circulations. For the relation between the southern Levant rainfall and NAO, they pointed out that weak correlation found between the rainfall and the NAO by Ben-Gai et al. (2001) and in this study is inconsistent with Eshel and Farrell (2000, 2001), who found that the rainfall variability over the EM is explained through the modulations of a North Atlantic/Mediterranean teleconnectivity. Other than rainfall variable, Ben-Gai et al. (2001) showed high correlations between the winter mode of the NAO and temperature and sea level pressure (SLP) in Israel with a magnitude of –0.8 and +0.9, respectively. However, none of these studies witnessed any significant correlation between NAO and the rainfall over the southern Levant.

Categorized rainfall intensity in different regions of Israel could be put on the table for searching the NAO signals. Yosef et al. (2009) examined rainfall intensity trends and searched NAO signals at the same time. Their results included that a positive significant correlation was found with the total rainfall in the centre and south and with some of the rainfall categories.

3 Conclusions

The findings in terms of consistent NAO signal for the eastern Mediterranean countries differ from one location to another. The following aspects are the articled highlights of the NAO signals in the eastern Mediterranean countries.

Highlights for Turkey;

1. The results of some lag correlations indicated a considerable prediction potential for precipitation and streamflow variables.
2. Correlation analysis revealed less sensitive characteristics of Turkish temperature patterns with the NAO.
3. The relation between the NAO index and streamflow PC1 scores (representing a basic streamflow anomaly pattern in the entire Turkey) was determined highly significant for annual average and various seasonal averages.
4. The NAO has quantifiable influences on the transition probabilities and expectation of the critical drought duration for annual flows at Göksu River.
5. Tuz, Sapanca, and Uluabat lakes demonstrated much stronger influences of the NAO than the other four lakes.
6. The NAO was shown as a factor affecting the form of hydrological regionalization depending on its positive and negative phases.

Highlights for Iran;

1. In the central part of Iran (Mashhad), the NAO provided a firm basis for a measurable teleconnection between crop yields and climate factors. It is likely to expect a higher wheat yield in Mashhad when the NAO index is negative.
2. Aridity index series compiled from Khorasan province in northeast of Iran revealed significant correlations between the NAO and NINO-3.4.
3. The NAO index alone could not be a predictor of the local climate in the Zagros Mountains, Iran. Similar to droughts, the wet winter seasons were not in association with the NAO in the same area.
4. In southwest Iran, the October–December NAO season was influenced most by the NAO in both dry and wet spells.
5. Significant relationship between the NAO index and winter surface air temperature was found in about 53% of 51 stations throughout Iran.

Highlights for Kuwait, Oman and Israel;

1. No significant correlation on a monthly or yearly time scale for the relationship between the NAO and Kuwait rainfall was noted.
2. During the period 1984–2007, the wet spells in Oman winter rainfall were negatively correlated with the NAO.
3. Extreme wet spells in northern Oman during winter corresponded to the extreme NAO negative index.

4. Poor correlation was found between the NAO and rainfall in the northern half of Israel.
5. High correlations were detected between the winter mode of the NAO and temperature and sea level pressure in Israel.
6. Positive significant correlation was found with the total rainfall in the centre and south and with some of the rainfall categories.

It will be a good practice to close this chapter by a rare NAO related indication for Iraq. Recalling preliminary research showing that the band of drying and cooling during the positive NAO phase extends from Spain through northern Africa and Italy to Turkey and the Middle East, this signal for temperature in Iraq was exceptionally strong, with the NAO explaining nearly 50% of the variance in wintertime temperatures according to the study by Sorooshian et al. (2003). Such a strong signal level is not common among climate studies, therefore it provides a potential for accurate seasonal forecasts.

Acknowledgements The author is grateful to the efforts of organizers of ESF-MedCLIVAR Workshop (Hydrological, socioeconomic and ecological impacts of the North Atlantic Oscillation in the Mediterranean region, Zaragoza, Spain, May 24–27, 2010) Drs. García-Herrera, Trigo and Vicente-Serrano. This work has been funded through support by TUBITAK Research Project, No: ÇAYDAG 108Y165.

References

Andy YK, Dorvlo AS, Kumar GTV (2009) Analysis of a 27-year rainfall data (1977–2003) in the Sultanate of Oman. Int J Climatol 29:605–617

Bannayan M, Lotfabadi SS, Sanjani S, Mohamadian A, Aghaalikhani M (2010a) Effects of precipitation and temperature on crop production variability in northeast Iran. Int J Biometeorol 55(3):387–401

Bannayan M, Sanjani S, Alizadeh A, Lotfabadi SS, Mohamadian A (2010b) Association between climate indices, aridity index, and rainfed crop yield in northeast of Iran. Field Crops Res 118:105–114

Ben-Gai T, Bitan A, Manes A, Alpert P, Kushnir Y (2001) Temperature and surface pressure anomalies in Israel and the North Atlantic Oscillation. Theor Appl Climatol 69:171–177

Charabi Y, Al-Hatrushi S (2010) Synoptic aspects of winter rainfall variability in Oman. Atmos Res 95:470–486

Cullen HM, Kaplan A, Arkin PA, Demenocal PB (2002) Impact of the north Atlantic oscillation on the Middle Eastern climate and streamflow. Clim Change 55:315–338

Dezfuli AK, Karamouz M, Araghinejad S (2010) On the relationship of regional meteorological drought with SOI and NAO over southwest Iran. Theor Appl Climatol 100:57–66

Eshel G, Farrell BF (2000) Mechanisms of Eastern Mediterranean rainfall variability. J Atmos Sci 57:3219–3232

Eshel G, Farrell BF (2001) Thermodynamics of Eastern Mediterranean rainfall variability. J Atmos Sci 58:87–92

Evans JP, Smith RB, Oglesby RJ (2004) Middle East climate simulation and dominant precipitation processes. Int J Climatol 24:1671–1694

Ghasemi R, Khalili D (2006) The influence of the Arctic oscillation on winter temperatures in Iran. Theor Appl Climatol 85:149–164

Hurrell JW (1996) Influence of variation in extratropical wintertime teleconnections on Northern Hemisphere temperature. Geophys Res Lett 23:665–668

Kalayci S, Kahya E (2006) Assessment of streamflow variability modes in Turkey: 1964–1994. J Hydrol 324:163–177

Karabörk MÇ, Kahya E, Karaca M (2005) The influences of the Southern and North Atlantic Oscillations on climatic surface variables in Turkey. Hydrol Process 19:1185–1211

Krichak SO, Kishcha P, Alpert P (2002) Decadal trends of main Eurasian oscillations and the Eastern Mediterranean precipitation. Theor Appl Climatol 72:209–220

Küçük M, Kahya E, Cengiz TM, Karaca M (2009) North Atlantic oscillation influences on Turkish lake levels. Hydrol Process 23:893–906

Lins HF (1985) Streamflow variability in the United States: 1931–1978. J Clim Appl Climatol 24:463–471

López-Moreno JI, Vicente-Serrano SM (2008) Positive and negative phases of the wintertime north Atlantic oscillation and drought occurrence over Europe: a multitemporal-scale approach. J Clim 21:1220–1242

Mann ME (2002) Large-scale climate variability and connections with the Middle East in the past centuries. Clim Change 55:287–314

Marcella MP, Eltahir EAB (2008) The hydroclimatology of Kuwait: explaining the variability of rainfall at seasonal and interannual time scales. J Hydrometeorol 9:1095–1105

Masih I, Uhlenbrook S, Maskey S, Smakhtin V (2010) Streamflow trends and climate linkages in the Zagros Mountains, Iran. Clim Change 104(2):317–338

Sabziparvar AA, Mirmasoudi SH, Tabari H, Nazemosadat MJ, Maryanajic Z (2010) ENSO teleconnection impacts on reference evapotranspiration variability in some warm climates of Iran. Int J Climatol 31(6)

Şarlak N, Kahya E, Bég AO (2009) Critical drought analysis: a case study of Göksu River (Turkey) and north Atlantic oscillation influences. J Hydrol Eng 14:795–802

Setoodeh P, Safavi A, Nazemosadat MJ (2004) Intelligent forecasting of rainfall and temperature of Shiraz city using neural networks. Iran J Sci Technol 28:165–174

Sorooshian S, Imam B, Mahani S, Pagano T, Whitaker M (2003) Hydrologic sciences and water resources management issues in a changing world. In: Alsharhan AS, Wood WW (eds) Water resources perspectives: evaluation, management and policy. Developments in water science, vol 50. Elsevier Science, Amsterdam, pp 83–92

Syed FS, Giorgi F, Pal JS, Keay K (2010) Regional climate model simulation of winter climate over Central–Southwest Asia, with emphasis on NAO and ENSO effects. Int J Climatol 30:220–235

Türkeş M (1996) Spatial and temporal analysis of annual rainfall variations in Turkey. Int J Climatol 16:1057–1076

Türkeş M, Erlat E (2003) Precipitation changes and variability in Turkey linked to the North Atlantic Oscillation during the period 1930–2000. Int J Climatol 23:1771–1796

Yosef Y, Saaroni H, Alpert P (2009) Trends in daily rainfall intensity over Israel 1950/1-2003/4. Open Atmos Sci J 3:196–203

Ziv B, Dayan U, Kushnir Y, Roth C, Enzel Y (2006) Regional and global atmospheric patterns governing rainfall in the southern Levant. Int J Climatol 26:55–73

Kalaycı S, Kahya E (2006) Assessment of streamflow variability modes in Turkey: 1964–1994. J Hydrol 324:163–177

Karabörk MÇ, Kahya E, Karaca M (2005) The influences of the Southern and North Atlantic Oscillations on climatic surface variables in Turkey. Hydrol Process 19(6):1185–1211

Krichak SO, Kishcha P, Alpert P (2002) Decadal trends of main Eurasian oscillations and the Eastern Mediterranean precipitation. Theor Appl Climatol 72(3):209–220

Küçük M, Kahya E, Cengiz TM, Karaca M (2009) North Atlantic Oscillation influences on Turkish lake levels. Hydrol Process 23:893–906

Lins HF (1985) Streamflow variability in the United States: 1931–1978. J Clim Appl Meteorol 24:463–471

López-Moreno JI, Vicente-Serrano SM (2008) Positive and negative phases of the wintertime north Atlantic oscillation and drought occurrence over Europe: a multitemporal-scale approach. J Clim 21(6):1220–1243

Mann ME (2002) Large-scale climate variability and connections with the Middle East in past centuries. Clim Change 55:287–314

Marcella MP, Eltahir EAB (2008) The hydroclimatology of Kuwait: explaining the variability of rainfall at seasonal and interannual time scales. J Hydrometeorol 9:1095–1105

Masih I, Uhlenbrook S, Maskey S, Smakhtin V (2010) Streamflow trends and climate linkages in the Zagros Mountains, Iran. Clim Change 104(2):317–338

Sabziparvar AA, [illegible] SH, Tabari H, Nazemosadat MJ, Maryanaji Z (2010) ENSO teleconnection impacts on reference evapotranspiration variability in some warm climates of Iran. Int J Climatol 31(6)

Sarlak N, Kahya E, Bég AO (2009) Critical drought analysis: a case study of Göksu River (Turkey) and North Atlantic Oscillation influences. J Hydrol Eng 14:795–802

[illegible], Safavi A, Nazemosadat MJ (2004) Intelligent forecasting of rainfall and temperature of Shiraz city using neural networks. Iran J Sci Technol 28:165–174

Sorooshian S, Imam B, Mahani S, [illegible], Whitaker M (2003) Hydrologic science and water resources management issues in a changing world. In: Alsharhan AS, Wood WW (eds) Water resources perspectives: evaluation, management and policy. Developments in water science, vol 50. Elsevier Science, Amsterdam, pp 83–92

Syed FS, Giorgi F, Pal JS, Keay K (2010) Regional climate model simulation of winter climate over Central–Southwest Asia, with emphasis on NAO and ENSO effects. Int J Climatol 30:220–235

Türkeş M (1996) Spatial and temporal analysis of annual rainfall variations in Turkey. Int J Climatol 16:1057–1076

Türkeş M, Erlat E (2003) Precipitation changes and variability in Turkey linked to the North Atlantic Oscillation during the period 1930–2000. Int J Climatol 23:1771–1796

Yosef Y, Saaroni H, Alpert P (2009) Trends in daily rainfall intensity over Israel 1950/1–2003/4. Open Atmos Sci J 3:196–203

Ziv B, Dayan U, Kushnir Y, Roth C, Enzel Y (2006) Regional and global atmospheric patterns governing rainfall in the southern Levant. Int J Climatol 26:55–73

Influence of Winter North Atlantic Oscillation Index (NAO) on Climate and Snow Accumulation in the Mediterranean Mountains

Juan I. López-Moreno, Sergio M. Vicente-Serrano, Enrique Morán-Tejeda, Jorge Lorenzo-Lacruz, Javier Zabalza, Ahmed El Kenawy, and Martin Beniston

Abstract This work analyses the influence of NAO on the interannual evolution of winter temperature, precipitation and snowpack in the Mediterranean mountains. Due to lack of snow data in many mountain areas, the occurrence of four different winter modes are used as a proxy of the amount of accumulated snow. Winter modes are defined on the basis of combined precipitation and temperature thresholds: warm and wet (WW), warm and dry (WD), cold and wet (CW), and cold and dry (CD). The study focuses on 15 relevant mountain areas located in the Mediterranean Europe, Morocco, Turkey and Lebanon. Moreover, we present the relationship between winter NAO and snow depth data in the Swiss Alps and the Spanish Pyrenees. It has been demonstrated that snowpack accumulation in a given year is closely related to the occurrence of these winter modes. However, such relationship is variable among mountain areas and also there are differences depending on elevation in a particular mountain range. Results show that occurrence of different winter modes is strongly related to the winter NAO for the majority of the mountain chains under study, although these relationships are weaker in the easternmost part of the Mediterranean basin. Moreover, it has also been proven that the snow cover response to winter NAO may differ spatially as a consequence of the different influence of winter NAO on precipitation and temperature. In Switzerland, NAO is correlated more with temperature than with precipitation. Therefore, the influence of NAO on snow is significant at the lowest elevation areas, where temperature is the main control on snowpack accumulation. On the other hand, over the Spanish Pyrenees the NAO mainly controls the interannual variability of precipitation. In this region, the highest correlation with snow is found at high elevations where the interannual variability of temperature does not significantly influences snowpack, whereas precipitation controls mainly the accumulation of snow.

Keywords Snow · North Atlantic Oscillation (NAO) · Precipitation · Temperature · The Mediterranean

J.I. López-Moreno (✉)
Instituto Pirenaico de Ecología, Spanish National Research Council (CSIC), Zaragoza, Spain
e-mail: nlopez@ipe.csic.es

S.M. Vicente-Serrano, R.M. Trigo (eds.), *Hydrological, Socioeconomic and Ecological Impacts of the North Atlantic Oscillation in the Mediterranean Region*, Advances in Global Change Research 46, DOI 10.1007/978-94-007-1372-7_6,

1 Introduction

In the Mediterranean mountains snow cover exerts a strong control over ecology, agriculture, availability of water resources, feasibility of a diverse range of economic activities, and associated risks in mountainous and high-latitude regions (Beniston, 2003; Breiling and Charamza, 1999; López-Moreno and García-Ruiz, 2004; López-Moreno, 2005; De Jong et al., 2009). The climate in the Mediterranean region is markedly characterized by a strong seasonality, with relatively low annual precipitation and high interannual variability (Lionello et al., 2006). Winter is the most humid season of the year, with a large proportion of the annual precipitation. Thus, climate anomalies during winter may have major implications for the annual water budget (López-Moreno et al., 2008).

Previous research has identified the North Atlantic Oscillation as one of the main atmospheric circulation modes that determines the temporal evolution of winter precipitation and temperature in the Mediterranean area (Hurrell, 1995; Corte-Real et al., 1995; Hurrell and van Loon, 1997). Thus, it can be expected to exert a strong control on snowpack variability in many mountain areas. However, such a hypothesis is difficult to verify owing to the scarcity of snow data in most of the Mediterranean mountain ranges. In this study, we investigated the possible effects of the NAO on winter precipitation and temperature in 15 mountain ranges in the Mediterranean Europe, Morocco, Turkey and Lebanon during the period 1950-2005 Moreover, snow data from the Swiss Alps and the Spanish Pyrenees has also enabled to analyze the direct influence of the NAO on the variability of snow depth in two contrasted regions in terms of winter NAO influences on climate. In this context, a special emphasis was given to analyze the effect of NAO on combined winter precipitation and temperature modes. Winter modes were identified according to possible combinations of cold-warm and dry-wet conditions. The occurrence of different winter modes is closely related to the amount of accumulated snow; although such a relation is highly determined by the elevation of the mountainous area. Thus, Beniston et al. (in press) concluded that cold and wet years in the Swiss Alps significantly favor the development of a thick snow pack in mountain areas, although during warm and wet years the potential for snow accumulation is essentially restricted to high altitude sites. On the other hand, the association between warm and dry years often results in low snow accumulation at all elevations, and particularly on the lowest slopes.

Although the effect of the NAO on precipitation and temperature has been widely analyzed in the Mediterranean area (e.g. Hurrell and van Loon, 1997; Krichak and Alpert, 2005; López-Bustins et al., 2008), there have not been previous studies focusing on the main mountain areas, and only a few studies have considered the local influence of the NAO on combined temperature and precipitation modes or snow depth.

2 Data and Methods

The CRU TS3.0 monthly gridded (latitude/longitude resolution = 0.5°) precipitation and temperature database was used for the entire Mediterranean region for the period 1950-2005. It was compiled by the Climate Research Unit at the University of East Anglia (http://www.cru.uea.ac.uk; Mitchell and Jones, 2005). In addition, climate and snow data was provided for the Alps by the national meteorological service (Meteoswiss). The climate data of the Spanish Pyrenees was also provided by the Spanish Meteorological Agency (AEMET). Snow depth data in the Pyrenees was obtained from 106 snow poles placed in the area since 1985. The poles and the collection of data are managed by the ERHIN Programme (*Estudio de los Recursos Hídricos INvernales*: Study of winter water resources), run by the Spanish Ministry of Environment. Measurements are taken regularly three times each year, although weather conditions slightly modify the measurement dates. In this study, only snow depths measured in March were considered. Snowpack at this time is determined by the climate of the previous winter (López-Moreno, 2005) and, hence, it adequately determines the accumulated role of the winter NAO conditions. The snow depth series for the period 1985–2005 showed a high correlation with the winter precipitation and temperature. These high correlations allowed the prediction of a regional series for snow pack in March between 1950 and 1999. More details about snow measurements and the methodology of building the snow depth series is found in López-Moreno (2005).

From the CRU database, only those series corresponding to the grid cells that cover the 15 selected mountain areas (see Fig. 1), were used in the study. All the mountain ranges selected for the current analysis experience some level of Mediterranean climatic influence, or drain to the Mediterranean Sea (or the Caspian Sea in the case of northern Turkey and the Caucasus). This included four mountain systems in the Iberian Peninsula (the Pyrenees, the Cantabrian Range, the Central System and the Sierra Nevada); the Atlas Mountains in Morocco; the Apennines in Italy; the Alps, the Carpathian, the Balkan and the Pindos mountains, and Dynaric Alps in the Balkan Peninsula; the Taurus and northern mountains in Turkey; the Caucasus in Armenia and Georgia and the Lebanese mountains.

We used the NAO index defined by Jones et al. (1997), which is based on the difference in anomalies of the sea level pressure measured at the Gibraltar (southwest Iberian Peninsula) and Reykjavik (southwest Iceland) stations. The NAO index was compiled by the Climatic Research Unit, University of East Anglia (http://www.cru.uea.ac.uk/cru/data/nao.htm) for the period 1821 to the present. Prior to correlating series of winter temperature and precipitation (from December to March; DJFM) with the winter NAO (DJFM), we linearly de-trended the time series for autocorrelation. This approach minimized the influence of time series trends and multi-year to decadal signal variability on the strength and significance of the computed correlations and the deduced NAO predictability.

To assess the impact of the winter NAO on combined precipitation and temperature, we used the joint quantile approach described by Beniston and Goyette (2007). New insights on statistics that integrate mutual feedbacks between temperature and

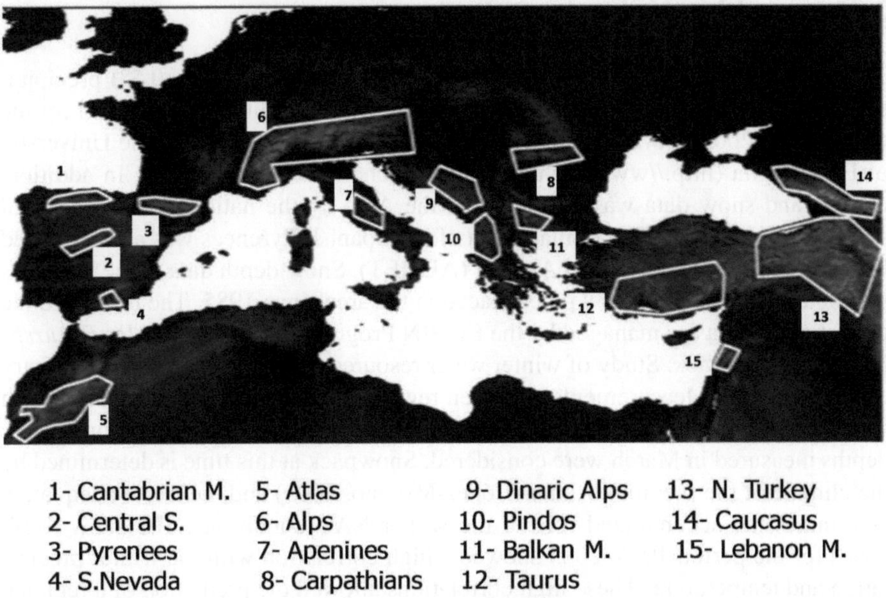

Fig. 1 Study area

moisture are possible when temperature and precipitation records are combined (Beniston, 2009). The 40 and 60% joint quantiles were used to define particular winter modes: (i) cold/dry (CD; when the joint quantiles were equal to or below $T_{40}P_{40}$; the subscript refers to the quantile threshold); (ii) cold/wet (CW; $T_{<40}P_{>60}$); (iii) warm/dry (WD; $T_{>60}P_{<40}$); and (iv) warm/wet (WW; $T_{>60}P_{>60}$). The winter NAO values for individual years belonging to each of the four winter modes were analyzed and compared with the other three modes. The 40 and 60 percentiles were selected, as opposed to the more restrictive percentiles used in previous research (Beniston, 2009). This was principally done to have a sufficient number of years for each winter mode to assess if differences in NAO among winter modes were statistically significant. The non-parametric Wilcoxon-Mann-Whitney rank test (Siegel and Castelan, 1988) was applied to determine whether differences in the NAO during the various winter modes were significant ($\alpha < 0.05$).

3 Results

3.1 *Effects of the Winter Modes on Snow Accumulation in Mountain Areas*

As an example, Fig. 2 shows how the occurrence of different winter modes affects the annual patterns of snow accumulation at four different locations in the Swiss

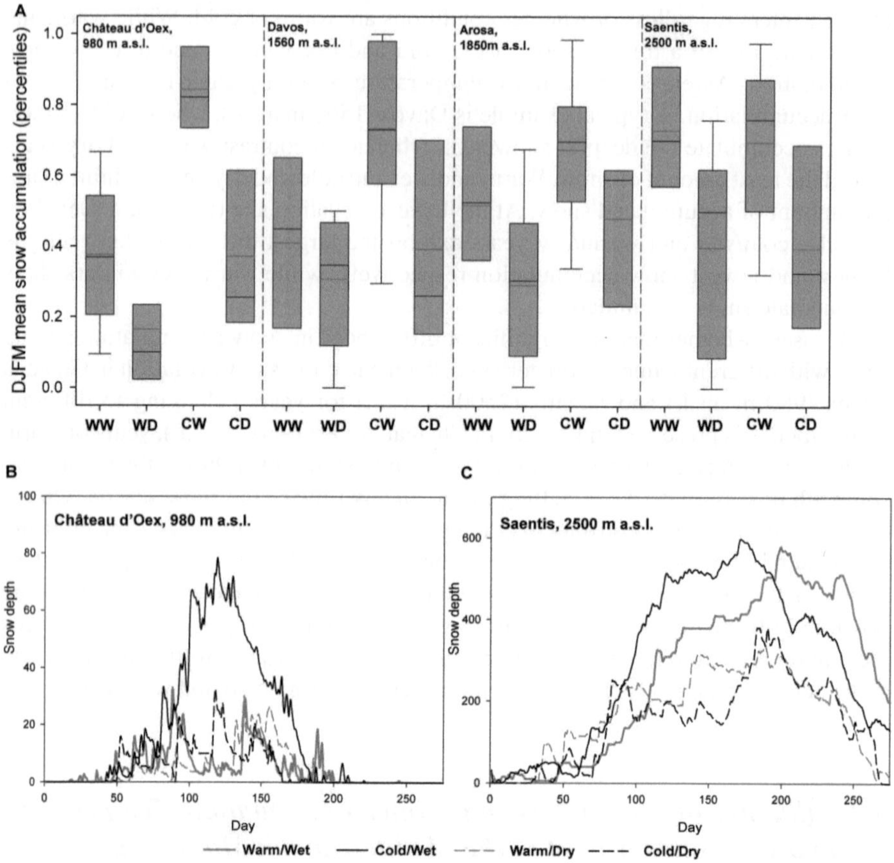

Fig. 2 (**a**) Mean snow accumulation (in cm) in four locations of the Swiss Alps during years that belong to different winter modes: warm and wet (WW), warm and dry (WD), cold and wet (CW) and cold and dry (CD). (**b**) Snow accumulation and melting patterns on Château d'Oex (**b**, 980 m a.s.l.) and Saentis (**c**, 2500 m a.s.l.). Day 1 is October 1st

Alps. Although this example is only based on data collected within the Swiss Alps, the results can be considered to be representative of the other mountainous areas because it depicts the absence of temperature influences on snow accumulation as elevation increases. This behavior can typically be expected in all mountain regions. Thus, the relation between winter modes and snow accumulation varies noticeably according to the elevation of the observatory. Figure 2a indicates the variability in mean snow accumulation from December to March according to different winter modes. Snow amounts have been standardized according to the value of their percentile, thereby enabling comparison between the impacts of the occurrence of winter modes on snow accumulation, independently of the magnitude of snowpack in each site. At high elevation sites (e.g. Saentis [2500 m a.s.l.] and Arosa [1850 m a.s.l.]), the occurrence of high accumulation is mainly related to the occurrence

of wet winters regardless of whether conditions are warm or cold. While warm and dry winters record a thinner snow pack, cold and dry winters record a low snow accumulation. As elevation declines, temperature plays a greater role in terms of snow accumulation. A typical example is Davos (1560 m a.s.l.), where cold and wet winters accumulate the deepest snowpack, whereas in contrast warm and dry years record the least amount of snow. Warm and wet and cold and dry years exhibit a similar amount of accumulated snow. At the lowest elevation site (Chateau d'Oex, 980 m a.s.l.), cold/wet and warm/dry years exhibit the largest differences between the highest and lowest snow accumulation respectively, while warm/wet winters show intermediate snow accumulation.

To assess whether there are significant differences in snow accumulation among years with different winter characteristics, the mean daily snow evolution in Château d'Oex (980 m a.s.l.) and Saentis (2500 m a.s.l.) for years belonging to different winter modes is presented in Fig. 2b. In Château d'Oex (980 m a.s.l.), years of warm winters are mainly characterized by very irregular snow cover during the whole season, with possibility to have melting events in any time of the snow season. On the contrary, snow remains constant over the ground for several months during cold and wet winters. This long period of snow accumulation is likely followed by a month of steady melting of snowpack. In Säntis (2500 m a.s.l.), the high elevation reduces the possible influence of winter type on snow accumulation. In particular, differences are mainly related to the amount of precipitation. Thus, during warm and wet years snow depth is practically the same to the observed during cold and wet years.

3.2 Influence of Winter NAO on Winter Precipitation, Temperature and Combined Modes in the Mediterranean Mountains

Figure 3 shows the correlation between the winter NAO (DJFM) and average temperature and precipitation for the 1950–2005 period in the Mediterranean region. It demonstrates that the NAO exerted marked control over both climate variables in the Mediterranean basin, with large areas with high and statistically significant correlation ($\alpha < 0.05$). Clearly, winter precipitation was negatively correlated with the NAO index over most of the Mediterranean basin. In particular, statistically significant negative correlations were found over large areas of Morocco and Tunisia, most of the Iberian Peninsula, southeastern France, Italy, the Balkan Peninsula, and large areas of central and northern Turkey. Nevertheless, exceptions were found along the coastal areas of Egypt and Libya, where the correlations were positive. With respect to temperature, a general positive correlation pattern was evident in the European Mediterranean. In contrast, a negative correlation was found in large areas of Turkey and the Middle East, whereas a non-significant correlation prevailed in north Africa (from Morocco to Egypt). The greatest positive correlations were found in northern Europe (r-values approximately 0.8); the correlations tended to decrease southward and eastward. Nonetheless, significant correlations were found over most of the Iberian Peninsula, France and the northernmost areas of Italy and the Balkan

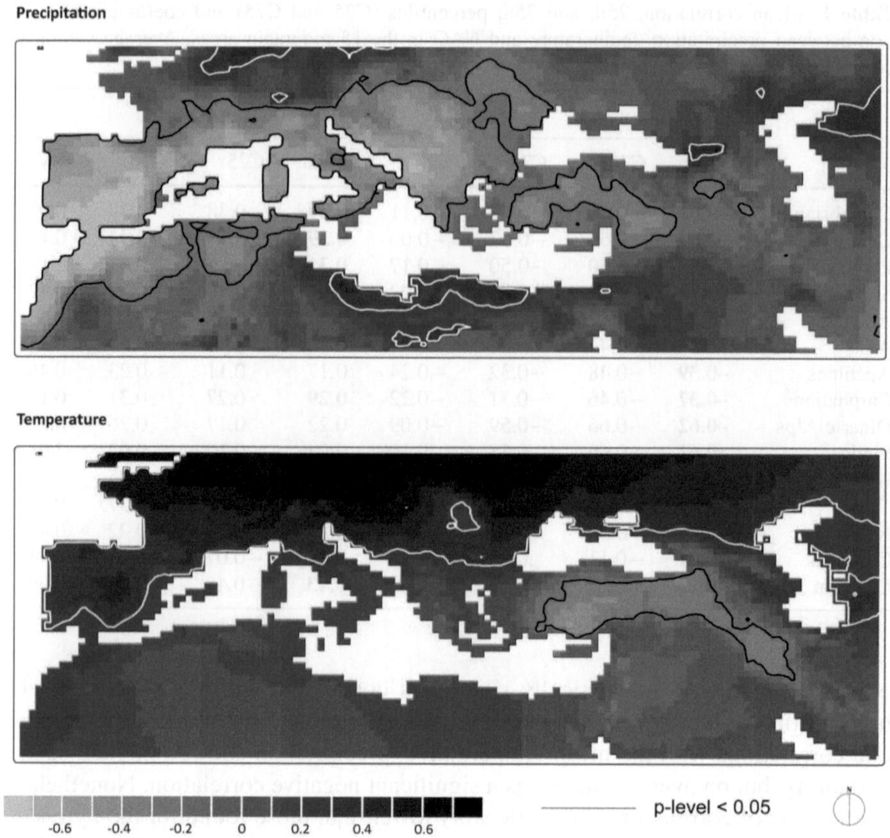

Fig. 3 Pearson's correlation coefficients between the winter (DJFM) NAO and precipitation and temperature for the period 1950–2005

Peninsula. In contrast, southern Italy, Greece and western Turkey exhibited low and non-significant correlations. These results are in good agreement with previous works describing the overall impact of NAO in European and Mediterranean temperatures (e.g. Trigo et al., 2002).

Table 1 presents the mean, the 25th and 75th percentiles and the coefficient of variation in the correlation of winter temperature and precipitation with the NAO index in each mountain area. In the Iberian Peninsula, the influence of the NAO on precipitation and temperature was evident for the four analyzed mountain areas. The correlation with mean temperature is statistically significant ($\alpha < 0.05$) in the Cantabrian Mountains, the Central System and over most of the Pyrenees. The Sierra Nevada, southeast the Iberian Peninsula, was the only mountain area with statistically non-significant correlation between the NAO and temperature. On the other hand, winter precipitation was negatively correlated with the NAO in the four mountain areas. Temperature in the Atlas Mountains in Morocco was negatively but not significantly correlated with the NAO. However, precipitation over most

Table 1 Mean correlation, 25th and 75th percentiles (C25 and C75) and coefficient of variation between precipitation, temperature and NAO in the 15 mountain areas. Average significant correlations ($\alpha < 0.05$) are in bold

	Precipitation				Temperature			
	Mean	C25	C75	CV	Mean	C25	C75	CV
Cantabrian M.	**−0.61**	−0.65	−0.54	−0.11	**0.39**	0.38	0.39	0.02
Central S.	**−0.63**	−0.66	−0.62	−0.06	**0.39**	0.34	0.44	0.17
Pyrenees	**−0.53**	−0.60	−0.50	−0.17	**0.31**	0.27	0.31	0.17
S. Nevada	**−0.72**	−0.72	−0.71	−0.02	0.14	0.10	0.17	0.42
Atlas	**−0.45**	−0.57	−0.37	−0.38	0.15	−0.28	−0.04	1.03
Alps	**−0.3**	−0.44	−0.24	−0.74	**0.52**	0.47	0.57	0.14
Apenines	**−0.39**	−0.48	−0.32	−0.24	0.17	0.11	0.23	0.46
Carphatians	**−0.37**	−0.46	−0.31	−0.22	**0.29**	0.27	0.31	0.1
Dinaric Alps	**−0.62**	−0.66	−0.59	−0.09	0.22	0.17	0.26	0.25
Pindos	**−0.61**	−0.66	−0.58	−0.10	0.04	−0.10	0.02	2.03
Balkan M.	**−0.49**	−0.57	−0.42	−0.18	0.20	0.15	0.25	0.25
Taurus	**−0.29**	−0.36	−0.23	−0.28	**0.41**	−0.45	−0.37	0.13
N. Turkey	−0.19	−0.26	−0.14	−0.60	0.25	−0.31	−0.22	0.45
Caucasus	0.01	−0.11	0.08	3.2	0.11	−0.07	0.18	0.94
Lebanon	−0.12	−0.15	−0.11	−0.43	**0.43**	−0.46	−0.40	0.09

of this area showed a statistically significant correlation with the NAO, but with high spatial variability. Although the Alps exhibited statistically significant positive correlations with the mean temperatures, precipitation showed marked spatial variability, but on average there was a significant negative correlation. Nonetheless, coefficients of correlation were noticeably lower than those found for the Pyrenees, which is a consequence to be an area between Southern and Northern Europe where NAO correlations with temperature goes increasingly south or north of the Alps. The Apennines exhibited a positive but non-significant relationship with temperature, and a negative correlation with precipitation. A similar pattern was evident for the Dynaric Alps and the Carpathians. However, for the Carpathians, the correlation with temperature was close to or exceeded the significance threshold. There was no clear correlation between the NAO and temperature in the Pindos and Balkan mountains although showed a marked influence on the interannual variability of winter precipitation. For the Taurus Mountains, the NAO was significantly and negatively correlated with both precipitation and temperature. The correlation for the mountains of northern Turkey was similar to that observed for the Taurus Mountains, but the α-coefficients were slightly above the statistical significance threshold. The NAO had not a significant influence on winter precipitation or temperature in the Caucasus. In the Lebanon Mountains, the NAO only showed a statistically significant negative correlation with temperature.

In the mountains of the Iberian Peninsula, the winter NAO values were clearly distributed according to the joint quantiles of precipitation and temperature. In particular, the WD winters were associated with the highest NAO values (Fig. 4), while the WW and CW winters accompanied to the lowest NAO values. Also, the CD

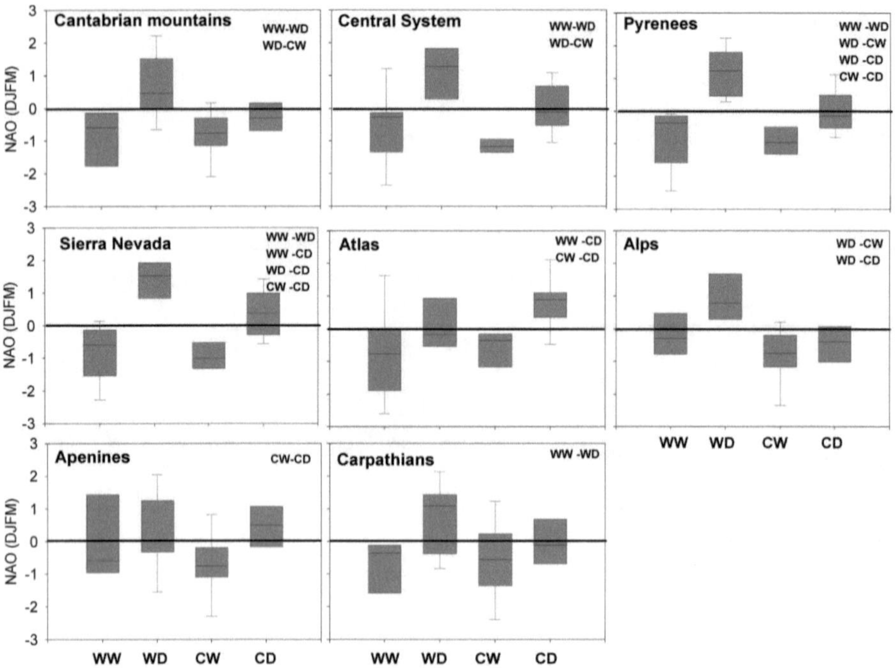

Fig. 4 Boxplots of the winter NAO values recorded during each winter mode in mountain areas of the Iberian Peninsula, Atlas, Alps, Apenines and the Carphatians. In each *panel* is depicted the pairs of winter modes in which NAO values has exhibited statistically significant differences according to the Mann-Whitney rank test

winters were related to the intermediate NAO values. The Mann-Whitney ranking indicated that the NAO values during the WD mode differed significantly from those observed during the wet modes (i.e. WW and CD).

Moreover, in the Sierra Nevada, the NAO values for the CD years were different from those of the CW and WD years. In the Pyrenees there were also significant differences between the NAO values during the WD and CD winters. In the Atlas Mountains (Morocco) precipitation is clearly affected by the NAO, which decreased from dry to wet winters. Thus, during the WW and CW winters, the NAO was much lower than those of the WD and CD winters. The differences between the CD winters and both the WW and CW winters were found statistically significant.

In the Alps the highest NAO values occurred during the WD winters. Significant differences in the NAO were identified between the CW and both CD and WD winters. For the Apennines there was no clear influence of the NAO on the winter modes; the only significant difference was between the CW and CD years.

In the Balkan Peninsula the NAO decreased from dry to wet years, with no clear pattern of influence on temperature. In the Carpathians (Fig. 4) the highest values of the NAO occurred during the WD years, which were statistically different from the WW years. In the Dinaric Alps (Fig. 5) there was a very similar pattern, but

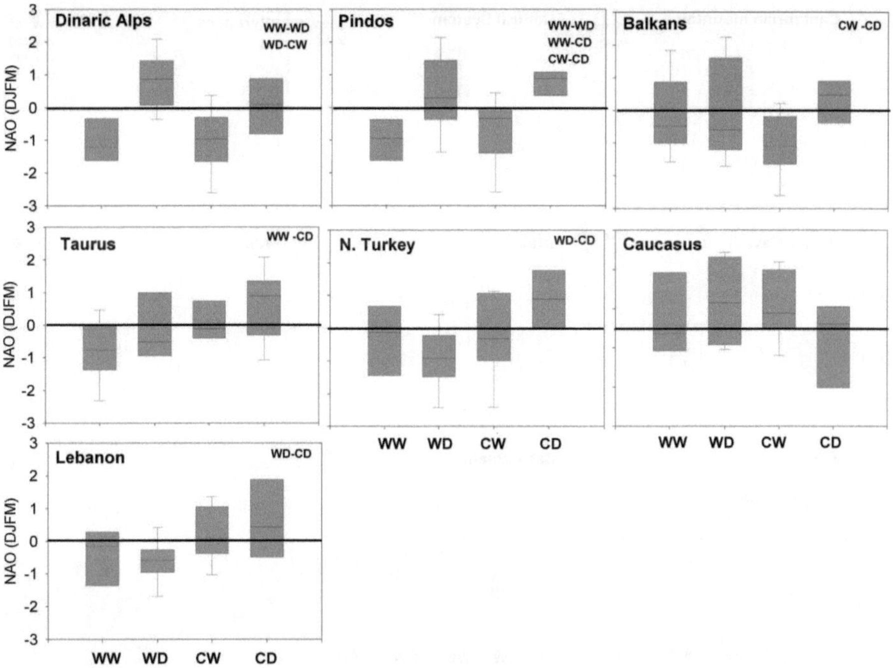

Fig. 5 Same as Fig. 4 but for the Dinaric Alps, Pindos, Balkans, mountain areas of Turkey and Lebanon

with statistically different WD years from both CW and WW winters. The Pindos Mountains had high NAO values during the WD and CD winters. However, in both cases the values were significantly different from those in the WW years. Moreover, the magnitude of the NAO during the CD years was significantly different to the CW years. Apart from the CW and CD years, the Balkan Mountains did not show clear differences in NAO values during the various winter modes.

The Taurus Mountains and the mountains of northern Turkey showed a negative NAO gradient as temperature increased. In the former there were significant differences between the WW and CD years, whereas in the later significant differences were found between the WD and CD years. In the Caucasus there was no evident signal in the magnitude of the NAO related to any of the four winter modes. In the Lebanese mountains, the NAO values varied with temperature, and significant differences were found between the WD and CD winters. Similarly, there were differences, but insignificant, in the mean NAO values between the WW and CW winters.

3.3 Influence of Winter NAO on Snow Accumulation in Switzerland

Table 2 shows the correlation between winter NAO and temperature, precipitation and snow indices (mean snow depth, days with snow over the ground and

Table 2 Correlation between winter NAO and winter temperature, precipitation, mean snow depth (Mean SD), days with snow (Days) and maximum snow accumulation (Max.snow)

	Elev. (m.a.s.l.)	Lat (°N)	Long (°E)	T_{max}	T_{min}	T_{avg}	Precip	Mean SD	Days	Max.snow
St Gallen	779	47.43	9.4	**0.52**	**0.52**	**0.52**	–0.15	**–0.49**	**–0.46**	**–0.40**
Saentis	2490	47.25	9.35	**0.51**	**0.48**	**0.51**	**0.45**	**0.25**	0.14	**0.24**
Neuchâtel	485	47.00	6.95	**0.60**	**0.50**	**0.55**	–0.19	**–0.4**	**–0.41**	**–0.41**
Lugano	273	46.00	8.97	**0.59**	**0.35**	**0.59**	**–0.37**	–0.19	–0.16	0.23
Geneva	420	46.25	6.13	**0.27**	**0.32**	**0.45**	**–0.24**	**–0.47**	**–0.49**	**–0.42**
Davos	1590	46.82	9.85	**0.41**	**0.33**	**0.55**	0.15	0.09	–0.04	0.13
Zurich	443	47.43	8.52	**0.55**	**0.52**	**0.55**	–0.05	**–0.41**	**–0.40**	**–0.40**
Arosa	1740	46.78	9.68	**0.43**	**0.46**	**0.47**	**0.25**	0.03	–0.10	0.05
Chateau D'Oex	981	46.28	7.08	**0.53**	**0.35**	**0.44**	0.00	–0.16	–0.10	–0.07

Bold numbers indicate significant correlations ($p < 0.05$).

maximum snow accumulation) in 9 locations across Switzerland. Following the results obtained for the Alps using the CRU dataset, average temperature exhibited a positive correlation with the NAO in all the considered locations, exceeding 0.44 in all cases. For precipitation, coefficients of correlation are markedly lower than for temperature; and they varied from negative (generally eastward) to positive (westward). In some cases contrasted correlations between the NAO and precipitation were found at very short distances. This was the case in St. Gallen (r = –015) and Saentis (r = 0.45), suggesting a more dominant influence of the complex topography of the region than an influence of large-scale atmospheric circulations on local precipitation. Given the response of temperature and precipitation to the NAO, interannual evolution of winter snow indices are only related to the NAO in those observatories located at low elevation (i.e. Zurich, Geneva, St Gallen and Neuchâtel). This is a consequence of the strong influence of the NAO on temperature, which determines not only the phase of precipitation (liquid or solid), but also the occurrence of melting events in those areas close to the 0°C winter isotherm. On the other hand, at highly elevated sites, temperature is always sufficiently cold for precipitation to fall essentially as snow during winter. In those areas, the interannual evolution of snowpack depends largely on winter precipitation amount, which is not generally controlled by the NAO in Switzerland. One exception is the Saentis weather station located at 2490 m a.s.l., where winter precipitation was clearly influenced by the NAO (r = 0.45), and snow indices accordingly exhibited a positive correlation with the NAO.

Figure 6 shows the temporal evolution of mean snow depth, days with snow cover and the NAO in Zurich (443 m a.s.l.) and Davos (1590 m a.s.l.). Both sites are located at a rather similar geographic position in the Swiss territory, but at a contrasted elevation. Both observatories have shown not only positive correlations between winter temperatures and the NAO, but also exhibited weak relationships between the NAO and precipitation. As previously mentioned, when the NAO only exerts a significant control on temperature as occurred in both sites, snow indices

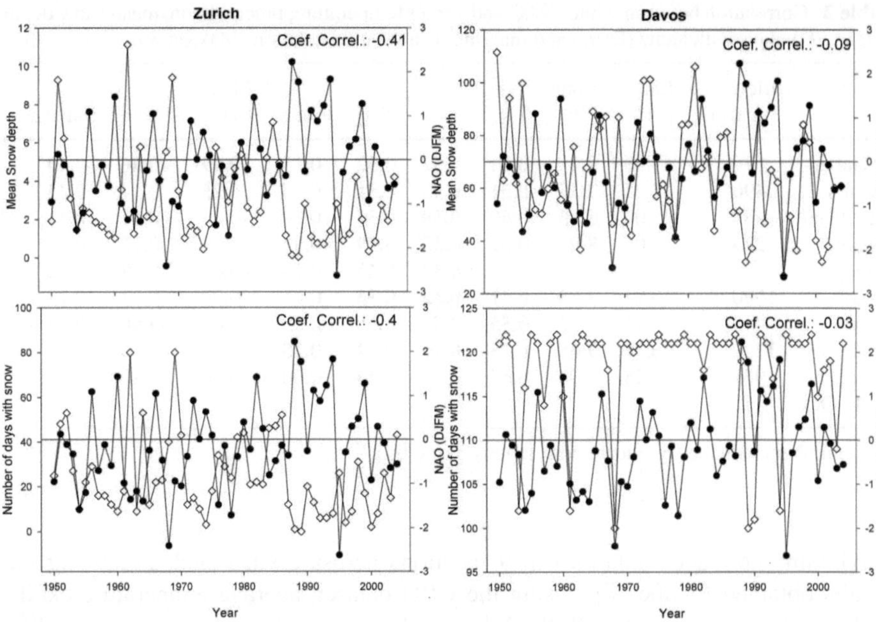

Fig. 6 Temporal evolution of mean snowdepth (in cm), days with snow cover and NAO in Zurich and Davos

were only correlated with the NAO at the lower elevation sites. Thus, mean snow depth and days with snow on the ground were significantly correlated with the NAO in Zurich (r = –0.41 and r = –0.40 respectively). On the other hand, both indices did not show any significant relationship in Davos. Indeed, we can observe how evolution of mean snow depth is independent of the winter NAO evolution. Likewise, the winter (DJFM) snow cover in Davos presents an inter-annual variability insensitive to the corresponding evolution of the winter NAO time series.

3.4 Influence of Winter NAO on Snow Accumulation in the Spanish Pyrenees

Previous studies using observational data in the Spanish Pyrenees (López-Moreno, 2005; López-Moreno and Vicente-Serrano, 2006) revealed that winter NAO had a positive but not significant correlation with the regional series of winter temperature (r = 0.19, $\alpha < 0.05$ for the 1950–2000 period), and a significant negative correlation with the regional precipitation (r = –0.6, $\alpha < 0.05$ for the 1950–2000 period). This strong relationship between the NAO and winter precipitation explains to large extent the negative correlation between the regional series of snow depth accumulated in March and winter NAO (Fig. 7).

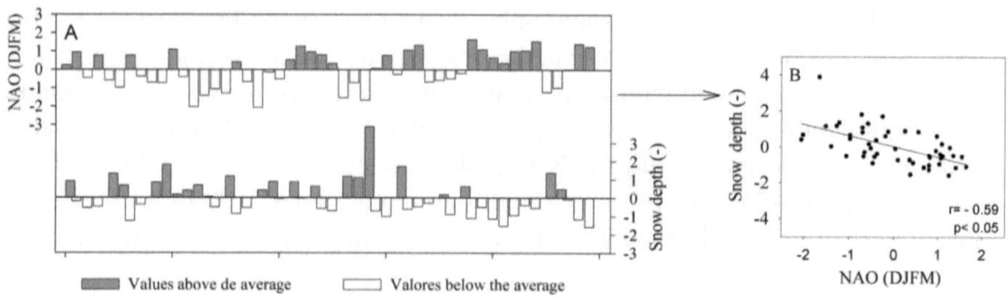

Fig. 7 (**a**) Temporal evolution of Winter (DJFM) NAO index and regional series of snow depth for the Pyrenees in March. (**b**) Correlation between NAO and snow depth series

Nonetheless, the weak relationship between winter NAO and temperature leads to a significant altitudinal gradient in the strength and significance of the correlation coefficients observed at individual sites in the Pyrenees. The differentiated influence of the NAO on snow depth depends on elevation. This is mainly a consequence of the necessity of low temperatures to ensure that precipitation falls as snow. At low elevation sites, interannual variability of winter precipitation is also controlled by the NAO. However, temperature determines the nature of precipitation (i.e. rain or snow) and also the occurrence of melting events during winter. In contrast, at high elevation sites, temperature is not the main control on snowpack and precipitation. Therefore, the NAO seems to be the principle factor governing the interannual variability of snow. Figure 8 shows the correlation coefficients between winter NAO

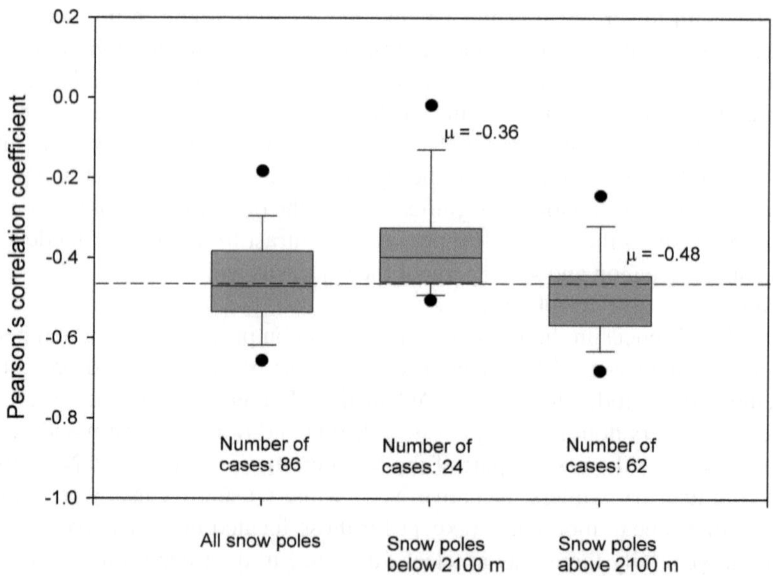

Fig. 8 Coefficient of correlation between winter NAO and snow depth measured in March in 86 locations at contrasted elevation. *Dashed line* indicate the threshold of statistical significance ($\alpha < 0.05$)

and snow depth measured at 86 snow poles located across the Pyrenees during the period 1986–2006. Depending on whether snow poles are located above or below 2100 m a.s.l., we can find statistical significant differences between both groups. Thus, snow depth measured at snow poles located below 2100 m a.s.l. showed a mean correlation of –0.36 with the NAO, which is statistically insignificant in the majority of cases. Conversely, snow depth measured above 2100 m a.s.l. exhibits a mean correlation of –0.48 with winter NAO, meeting the statistical significance threshold in most cases.

4 Discussion and Conclusions

The results of this study confirm the major influence of winter NAO on climate and snowpack in most of the Mediterranean mountains. Following the results obtained ion previous works we have identified large scale negative (positive) correlation with precipitation (temperature) (e.g., Trigo et al., 2002). However, there are noticeable spatial differences in the magnitude or sign of the correlations (Dunkeloh and Jacobeit, 2003). Thus, the negative correlation with precipitation tended to be higher in the western part of the Mediterranean basin (i.e. the Iberian Peninsula and northwestern Morocco), northern Italy, the western part of the Balkan Peninsula and some parts of central Turkey. Winter precipitation exhibited a close relationship with the NAO in the mountains of the Iberian Peninsula (the Cantabrian Mountains, Central System, Pyrenees and Sierra Nevada), the Atlas, the Pindos and the Balkan mountains, and the Dinaric Alps. Other mountains showed negative correlation with the NAO, but with lower coefficients. The exceptions were found in northern Turkey, the Caucasus and the Lebanese mountains, where no significant correlations were exhibited. The positive correlation between temperature and winter NAO tended to decrease southward in the Mediterranean basin. Only the Alps, the Cantabrian Mountains, the Central System, the Pyrenees and the Carpathians showed significant positive correlations. In the case of the Pyrenees, the correlation with temperature was positive but not statistically significant when the observational data was considered for analysis in the 1950–2000 period. In contrast to the general tendency, the mountains of Lebanon and some parts of the Caucasus and northern Turkey showed a significant negative correlation.

Spatial differences in the influence of the NAO on precipitation and temperature explain why there were differences in the response of the combined precipitation and temperature modes to winter NAO in the Mediterranean mountain areas. In most cases winters dominated by positive NAO conditions had warm and dry conditions, except for the eastern part of the Mediterranean basin, where NAO usually yields cold and dry winters. Negative NAO winters usually cause wet conditions in the Mediterranean mountains (except for those located in the eastern part of the basin), but generally there was no clear difference in their occurrence under warm or cold modes. Irrespective of the geographical location, the NAO had a statistically significant influence on the occurrence of winter modes in all the mountains, except the Caucasus and the Lebanese mountains. The NAO had a major influence

on the occurrence of various winter modes in the Pyrenees, the Sierra Nevada and the Pindos Mountains.

The occurrence of different winter modes has a major influence on the accumulation of snow in the mountain areas. Using snow depth data from the Swiss Alps, we have confirmed that snowpack varies noticeably depending on the occurrence of a given winter mode. However, the elevation of each specific mountain area or altitudinal gradients within a particular mountain sector often produces a contrasted response of snowpack to winter modes. In particular, the wet modes in the highly elevated sites of the Swiss Alps are associated with a deep snowpack, independently of whether the winters are warm or cold. As elevation decreases, the role of temperature increases. At the lowest altitudes, the cold and wet modes are exclusively associated to high snow accumulation. In all cases warm and dry modes are associated with winters that have an insufficient snowpack.

Given the relationship between the winter modes and snow pack, and also considering the large influence of the NAO on winter modes, it is not surprising that previous research has suggested the NAO pattern as an important driver of interannual variability of snowpack in mountain areas including: the Pyrenees (López-Moreno and Vicente-Serrano, 2006; García et al., 2009), the Alps (Beniston, 1997; Scherrer and Appenzeller, 2006; Schöner et al., 2009), the Romanian mountains (Micu, 2009), the Bulgarian Carpathians (Petkova et al., 2004). Following the results of this study, it can be concluded that the snow in mountain areas of the Iberian Peninsula (the Sierra Nevada, the Cantabrian Mountains, the Central System), the Atlas, the Pindos and Turkish mountains could also be markedly affected by the NAO in winter.

This research provides clear evidence of the role of elevation as an important variable in explaining the spatial variability of the impacts of the NAO on snowpack evolution within a specific mountain region. The Swiss and Spanish case studies have enabled to assess how coefficients of correlation between NAO and snow indices may exhibit opposite dependence with elevation. More specifically, winter NAO in Switzerland controls mainly temperature, with a variable influence on precipitation. Accordingly, the NAO does not explain interannual variability of snow at highly elevated areas, where temperature is generally below the freezing level. Similarly, snowpack is significantly influenced by the NAO in sites located at low altitudes. In the valley bottoms, temperature determines whether precipitation is solid or liquid, and the NAO controls the interannual characteristics of winter temperature. An opposite example is the Spanish Pyrenees, where the NAO mainly controls the interannual evolution of winter precipitation rather than temperature. Thus, at high elevation sites, where temperature is usually below the freezing point, the NAO explains to a larger extent the interannual variability of the snowpack. However, when elevation decreases, the influence of the precipitation on snowpack weakens, and temperature becomes the major control on snow pack duration and magnitude.

Nonetheless, under global warming conditions expected in the Mediterranean region during the next decades (Giorgi, 2006), the relationships obtained in this study may be different from those for past and current climates. For instance, the

frequency of cold modes is expected to dramatically decrease in favor of both the WD and WW modes in the Swiss Alps (Beniston et al., in press). Also, a higher frequency of warm modes is likely to occur in all mountain areas under study owing to the expected increase in the Mediterranean temperature. In large areas of the Mediterranean region winter precipitation is expected to decrease (López-Moreno et al., 2009). A higher frequency of warm and dry winters in those areas could negatively affect snow cover. In other areas where an increase in winter precipitation is expected, an increase in the WW winters could likely occur in several mountain areas. An increase in the WW winters will be favorable for snow accumulation only at very high altitudes, where precipitation falling as snow can remain in the solid state over a number of consecutive days, even under generally warmer conditions. This hypothesis coincides with numerous studies that have simulated future snowpack and predicted a general decrease in the Mediterranean mountains, especially for sites at low elevations (López-Moreno et al., 2008; Uhlmann et al., 2009).

Acknowledgements This work was supported by the research projects CGL2006-11619/HID, CGL2008-01189/BTE, and CGL2008-1083/CLI, financed by the Spanish Commission of Science and Technology and FEDER, EUROGEOSS (FP7-ENV-2008-1-226487) and ACQWA (FP7-ENV-2007-1-212250), financed by the VII Framework Programme of the European Commission; and "La nieve en el Pirineo Aragonés y su respuesta a la variabilidad climática", financed by "Obra Social La Caixa" and the Aragón Government.

References

Beniston M (1997) Variations of snow depth and duration in the Swiss Alps over the last 50 years: links to changes in large-scale climatic forcings. Clim Change 36:281–300

Beniston M (2003) Climatic change in mountain regions: a review of possible impacts. Clim Change 59:5–31

Beniston M (2009) Trends in joint quantiles of temperature and precipitation in Europe since 1901 and projected for 2100. Geophys Res Lett 36:L07707

Beniston M, Goyette S (2007) Changes in variability and persistence of climate in Switzerland: exploring 20th century observations and 21st century simulations. Glob Planet Change 57:1–20

Beniston M, Uhlmann B, Goyette S, Lopez-Moreno JI (in press) Will snow-abundant winters still exist in the Swiss Alps in a greenhouse climate? Int J Climatol. doi:10.1002/joc.2151

Breiling M, Charamza P (1999) The impact of global warming on winter tourism and skiing: a regionalised model for Austrian snow conditions. Reg Environ Change 1:4–14

Corte-Real J, Zhang X, Wang X (1995) Large-scale circulation regimes and surface climatic anomalies over the Mediterranean. Int J Climatol 15:1135–1150

De Jong C, Lawler D, Essery R (2009) Mountain hydroclimatology and snow seasonality and hydrological change in mountain environments. Hydrol Process 23:955–961

Dunkeloh A, Jacobeit J (2003) Circulation dynamics of Mediterranean precipitation variability 1948–98. Int J Climatol 23:1843–1866

García C, Martí G, Oller P, Moner I, Gavalda J, Martinez P, Peña JC (2009) Major avalanches occurrence at regional scale and related atmospheric circulation patterns in the Eastern Pyrenees. Cold Regions Sci Technol 59:106–118

Giorgi F (2006) Climate change hot-spots. Geophys Res Lett 33:L08707. doi:10.1029/2006GL025734

Hurrell J (1995) Decadal trends in the North-Atlantic Oscillation. Regional temperatures and precipitation. Science 269:676–679

Hurrell J, van Loon H (1997) Decadal variations in climate associated with the North Atlantic Oscillation. Clim Change 36:301–336

Jones PD, Jónsson T, Wheeler D (1997) Extensions to the North Atlantic oscillation using early instrumental pressure observations from Gibraltar and southwest Iceland. Int J Climatol 17:1433–1450

Krichak SO, Alpert P (2005) Decadal trends in the East Atlantic–West Russia pattern and Mediterranean precipitation. Int J Climatol 25:183–192

Lionello P, Malanotte-Rizzoli P, Boscolo R, Alpert P, Artale V, Li L, Luterbacher J, May W, Trigo R, Tsimplis M, Ulbrich U, Xoplaki E (2006) The Mediterranean climate: an overview of the main characteristics and issues. In: Lionello P, Malanotte-Rizzoli P, Boscolo R (eds) Mediterranean climate variability. Elsevier, Amsterdam, pp 1–26

López-Bustins J-A, Martin-Vide J, Sanchez-Lorenzo A (2008) Iberia winter rainfall trends based upon changes in teleconnection and circulation patterns. Glob Planet Change 63:171–176

López-Moreno JI (2005) Recent variations of snowpack depth in the Central Spanish Pyrenees. Arct Antarct Alpine Res 37:253–260

López-Moreno JI, García-Ruiz JM (2004) Influence of snow accumulation and snowmelt on streamflow in the Central Spanish Pyrenees. Int J Hydrol Sci 49:787–802

López-Moreno JI, García-Ruiz JM, Beniston M (2008) Environmental change and water management in the Pyrenees. Facts and future perspectives for Mediterranean mountains. Glob Planet Change 66:300–312

López-Moreno JI, Vicente-Serrano SM (2006) Atmospheric circulation influence on the interannual variability of snowpack in the Spanish Pyrenees during the second half of the twentieth century. Nordic Hydrol 38:38–44

López-Moreno JI, Vicente-Serrano SM, Gimeno L, Nieto R (2009) The stability of precipitation regimes in the Mediterranean region: observations since 1950 and projections for the twentieth-one century. Geophys Res Lett 36:L10703. doi:10.1029/2009GL037956

Micu D (2009) Snow pack in the Romanian Carpathians under changing climatic conditions. Meteorol Atmos Phys 105:1–16

Mitchell TD, Jones PD (2005) An improved method of constructing a database of monthly climate observations and associated higher resolution grids. Int J Climatol 25:693–712

Petkova N, Koleva E, Alexandrov V (2004) Snow cover variability and change in mountainous regions of Bulgaria, 1931–2000. Meteorologische Zeitschrift 13:19–23

Scherrer SC, Appenzeller C (2006) Swiss Alpine snow pack variability: major patterns and links to local climate and large-scale flow. Clim Res 32:187–199

Schöner W, Auer I, Böhm R (2009) Long term trend of snow depth at Sonnblick (Austrian Alps) and its relation to climate change. Hydrol Process 23:1052–1073

Siegel S, Castelan NJ (1988) Nonparametric statistics for the behavioral sciences. McGraw-Hill, New York

Trigo RM, Osborn TJ, Corte-Real JM (2002) The North Atlantic Oscillation influence on Europe: climate impacts and associated physical mechanisms. Clim Res 20:9–17

Uhlmann B, Gollete S, Beniston M (2009) Sensitivity analysis of snow patterns in Swiss ski resorts to shifts in temperature precipitation and humidity under condition of climate change. Int J Climatol 29:1048–1055

Hurrell J, van Loon H (1997) Decadal variations in climate associated with the North Atlantic Oscillation. Clim Change 36:301–326
Jones PD, Jonsson T, Wheeler D (1997) Extensions to the North Atlantic oscillation using early instrumental pressure observations from Gibraltar and southwest Iceland. Int J Climatol 17:1433–1450
Krichak SO, Alpert P (2005) Decadal trends in the East Atlantic–West Russia pattern and Mediterranean precipitation. Int J Climatol 25:183–192
Lionello P, Malanotte-Rizzoli P, Boscolo R, Alpert P, Artale V, Li L, Luterbacher J, May W, Trigo R, Tsimplis M, Ulbrich U, Xoplaki E (2006) The Mediterranean climate: an overview of the main characteristics and issues. In: Lionello P, Malanotte-Rizzoli P, Boscolo R (eds) Mediterranean climate variability. Elsevier, Amsterdam, pp 1–26
López Bustins JA, Martin-Vide J, Sanchez-Lorenzo A (2008) Iberia winter rainfall trends based upon changes in teleconnection and circulation patterns. Glob Planet Change 63:171–176
López-Moreno JI (2005) Recent variations of snowpack depth in the Central Spanish Pyrenees. Arct Antarct Alpine Res 37:253–260
López-Moreno JI, García-Ruiz JM (2004) Influence of snow accumulation and snowmelt on streamflow in the Central Spanish Pyrenees. Int J Hydrol Sci 49:787–802
López-Moreno JI, García-Ruiz JM, Beniston M (2008) Environmental change and water management in the Pyrenees. Facts and future perspectives for Mediterranean mountains. Glob Planet Change 66:300–312
López-Moreno JI, Vicente-Serrano SM (2006) Atmospheric circulation influence on the interannual variability of snowpack in the Spanish Pyrenees during the second half of the twentieth century. Nordic Hydrol 38:38–44
López-Moreno JI, Vicente-Serrano SM, Gimeno L, Nieto R (2009) The stability of precipitation regimes in the Mediterranean region: observations since 1950 and projections for the twenty-one century. Geophys Res Lett 36:L10703. doi:10.1029/2009GL037956
Micu D (2009) Snow pack in the Romanian Carpathians under changing climatic conditions. Meteorol Atmos Phys 105:1–16
Mitchell TD, Jones PD (2005) An improved method of constructing a database of monthly climate observations and associated high-resolution grids. Int J Climatol 25:693–712
Petkova N, Koleva E, Alexandrov V (2004) Snow cover variability and change in mountainous regions of Bulgaria, 1931–2000. Meteorologische Zeitschrift 13:19–23
Scherrer SC, Appenzeller C (2006) Swiss Alpine snow pack variability: major patterns and links to local climate and large-scale flow. Clim Res 32:187–199
Schöner W, Auer I, Böhm R (2009) Long term trend of snow depth at Sonnblick (Austrian Alps) and its relation to climate change. Hydrol Process 23:1052–1063
Siegel S, Castellan NJ (1988) Nonparametric statistics for the behavioral sciences. McGraw-Hill, New York
Trigo RM, Osborn TJ, Corte-Real JM (2002) The North Atlantic Oscillation influence on Europe: climate impacts and associated physical mechanisms. Clim Res 20:9–17
Uhlmann B, Goyette S, Beniston M (2009) Sensitivity analysis of snow patterns in Swiss ski resorts to shifts in temperature, precipitation and humidity under condition of climate change. Int J Climatol 29:1048–1055

Impact of NAO on Mediterranean Fisheries

Francesc Maynou

Abstract The effect of the North Atlantic Oscillation (NAO) on fisheries production has received attention in the last 2 decades, especially in relation to the inter-annual fluctuation of the important North Atlantic cod fisheries. Results of these studies show that the effects of NAO on cod stocks vary geographically, with opposite patterns in the western and eastern North Atlantic and that NAO affects mainly the recruitment success of cod, especially on heavily fished stocks. In the Mediterranean sea the number of studies is much more limited, due in part to the paucity of sufficiently long and reliable data series for Mediterranean fisheries. A recent study on the interannual variation of landings of the red shrimp *Aristeus antennatus* in Catalonia shows that the NAO explains a large part of the variability in population abundance with a lag of 2–3 years. New results presented here show that, at the scale of the entire Mediterranean, stocks of hake (*Merluccius merluccius*) respond to the NAO with different trends in the northwestern and southeastern stocks. Analysis of detailed life history data of this species in the Spanish Mediterranean coast allows to postulate the hypothesis that positive NAO years enhance hake fishery production through increasing the individual size of recruits, as well as the individual weight and abundance of adult hake. No effect on other demographic parameters such as recruitment strength or natural mortality could be demonstrated.

Keywords Aristeus antennatus · Fisheries · Mediterranean · Merluccius merluccius · NAO

1 Introduction

Fisheries production depends on the interaction between human inputs (*fishing effort*) and the biomass of fish. While fishing effort may change over time due to socioeconomic factors, such as market interests, changing costs of effort or other reasons, fish biomass fluctuates over time due to environmental factors, such as biological productivity of its prey resources, or abiotic factors such as

F. Maynou (✉)
Institut de Ciències del Mar, Spanish National Research Council (CSIC), Barcelona, Spain
e-mail: maynouf@icm.csic.es

S.M. Vicente-Serrano, R.M. Trigo (eds.), *Hydrological, Socioeconomic and Ecological Impacts of the North Atlantic Oscillation in the Mediterranean Region*, Advances in Global Change Research 46, DOI 10.1007/978-94-007-1372-7_7,

changes in temperature or salinity (Bakun, 1996; Cushing, 1996; Ottersen et al., 2004).

The abundance of marine fish is difficult to determine and many different methods have been developed over time (Hilborn and Walters, 1992). Briefly, fish abundance can be estimated from two basic sources: fisheries dependent data (commercial catches) or fisheries-independent data. In the first case, data series of catches are relatively easy to obtain because most countries maintain relatively long data bases on fish production by species (Evans and Grainger, 2001). However, to interpret catch data it is often necessary to know the evolution of fishing effort and catchability, which are often not available or difficult to estimate (Hilborn and Walters, 1992). Fisheries-independent programmes are more reliable but expensive to maintain and rarely have the long time spans required to assess the relationship between environmental variations and fish abundance (Sparre and Venema, 1998). In the European North Atlantic, there exist data series derived from scientific trawl surveys longer than 40 years (International Bottom Trawl Survey: http://datras.ices.dk), but in the Mediterranean, the only continuous fisheries monitoring programme (MEDITS: Mediterranean International Trawl Surveys: http://www.ifremer.fr/Medits_indices) to cover all EU member states started in 1994 and the data series is thus considerably shorter than in the Atlantic (Bertrand et al., 2002). In any case both types of data sources, commercial and trawl surveys, have advantages and disadvantages and it is convenient to use the two types of data concurrently when possible.

It is known that the changes in total biomass of a fishery target species can be due to individual changes in (one or more) individual demographic parameters. For instance, the abundance of a given year class may fluctuate from year to year due to environmental factors that cause enhanced survival (or conversely, increased mortality) of the individuals of that particular year class (Cushing, 1996; Myers, 2002; Watanabe et al., 2003). This is specially important in the first age class recruited to the fishery (known as recruitment), and several studies have shown the relationship between interannual fluctuations in recruitment and environmental variability (e.g. Mendelsohn and Mendo, 1987; Watanabe et al., 1996; Stige et al., 2006). Other key life history parameters affected by environmental variability are the individual growth rate, the natural mortality or the reproductive success (Hänninen et al., 1999; Jobling, 2002).

It is often difficult to determine the causes of the relationship between large scale climatic indices and fisheries due to the paucity of data, but also due to the complexity of the mechanisms involved (Cushing, 1996). In recent years, the usefulness of employing large scale climatic indices such as the NAO index as a proxy for complex space-time variations in ecological variables has been suggested (Hallett et al., 2004; Stenseth and Mysterud, 2005) and indeed the relationship between climatic indices and fisheries is well documented in non-Mediterranean settings, particularly in upwelling ecosystems (e.g. anchovy/sardine substitutions in the Pacific Kuroshio and Humboldt Current ecosystems and teleconnections among fish populations, reviewed in Alheit and Bakun, 2010).

The effects of NAO on North Atlantic fisheries has received attention in recent decades, especially in the last 10 years (Reid et al., 2001; Beaugrand, 2004), with studies focused particularly on cod (*Gadus morhua*, Mann and Drinkwater, 1994; Beaugrand et al., 2003; Stige et al., 2006). The results of these studies show that the NAO affects cod fisheries mainly by modulating the strength of cod recruitment with a time lag of 2 years (Brodziak and O'Brien, 2005). It has been shown also that the effect is stronger when spawning stock biomass (SSB) is low (Brander, 2005). Most significantly, some of these studies showed that the effect of the NAO varies spatially across North Atlantic cod stocks (synthesized in the model by Stige et al., 2006): while cod recruitment is enhanced in Newfoundland, Greenland, Iceland, northern Norway and the Faroes during positive NAO years, it is decreased in the eastern US, around the British Isles, in the North sea and in the Baltic sea (figure 4 in Stige et al., 2006).

In many cases the relationship between the NAO and fisheries production has been shown to be mediated by changes in the composition and abundance of zooplankton (Fromentin and Planque, 1996; Ottersen and Stenseth, 2001; Reid et al., 2003; Molinero et al., 2008).

In the Mediterranean, the relationship between the NAO and fisheries has received less attention, probably because data series are generally shorter than in Atlantic fisheries. Lloret et al. (2001) assessed the relationship between river runoff in 13 coastal fishery species of the Catalan coasts. They found that catches per unit of effort were positively correlated with the wind mixing index in the Gulf of Lions and the runoff of local rivers, which varied synchronously with the NAO index: negative NAO indices were associated with higher river runoff (probably because during negative NAO years precipitation increases in the NW Mediterranean, Mariotti et al., 2002). Maynou (2008) showed that the detrended landings of the deepwater red shrimp *Aristeus antennatus* were positively correlated with the NAO index with time lags of 2 and 3 years. An ecological mechanism to explain this correlation was produced, based on current knowledge of the biology of this species and the water mass dynamics in the study area (NW Mediterranean): positive NAO years are correlated with decreased rainfall in the NW Mediterranean sea, especially in the Gulf of Lions, the Ligurian sea and adjacent hinterland (Mariotti et al., 2002). Decreased rainfall is linked to enhanced vertical mixing of the water masses in the Gulf of Lions (Demirov and Pinardi, 2002; Jordi and Hameed, 2009), due to the saltier (hence, more dense) surface water mass in low rainfall years. Knowing that food supply in winter is key to the reproductive success of *A. antennatus*, because during this period females switch from their normal generalist diet to a high-energy content diet based on zooplankton (Cartes et al., 2008; Maynou, 2008) postulated that positive NAO years would enhance egg production and higher abundance of recruits to the fishery.

The objective of this contribution is to assess the effects of NAO in another important Mediterranean demersal fishery resource, the hake *Merluccius merluccius* at the scale of the entire Mediterranean basin, in order to investigate whether ecological mechanisms linking the NAO with fisheries productivity demonstrated in local areas (Catalan sea, Ligurian sea) can be extrapolated to wider areas.

2 Results

European hake (*Merluccius merluccius*) is an important resource of the Mediterranean demersal fishery, with catches fluctuating between 20 and 50,000 t in recent decades (FAO Fisheries Department, 2000). The biology of this species is well-known (e.g. the review by Oliver and Massutí, 1995; Recasens et al., 1998). Hake is distributed mainly over the continental shelf between 50 and 200 m depth, although occasional large individuals are caught down to 800 m. It is a heavily exploited species and catches are at present mainly composed of juvenile individuals of age classes 0–2. The recruitment period of hake varies across Mediterranean subareas and is relatively protracted, but it occurs mainly in early summer (June–July) in the NW Mediterranean (Recasens et al., 1998).

I analyzed the annual data series of hake catches for the period 1970–2005 in the seven geographical areas of the Mediterranean used by the FAO for fisheries statistics purposes (Fig. 1) based on the FISHSTAT database (FAO Fisheries Department, 2000).

As shown in Fig. 2, the data series of hake catches are very noisy, although some general patterns emerge. First, hake catches increased from 1970 until the early 1990s in most areas (Gulf of Lions, Adriatic, Ionian and Aegean). In these areas, and also in the Levant, hake production decreased since the early 1990s to the present due probably to overfishing (Lleonart, 2008). In the Balearic area no overall trend can be distinguished, but periods of high catches and low catches can be discerned, while in the Sardinia area hake catches have been decreasing since the beginning of the time series (Fig. 2).

To analyze these data sets, a Generalized Additive Model (GAM) was set up for each statistical area, using *year* as the main effect (because it is expected that long-term trends in interannual variations in catches are mainly due to the observed

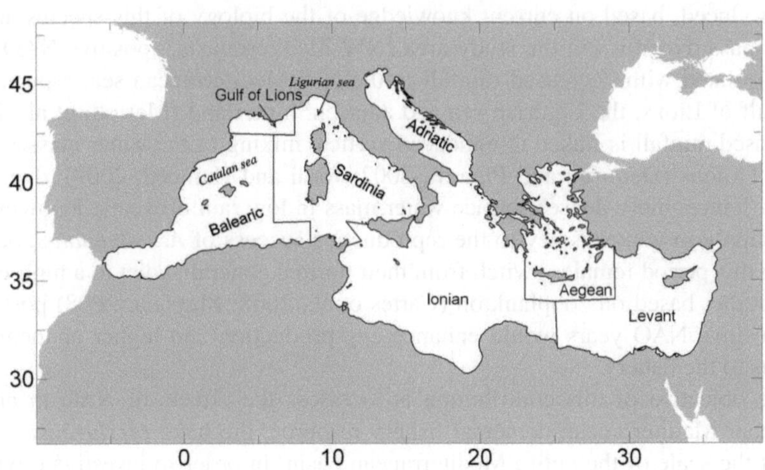

Fig. 1 FAO fisheries divisions in the Mediterranean and geographic names mentioned in the text (italics)

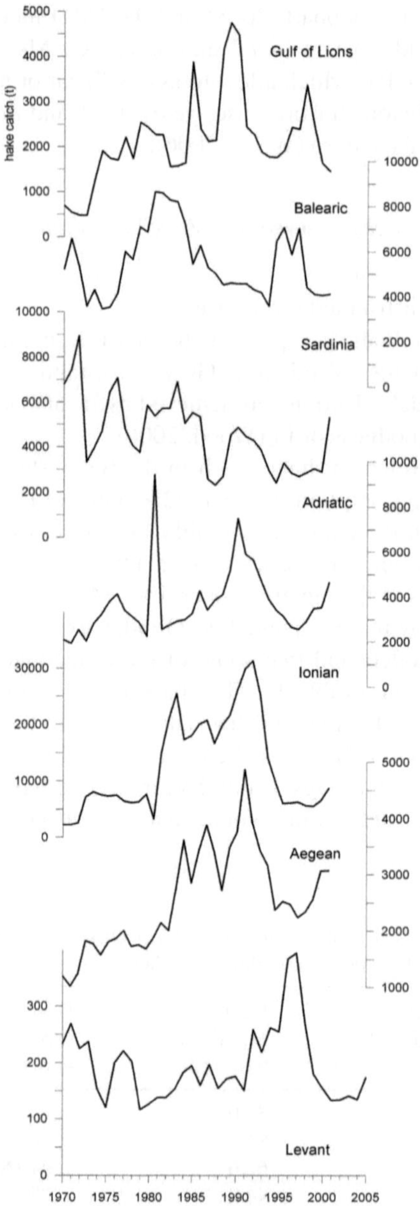

Fig. 2 Data series of hake (*Merluccius merluccius*) landings in the seven Mediterranean FAO fisheries divisions for the period 1970–2005 (data extracted from FISHSTAT, FAO Fisheries Department, 2000)

increase in fishing effort, Lleonart, 2008) and the NAO index as additional effect with time lags of 1 and 2 years (*nao1* and *nao2*). GAMs are a flexible class of statistical regression models which allow to assess linear or nonlinear relationships between a set of predictors (in our case, *year*, *nao1* and *nao2*) and a dependent variable (in our case hake *catch*) (Wood, 2006):

$$\log(catch) = \mathrm{s}(year) + \mathrm{s}(nao1) + \mathrm{s}(nao2) + \varepsilon,$$

where ε is a normally distributed error term.

I used smoothing splines to represent the possibly nonlinear effect of predictors and Generalized Cross Validation (GCV) to establish the optimal degree of the smooth terms. Model selection was achieved by means of Akaike's Information Criteria to assess the goodness of fit (Wood, 2006).

The GAM models fitted for hake catch in the seven statistical subareas for the period 1970–2005 helped explain between 62% (Gulf of Lions) and 96% (Aegean sea) of the deviance in the data series (Table 1). These results show that most of the variability in the catch data series can be attributed to the year effect, which mostly reflects changes in fishing effort over time. More interestingly, the results of the GAM analysis show that the lagged NAO index explains between 2 and 8% of the deviance of the models and that some of the functional non-parametric terms are statistically significant (Table 1). The shape of these nonparametric terms can be examined in Fig. 3, which shows that hake stocks in the western and northern Mediterranean areas (Gulf of Lions, Balearic and Adriatic) are positively correlated with the NAO index at lags 1 and 2 years, while hake stocks in the eastern Mediterranean (Aegean and Levant) would be negatively correlated with these NAO

Table 1 Results of the GAM analysis of hake (*Merluccius merluccius*) landings in the seven Mediterranean areas for fishery purposes defined by FAO

Area	Deviance explained by full model (%)	Deviance explained by term *year* (%)	Deviance explained by term *nao1*	Deviance explained by term *nao2*
Gulf of Lions	62.3	55.0	4.3%	3.0%
Balearic	87.5	81	5.7%	0.8% (NS)
Sardinia	69.1	62.6	2.1% (NS)	4.4%
Adriatic	66.8	44.9	1.2%	0.7% (NS)
Ionian	95.0	93.9	0.1% (NS)	1%
Aegean	96.2	89.7	4.2%	2.0%
Levant	81.5	77.2	3.7%	0.6% (NS)

Data series from FAO Fisheries Department (2000).
nao1 and *nao2* are the NAO index lagged 1 and 2 years, respectively.
NS: non-significant term, otherwise significant terms at $p < 0.1$ level.

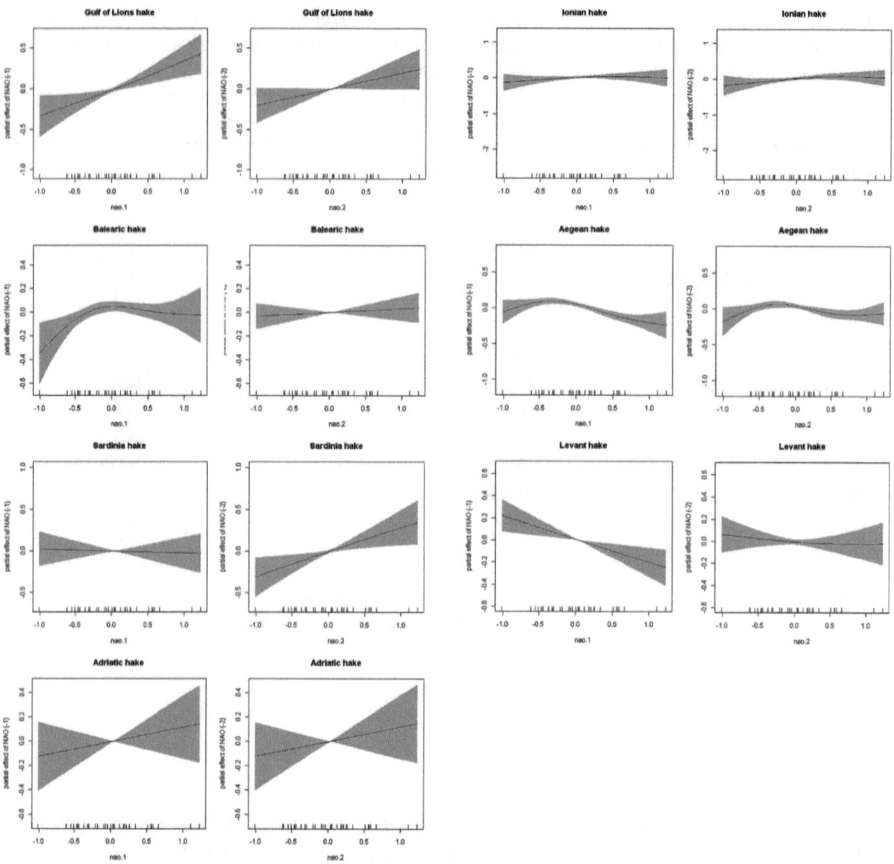

Fig. 3 Non-parametric smooth terms determined by the GAM analysis for hake (*Merluccius merluccius*) landings in the seven Mediterranean FAO fisheries divisions. *Left column*: effect of NAO index lagged 1 year. *Right column*: effect of NAO index lagged 2 years

indices (Fig. 3). Hake stocks in the central Mediterranean show weak correlations (Ionian) or of opposite sign (Sardinian).

To investigate the causes linking NAO with hake fisheries variability I analyzed the correlation between several life-history indicators derived from fisheries independent trawl surveys in the western Mediterranean (MEDITS surveys carried out along the continental coast of Spain, Bertrand et al., 2002) by means of autocorrelation function analysis (Venables and Ripley, 2002). The span of this data set (1994–2009) is shorter than the FAO production statistics, but includes a critical transitional period from positive values of the NAO index in the 1990s to negative or zero values later on. I extracted several demographic indicators from the

Table 2 Demographic indicators of hake (*Merluccius merluccius*) analyzed by autocorrelation functions. Indicators were derived from the annual MEDITS trawl surveys carried out along the Mediterranean continental coast of Spain (1994–2009)

	Age class	Result
Abundance indices	0 (recruits)	NS
No of individuals/km^2	1	NS
	2	NS
	Average for ages 3–6	S (lags 1 and 2)
Individual length (cm TL)	0 (recruits)	S (lag 1)
	1	NS
	2	NS
	Average for ages 3–6	NS
Individual weight (g)	0 (recruits)	NS
	1	NS
	2	NS
	Average for ages 3–6	S (lag 1)
Natural mortality (year^{-1})	0 (recruits)	NS
	1	NS
	2	NS
	Average for ages 3–6	NS
Biomass	Total population biomass	NS
	Spawning stock biomass (SSB)	NS

Data elaborated from http://www.ifremer.fr
S: significant; NS: Not Significant at $p<0.05$ level.

MEDITS database (http://www.ifremer.fr/Medits_indices): abundance by age class, natural mortality by age class, total biomass, spawning stock biomass, and average individual length and weight (Table 2).

The results of autocorrelation function analysis (Fig. 4) show that the mean length of recruits (hake of age class 0) is positively correlated with NAO at lag 1, i.e. recruits are larger than average after positive NAO years. Also the abundance of adults (age classes 3–6) of hake is increased 1 and 2 years after positive NAO years, while the average weight of adults is positively correlated with NAO at lag 1. Despite this increase in abundance and mean weight of adults, no increase in spawning stock biomass could be shown in positive NAO years, probably because the biomass of the adult fraction of the population is much lower than the juvenile fraction (age classes 0–2) in this heavily exploited population. Natural mortality was not correlated to the NAO. These results show that in the Balearic subarea, and probably also in the Gulf of Lions and Adriatic, positive NAO years have a positive effect in at least 3 critical indicators of the hake life history (length of recruits, abundance of adults and weight of adults) suggesting that the positive effects on the catches shown earlier from the FAO production data series can be related to increase size and weight of the hake population and not, for instance, to decreased natural mortality or increased reproductive success, as in the red shrimp (Maynou, 2008).

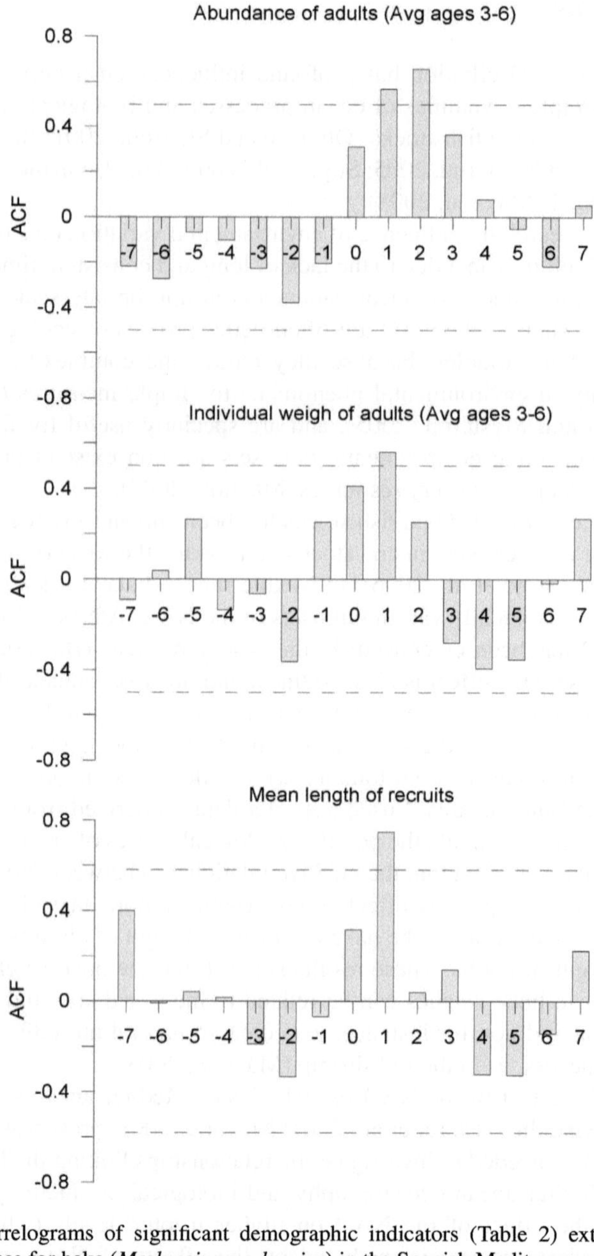

Fig. 4 Autocorrelograms of significant demographic indicators (Table 2) extracted from the MEDITS database for hake (*Merluccius merluccius*) in the Spanish Mediterranean coast (subarea 6). *Dotted lines* indicate significant correlation at the $p < 0.05$ level

3 Conclusions

The North Atlantic Oscillation has profound influences on fisheries productivity, mediated by complex dynamics of ocean processes and biological productivity, as shown in North Atlantic fish stocks (Ottersen and Stenseth, 2001; Reid et al., 2001, 2003; Stenseth and Mysterud, 2005; Stige et al., 2006), but also in the Mediterranean (Lloret et al., 2001; Maynou, 2008).

In fisheries research the influence of environmental variation on stock productivity may be difficult to detect due to the lack of long and consistent time series of the main oceanographic and ecological parameters at appropriate scales. Large-scale climatic indices such as the NAO are often better proxies of ecological processes than local weather variables, because they reduce the complexities of time and space variability in environmental phenomena to simple measures (Hallett et al., 2004; Stenseth and Mysterud, 2005), and are specially useful for fish abundance data where comparable environmental data sets are non existent or expensive to obtain (e.g. for deep-sea fishery resources, Maynou, 2008).

The influence of the NAO on fish stocks has been shown to vary geographically, with impacts of different sign in cod stocks of, broadly, the western Atlantic and the eastern Atlantic (Stige et al., 2006). In Mediterranean hake stocks, present results suggest also geographically varying impacts of the NAO. A N/S – W/E trend in the effects of NAO has been evidenced: in the N and W areas (Balearic sea, Gulf of Lions, Adriatic sea) the effect is close to linear and positive (enhanced hake catches at positive NAO values). In the E and S subareas (Aegean sea, Levantine sea) the effects is negative (enhanced catches at negative NAO values), while in the two central subareas (Sardinian seas and Ionian sea) the effect is weak but mainly positive. Focusing on the Balearic area, I used a shorter data set derived from fisheries independent surveys to investigate the possible biological causes of enhanced catches by NAO in this area. I showed that the positive relationship between NAO and catches of hake is related to the positive effect of NAO (with temporal lags of 1 or 2 years) on at least 3 critical indicators of the hake life history: length of recruits, abundance of adults and weight of adults. These results suggest that the positive effects of NAO on Mediterranean hake catches can be related to increased size and weight of the hake population and not, for instance, to decreased natural mortality nor increased reproductive success, as in the red shrimp (Maynou, 2008).

Although the effect of the NAO on fisheries of Mediterranean stocks has been shown in a few studies (Lloret et al., 2001; Maynou, 2008; present work for hake), further research is needed to investigate the relationships linking the North Atlantic Oscillation, Mediterranean oceanography and biological productivity. Although it is likely that the effect of the NAO on higher trophic levels (which constitute the fisheries target species) is mediated by the effects of NAO on zooplankton (Fromentin and Planque, 1996; Maynou, 2008; Molinero et al., 2008) the details of these relationships are often mere working hypotheses that need to be refined and tested.

References

Alheit J, Bakun A (2010) Population synchronies within and between ocean basins: apparent teleconnections and implications as to physical-biological linkage mechanisms. J Mar Syst 79:267–285

Bakun A (1996) Patterns in the ocean: ocean processes and marine population dynamics. California Sea Grant College System/Centro de Investigaciones Biológicas del Noroeste, La Paz, 323 pp

Beaugrand G (2004) The North Sea regime shift: evidence, mechanisms and consequences. Prog Oceanogr 60:245–262

Beaugrand G, Brander KM, Lindley JA, Souissi S, Reid PC (2003) Plankton effect on cod recruitment. Science 426:661–664

Bertrand JA, Gil de Sola L, Papaconstantinou C, Relini G, Souplet A (2002) The general specifications of the MEDITS surveys. Sci Marina 66:9–17

Brander K (2005) Cod recruitment is strongly affected by climate when stock biomass is low. ICES J Mar Sci 62:339–343

Brodziak J, O'Brien L (2005) Do environmental factors affect recruits per spawner anomalies of New England groundfish? ICES J Mar Sci 62:1394–1407

Cartes JE, Madurell T, Fanelli E, López-Jurado JL (2008) Dynamics of suprabenthos-zooplankton communities around the Balearic Islands (western Mediterranean): influence of environmental variables and effects on the biological cycle of *Aristeus antennatus*. J Mar Syst 71:316–335

Cushing DH (1996) Towards a science of recruitment in fish populations. International Ecology Institute, Oldendorf/Luhe

Demirov E, Pinardi N (2002) Simulation of the Mediterranean Sea circulation from 1979 to 1993: Part I. The interannual variability. J Mar Syst 33–34:23–50

Evans D, Grainger R (2001) Gathering data for resource monitoring and fisheries management. In: Hart PJB, Reynolds JD (eds) Handbook of fish biology and fisheries, vol 2, chapter 5. Blackwell Publishing, Oxford

FAO Fisheries Department (2000) *FISHSTAT plus: universal software for fishery statistical time series*. Version 2.30. FAO Fisheries Department, Fishery Information, Data and Statistics Unit

Fromentin JM, Planque B (1996) *Calanus* and environment in the eastern North Atlantic. II. Influence of the North Atlantic Oscillation on *C. finmarchicus* and *C. helgolandicus*. Mar Ecol Prog Ser 134:111–118

Hallett TB, Coulson T, Pilkington JG, Clutton-Brock TH, Pemberton JM, Grenfell BT (2004) Why large-scale climate indices seem to predict ecological processes better than local weather. Nature 430:71–75

Hänninen J, Vuorinen I, Hjelt P (1999) Climatic factors in the Atlantic control the oceanographic and ecological changes in the Baltic Sea. Limnol Oceanogr 45:703–710

Hilborn R, Walters CJ (1992) Quantitative fish stock assessment: choice, dynamics and uncertainty. Chapman and Hall, London

Jobling M (2002) Environmental factors and rates of development and growth. In: Hart PJB, Reynolds JD (eds) Handbook of fish biology and fisheries, vol 1, chapter 5. Blackwell Publishing, Oxford

Jordi A, Hameed S (2009) Influence of the Icelandic low on the variability of surface air temperature in the Gulf of Lion: implications for intermediate water formation. J Phys Oceanogr 39:3228–3232

Lleonart J (2008) Review of the state of Mediterranean and Black Sea fishery resources. In: Basurco B (ed) The Mediterranean fisheries sector. *Options méditerranéennes, Série B: Études et recherches*, n. 62. CIHEAM/FAO/GFCM, Rome, 179 pp

Lloret J, Lleonart J, Solé I, Fromentin J-M (2001) Fluctuations of landings and environmental conditions in the north-western Mediterranean Sea. Fish Oceanogr 10:33–50

Mann KH, Drinkwater KF (1994) Environmental influences on fish and shellfish production in the Northwest Atlantic. Environ Rev 2:16–32

Mariotti A, Struglia MV, Zeng N, Lau KM (2002) The hydrological cycle in the Mediterranean region and implications for the water budget of the Mediterranean Sea. J Clim 15:1674–1690

Maynou F (2008) Influence of the North Atlantic Oscillation on Mediterranean deep-sea shrimp landings. Clim Res 36:253–257

Mendelsohn R, Mendo J (1987) Exploratory analysis of anchoveta recruitment off Peru and related environmental series. In: Pauly D, Tsukuyama I (eds) The Peruvian anchoveta and its upwelling ecosystem: three decades of change. ICLARM studies and reviews, Instituto del Mar de Peru, Callao (Peru), vol 15. pp 109–116

Molinero JC, Ibanez F, Souissi S, Buecher E, Dallot S, Nival P (2008) Climate control on the long-term anomalous changes of zooplankton communities in the northwestern Mediterranean. Glob Change Biol 14:11–26

Myers R (2002) Recruitment: understanding density-dependence in fish populations. In: Hart PJB, Reynolds JD (eds) Handbook of fish biology and fisheries, vol 1, chapter 6. Blackwell Publishing, London

Oliver P, Massutí E (1995) Biology and fisheries of western Mediterranean hake (*M. merluccius*). In: Alheit J, Pitcher TJ (eds) Hake: biology, fisheries and markets. Chapman and Hall, London, pp 181–202

Ottersen G, Alheit J, Drinkwater K, Friedland K, Hagen E, Stenseth NC (2004) The responses of fish populations to ocean climate fluctuations. In: Stenseth NC, Ottersen G, Hurrell J, Belgrano A (eds) Marine ecosystems and climate variation: the North Atlantic. Oxford University Press, Oxford, pp 73–94

Ottersen G, Stenseth NC (2001) Atlantic climate governs oceanographic and ecological variability in the Barents Sea. Limnol Oceanogr 46:1774–1780

Recasens L, Lombarte A, Morales-Nin B, Torres GJ (1998) Spatiotemporal variation in the population structure of the European hake in the NW Mediterranean. J Fish Biol 53:387–401

Reid PC, Borges MF, Svendsen E (2001) A regime shift in the North Sea circa 1988 linked to changes in the North Sea horse mackerel fishery. Fish Res 50:163–171

Reid PC, Edwards M, Beaugrand G, Skogen M, Stevens D (2003) Periodic changes in the zooplankton of the North Sea during the 20th century linked to oceanic inflow. Fish Oceanogr 12:260–269

Sparre P, Venema SC (1998) Introduction to tropical fish stock assessment – Part 1: Manual. FAO Fisheries Technical Paper 306/1

Stenseth NC, Mysterud A (2005) Weather packages: finding the right scale and composition of climate in ecology. J Animal Ecol 74:1195–1198

Stige LC, Ottersen G, Brander K, Chan KS, Stenseth NC (2006) Cod and climate: effect of the Northa Atlantic Oscillation on recruitment in the North Atlantic. Mar Ecol Prog Ser 325:227–241

Venables WN, Ripley BD (2002) Modern applied statistics with S, 4th edn. Springer, Berlin

Watanabe Y, Kurita Y, Noto M, Oozeki Y, Kitagawa D (2003) Growth and survival of Pacific saury (*Cololabis saira*) in the Kuroshio-Oyashio transitional waters. J Oceanogr 59:403–414

Watanabe Y, Zenitani H, Kimura R (1996) Causes of population decline in the Japanese sardine (*Sardinops melanostictus*): overfishing or early mortality? In: Watanabe Y, Yamashita Y, Oozeki Y (eds) Survival strategies in early life stages of marine resources. Balkema, Rotterdam, pp 83–94

Wood S (2006) Generalized additive models: an introduction with R. Chapman and Hall/CRC, London, p 410

Impacts of the NAO on Mediterranean Crop Production

Simone Orlandini, Anna Dalla Marta, Marco Mancini, and Daniele Grifoni

Abstract Meteorological conditions can influence the main biological processes responsible of plant growth and development affecting vegetative growing, setting and ripening of different plant organs, onset and duration of phenological stages and the final production, exerting a strong impact on agricultural activities. In particular, the study of the variability of timing and length of the plants growing season is gaining importance because plant phenology is a sensitive indicator of climate change and has large impacts on terrestrial ecosystems through changes in productivity and in the annual carbon and water cycles. For many years, starting from meteorological information, mathematical descriptions of these effects have been formulated in order to provide users (farmers, technicians, extension services, researchers, etc.) with operational tools for improving management and planning activities. Besides common meteorological information supplied by local or synoptic weather stations, the use of large-scale climatic variables has also been investigated in order to forecast agricultural yields and production quality in several regions of the world. In Europe, the North Atlantic Oscillation (NAO) seems to be responsible for a large component of climate variability, particularly influencing winter and, through climate interactions on plants, crop production quality and quantity characteristics. The use of large-scale meteorological information showed great potential, particularly for the development of crop production forecasting systems.

Keywords Wheat · Grapevine · Growth · Development · Yield

1 Introduction

Observations of the physical and biological variables in the environment are essential in agricultural meteorology. Meteorological considerations enter in assessing the performance of a plant or animal because their development is a result of the combined effect of genetic characteristics and their response to environment. Without quantitative data, agrometeorological planning, forecasting, research and services by agrometeorologists cannot properly assist agricultural producers to meet

S. Orlandini (✉)
Department of Plant, Soil and Environmental Science, University of Florence, Florence, Italy
e-mail: simone.orlandini@unifi.it

S.M. Vicente-Serrano, R.M. Trigo (eds.), *Hydrological, Socioeconomic and Ecological Impacts of the North Atlantic Oscillation in the Mediterranean Region*, Advances in Global Change Research 46, DOI 10.1007/978-94-007-1372-7_8,

the ever-increasing demands for food and agricultural products. Such data are also needed to assess the impacts of agricultural activities and processes on the environment and climate.

Crop production is the result of a complex interaction among environmental conditions, agronomic management and plant characteristics. Soil, described by its chemical, biological and physical properties, and climate (i.e. degree days accumulation, rainfall, total solar radiation) are the main environmental factors affecting plant responses.

Crop production is influenced by meteorological conditions such as temperature, precipitation, solar radiation (intensity, photoperiod, and quality) and wind. As these variables correspond to the main driving factors of crops production spatial and temporal variability, they are commonly used for the analysis and simulation of crop yield.

A number of approaches have been used to describe crop yield based on weather parameters. In most statistical approaches, regressions are obtained correlating environmental variables and yield. The main problems with statistical approaches are that yield is more complex than growth simulation and that there is little confidence in extrapolating the results beyond the original limits of the data-sets. On the other hand, in using mechanistic approaches, difficult arises from numerous assumptions. Intermediate approaches between statistical and complex models are to use simplified mechanistic models which define crop behaviour by only few relationships (Bindi et al., 1997).

Agrometeorological stations can provide data to feed crop yield simulators. The quality of agrometeorological variables is the basis for a precise simulation and consequently for the processing of information that meets the needs of the end-users. However, ground weather stations often fail to satisfy these quality standards, both in terms of accuracy and precision of measured data and of spatial and temporal representativeness (WMO, 1981). The implementation of networks of weather stations could partly solve these problems, but would involve high installation and maintenance costs (Vose and Menne, 2004).

Moreover meteorological observations do not inform us about the state of the local atmosphere unless we know how the observations were made – the instrument, its installation height and exposure, used sampling and averaging times – and the way in which the measurements were processed. Specifications of all these links of the observation chain are called metadata, and their availability determines the value of measurements. Unfortunately they are not commonly available, making the use of station data unable to provide a realistic description of crop responses, particularly for application at meso or macroscale level.

The internet has increasingly been used to disseminate climate and meteorological data, given its rapid manipulation and displays of information, the interaction and feedback with the users, and the reduction of cost. The internet allows for finding suitable applications providing free access to meteorological information with different spatial domains and temporal resolutions.

Besides common meteorological information supplied by local or synoptic weather stations, the use of large-scale climatic variables has also been investigated

in order to forecast agricultural yields and production quality in several regions of the world (Atkinson et al., 2005). Among these indices that identify the large-scale distribution of pressures and temperatures over defined geographical areas, the North Atlantic Oscillation (NAO) is one of the most studied due to its significant impact on the climate over important production regions worldwide (Tables 1 and 2). In Spain, Gimeno et al. (2002) found that the yield of many crops (lemon, orange, tangerine, wheat, rye and olive) is affected by NAO, while in Italy the effect of NAO was related to the phenology of grapevine (Dalla Marta et al., 2010a) and quality of wine (Grifoni et al., 2006). Many studies demonstrated a relationship between the wheat yield and grain quality (Hagber falling number, specific weight and protein concentration) and the winter NAO in Italy (Dalla Marta et al., 2010b), Portugal (Gouveia and Trigo, 2008) and UK (Kettlewell et al., 1999; Wanner et al., 2001; Hurrell et al., 2003; Atkinson et al., 2005, 2008). According to Kettlewell et al. (2003) the winter NAO has an effect on cumulated precipitation in summer in England and Wales and, consequently, on wheat quality, while Colman (1997) suggested an association between Sea Surface Temperature (SST) and NAO in the UK.

Large-scale climate variability affects average conditions considerably and is therefore likely to influence the phenology of plants. In Europe, the North Atlantic Oscillation (NAO) seems to be responsible for a large component of climate variability (Post and Stenseth, 1999; Benestad, 2001; Wanner et al., 2001; Sivle, 2005), particularly influencing winter weather and, through climate interactions on plants, flowering time of many tree species (Klaveness and Wielgolaski, 1996; Gormsen et al., 2005; Avolio et al., 2008).

At the same time, winter meteorological conditions play an important role in processes controlling the termination of dormancy and bud-break, strongly affecting the onset of the growing season, spring flowering phenophases and pollen season of different woody plants in North and southern Europe (Fitter et al., 1995; Klaveness and Wielgolaski, 1996; D'Odorico et al., 2002; Gormsen et al., 2005; Avolio et al., 2008; Nordli et al., 2008).

Table 1 Example of NAO effect on vegetation, forest and trees

Ecological descriptor	Parameter	Correlation	Period	Variable affected	References
Europe vegetation	Timing of spring phenology	Negative	Jan–Mar	Temperature, precipitation	D'Odorico et al. (2002)
Europe vegetation	Timing of phenology	Negative	Winter	Temperature, precipitation	Gouveia et al. (2008)
Europe forests	Timing of phenology	Negative	N.A.	Temperature	Cook et al. (2005)
Norway trees	Timing of phenology	Negative	Winter	Temperature	Post et al. (1999)
Norway trees	Timing of phenology	Negative	Winter	Temperature	Nordli et al. (2008)
Germany trees	Timing of spring/ summer phenology	Negative	Jan–Mar Feb–Mar	Temperature	Menzel (2003)

Table 2 Example of NAO effect on crop responses

Ecological descriptor	Parameter	Effect	Period	Variable affected	References
Spain wheat, orange, lemon	Yield	Positive	Yearly	Temperature, precipitation	Gimeno et al. (2002)
Portugal wheat	Yield	N.A.	Feb–Mar	Temperature, precipitation	Gouveia et al. (2008)
UK wheat	Quality (specific weight)	Positive	Winter	Radiation, precipitation	Atkinson et al. (2005)
UK wheat	Quality (specific weight)	Positive	Jan	Radiation, precipitation	Atkinson et al. (2008)
UK wheat	Quality (specific weight, protein)	Positive	Jan–Feb	Radiation, temperature	Kettlewell et al. (1999)
UK wheat	Quality (specific weight)	Positive	Winter	Summer precipitation	Kettlewell et al. (2003)
Italy durum wheat	Quality (protein content)	Positive	Nov–Feb	Winter precipitation	Dalla Marta et al. (2010b)
Italy grapevine	Timing of phenology	Negative	Feb–Mar	Radiation and temperature	Dalla Marta et al. (2010a)
Italy grapevine	Quality (rating)	Negative	Apr–Jul	Temperature, precipitation	Grifoni et al. (2006)
Italy olives	Pollen season	Negative	Mar–Jun	Temperature	Avolio et al. (2008)

1.1 Objectives

The main objective of this chapter was to analyze the correlations between NAO index (Jones et al., 1997) and several Italian crop productions. Different case studies were considered, all taking into consideration high quality Mediterranean crops, such as wheat and grapevine. Local ground weather stations data were also analysed, to compare obtained results by using large scale indices. Different temporal scales were compared in order to propose a crop response analysis and forecasting method based on meteorological conditions observed several months before harvest. Results are discussed with the aim of evaluating the possibility of assessing crop production with the freely available data, solving the problems related to the use of ground weather stations.

2 Results

2.1 NAO and Grapevine Phenology

In this study, data series of phenological stages of Sangiovese grapevine were analysed for the productive area of Montepulciano wine. Bud-break (1979–2006),

flowering and harvest (1970–2006) dates were used for the analysis and related to the NAO index. The phenology is the scientific study of periodic biological phenomena such as flowering, breeding, and migration in relation to climatic conditions. Phenological data, expressed as the day of the year in which the phenological stage (bud-break, flowering, fruit set, etc.) was reached (Balloid and Baggiolini, 1993), were collected by skilled field operators and considered as the mean value of 20 vines.

Bud-break date was negatively correlated with NAO index for February ($r = -0.402$, $p \leq 0.05$) and for the period February–March ($r = -0.377$, $p \leq 0.05$) (Fig. 1), even if the month of March did not show a significant correlation when considered independently.

The negative correlation between February and March NAO index and bud-break data was due to the relation between NAO and weather conditions in the wine production area at the beginning of the grapevine vegetative season. Winter and early spring NAO has a strong and well-documented effect on precipitation and consequently on the cloudiness of the same period. The role of NAO, therefore, could be mainly due to a reduction in cloud cover and consequently greater insolation than average for the positive NAO index phase. Greater sunshine probably leads to an increased surface temperature and thus the heat requirement for various stages of plant development could be reached earlier (Avolio et al., 2008).

Flowering of grapevine in the study area usually occurs between mid-May and early June. The NAO index of single months showed no correlation with flowering stage (data not shown) but a significant negative correlation was found between the NAO index for February–March ($r = -0.397$, $p \leq 0.05$) (Fig. 2) and the phenological stage date. The effect of winter NAO on flowering stage was probably an indirect result beginning at the bud-break stage. In fact, the effect of NAO on meteorological conditions is strong during winter and decreases in the spring and early summer months. However, the two stages of bud-break and flowering are partially

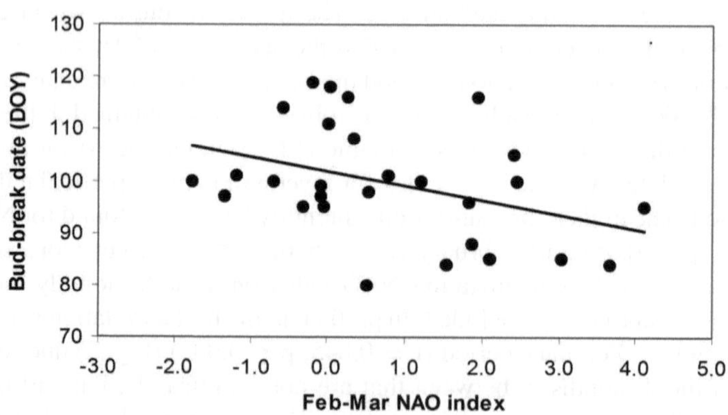

Fig. 1 Linear correlation between NAO index of the period February–March and grapevine bud-break phase date

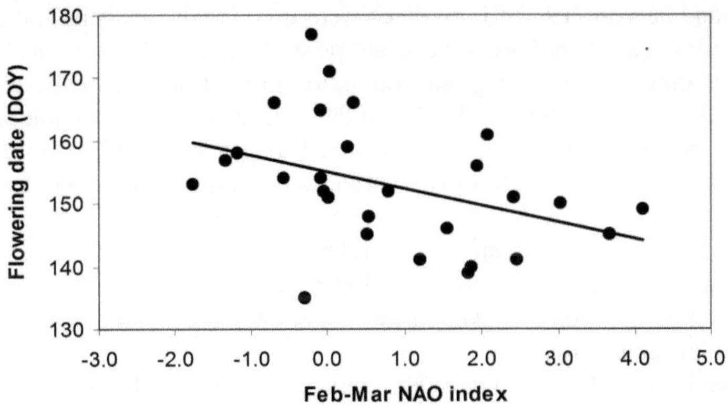

Fig. 2 Linear correlation between NAO index of the period February–March and grapevine flowering phase date

auto-correlated and the period in between is mainly determined by a thermal summation. Therefore, the anticipation of bud-break due to high values of winter NAO also has the indirect effect of anticipating flowering in late spring.

Harvest is usually conducted from mid-September to mid-October, depending on different environmental and physiological factors. The NAO index did not significantly affect harvest date, confirming that its effect on meteorological conditions is really evident in winter and, less strongly, in early spring.

2.2 NAO and Wheat Quality

The research was carried in Val d'Orcia (Siena, Italy) where durum wheat is considered an important quality production with a cultivated area of almost 135,000 ha, about 15% of the total cultivated surface. The quality of durum wheat grain was described by its protein content, expressed as the percentage of dry matter, due to its direct relation with the production of good quality pasta. Data were analyzed for the period 1997–2009. These results are not as robust as those obtained in the previous section due to the short time-series available (13-year), but the wheat quality was positively correlated with the NAO index for several months, in particular during the winter period. The higher correlations on a monthly basis were found for November ($r = 0.580$, $p \leq 0.05$) and February ($r = 0.546$, $p \leq 0.05$), but more significant results were obtained by aggregating NAO index on a multi-monthly basis when November and February are included. In particular, the best correlation was obtained for the November–February period ($r = 0.845$, $p \leq 0.001$) (Fig. 3) due to the contribution of the 2 months in between that are not significantly, but still positively, correlated to the final grain protein content. This result can be explained by the winter NAO index defining a specific atmospheric synoptic configuration over the Atlantic Ocean, strongly affecting winter precipitation over the Mediterranean sea and consequently over the study area (Hurrell and Van Loon, 1997; Bartolini et al.,

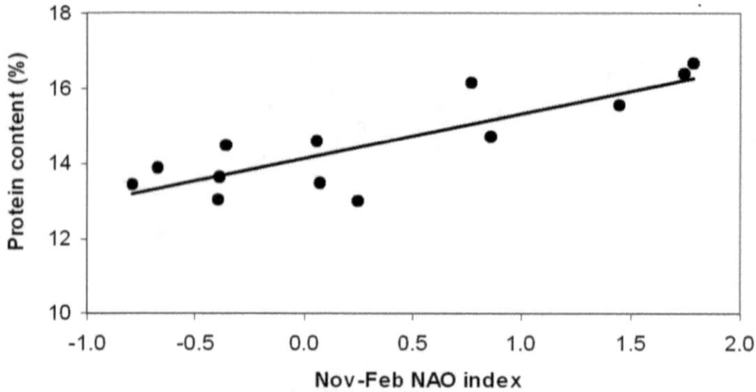

Fig. 3 Linear correlation between NAO index of the period November–February and durum wheat quality

2009). In turn, winter precipitation is negatively correlated with protein content. The negative effect of rainfall on grain quality is probably related to problems concerning crop management. In fact, abundant precipitations in November can create serious problems for field operations, even causing a significant delay in the sowing date. Nevertheless, even if wheat can be sown in time, a too high soil water content may give rise to asphyxia phenomena leading to further problems during crop development. Moreover, the negative effect of rainfall may be due to the losses of nitrogen applied with fertilization. Anyway, wheat protein resulted slightly more correlated with the NAO index than with cumulated precipitation, probably because the former is able to better describe the environmental conditions (being a consequence of atmospheric circulation) rather than just the precipitation.

To investigate the effect of NAO and common meteorological variables as a whole, a stepwise linear multiple regression analysis was performed. The model, based on an automatic procedure of variable inclusion/exclusion ($p \leq 0.05$ and $p \geq 0.10$ respectively), was tested by using the multi-monthly period for which the best correlation coefficients were obtained for each single meteorological variable. In particular, the variables used were February–June air temperature ($r = 0.609$, $p \leq 0.05$), November–May precipitations ($r = 0.732$, $p \leq 0.01$) and November–February NAO ($r = 0.845$, $p \leq 0.001$). Within the stepwise regression results, the simplest model was obtained using NAO index ($r = 0.845$, $p \leq 0.001$), while precipitation and air temperature were excluded since their addition to the model failed to improve its performance. This result is consistent with the previous considerations concerning the dependence of these variables on NAO, which already contains their partial contribution to the determination of wheat quality.

2.3 NAO and Wine Quality

This case study analyzed the relationships between NAO index and the quality of different important Italian wines for the period 1970–2002. The quality data

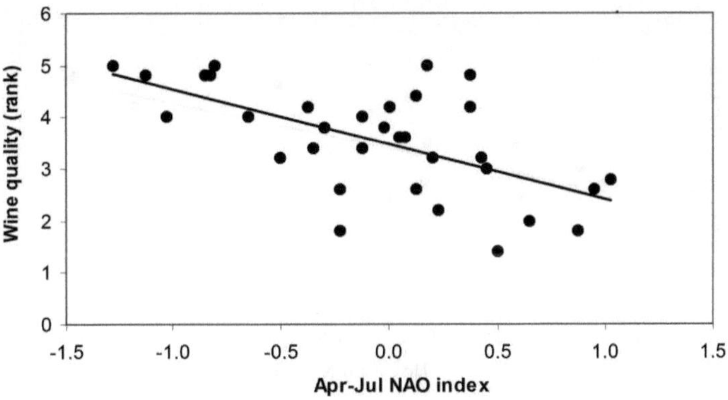

Fig. 4 Linear correlation between NAO index of the period April–July and Italian wine quality

series of the following six wines were used: Brunello di Montalcino, Nobile di Montepulciano and Chianti Classico (Tuscany region, central Italy), Barolo and Barbaresco (Piedmont region, north-western Italy), and Amarone (Veneto region, north-eastern Italy). Vintage ratings were used to define wine quality ranking. The rating of a given vintage is conducted in a single blind tasting of the individual varietals by a panel of experts. Although this rating does not consider variations in quality among the individual vineyards, it is the most comprehensive overall rating compiled for high-quality Italian wines (Corsi and Ashenfelter, 2001). This analysis allows for verifying whether the wine meets market expectations and represents a composite of all the quality elements of the wine. In production area, this ranking is based upon the collection of estimates from 1 to 5 classes (from insufficient to excellent) (Riou, 1994). The relationships between quality data of wines and NAO index were investigated through regression analysis on monthly and multi-monthly bases.

Wine quality rankings were inversely correlated with the NAO Index of several previous months (included in the growing season) and in particular, the higher negative correlations were obtained for April ($r = -0.439$, $p \leq 0.05$) and May ($r = -0.359$, $p \leq 0.05$).

More significant results were obtained by developing regression of the wine quality versus NAO Index aggregated on a multi-monthly basis including April and May. The best result, for some selected aggregations, was obtained for the April–July period ($r = 0.607$, $p \leq 0.001$) (Fig. 4). These results could be explained by the fact that the NAO index defines a specific atmospheric synoptic configuration over the Atlantic Ocean, affecting the regional rainfall and air temperature.

3 Conclusions

Obtained results are useful for agricultural communities when used to modify management decisions and to adopt practices that are better adapted to the expected

conditions. However the possible use of seasonal climate information for agricultural aims has been slower than expected for a number of reasons. The most significant is that the deterministic information requested by the farmers is beyond that elaborated by the NAO and other indices (ENSO, AO, etc.). As an example, the quality of short term forecast is not comparable with the information available from seasonal index. Recent development, however, provide some evidence of progress toward meeting user needs and forecast capabilities. This is due to the consideration that reliable sources of daily rainfall and temperature data are becoming more widely available for many parts of the world. Secondly, climate scientists are beginning to use these data to examine the observed statistics of climatic variables over the course of the season and how these statistics influence climatic indices. Further, investigations are being initiated to examine the ability of numerical weather models to replicate these statistics in analyses and forecasts (Baethgen et al., 2009).

Acknowledgements The authors wish to thank Fondazione Monte dei Paschi di Siena, Consorzio Agrario di Siena and COST734 Action for their support of this research.

References

Atkinson MD, Kettlewell PS, Hollins PD, Stephenson DB, Hardwik NV (2005) Summer climate mediates UK wheat quality response to winter North Atlantic Oscillation. Agri Forest Meteorol 130:27–37

Atkinson MD, Kettlewell PS, Poulton PR, Hollins PD (2008) Grain quality in the Broadbalk Wheat Experiment and the winter North Atlantic Oscillation. J Agric Sci 146:541–549

Avolio E, Pasqualoni L, Federico S, Fornaciari M, Bonofiglio T, Orlandi F, Bellocci C, Romano B (2008) Correlation between large-scale atmospheric fields and the olive pollen season in Central Italy. Int J Biometeorol 52:787–796

Baethgen WE, Carriquiry M, Ropelewski C (2009) Tilting the odds in maize yields. Bull Am Meteorol Soc 2:179–183

Balloid M, Baggiolini M (1993) Les stades repères de la vigne. Revue Suisse de Viticulture Arboriculture Horticulture 25:7–9

Bartolini E, Claps P, D'Odorico P (2009) Interannual variability of winter precipitation in the European Alps: relations with the North Atlantic Oscillation. Hydrol Earth Syst Sci 13:17–25

Benestad RE (2001) The cause of warming over Norway in the ECHAM4/OPYC3 GHG integration. Int J Climatol 21:371–387

Bindi M, Miglietta F, Gozzini B, Orlandini S, Seghi L (1997) A simple model for simulation of growth and development in grapevine (Vitis vinifera L.). I Model description. Vitis 36:67–71

Colman A (1997) Prediction of summer central England temperature from preceding North Atlantic winter sea surface temperature. Int J Climatol 17:1285–1300

Cook BI, Smith TM, Mann ME (2005) The North Atlantic Oscillation and regional phenology prediction over Europe. Glob Change Biol 11:919–926

Corsi A, Ashenfelter O (2001) Predicting Italian wines quality from weather data and experts' ratings, 18 pp. Cahier Scientifique de l'OCVE, n 4

Dalla Marta A, Grifoni D, Mancini M, Storchi P, Zipoli G, Orlandini S (2010a) Analysis of the relationships between climate variability and grapevine phenology in the Nobile di Montepulciano wine production area. J Agric Sci 148:657–666

Dalla Marta A, Grifoni D, Mancini M, Zipoli G, Orlandini S (2010b) The influence of climate on durum wheat quality in Tuscany, Central Italy. Int J Biometeorol 55:87–96

D'Odorico P, Yoo JC, Jaeger S (2002) Changing seasons: an effect of the North Atlantic Oscillation? J Clim 15:435–445

Fitter AH, Fitter RSR, Harris ITB, Williamson MH (1995) Relationship between first flowering date and temperature in the flora of a locality in Central England. Funct Ecol 9:55–60

Gimeno L, Ribera P, Iglesias R, de la Torre L, Garcia R, Hernandez E (2002) Identification of empirical relationships between indices of ENSO and NAO and agricultural yields in Spain. Clim Res 21:165–172

Gormsen AK, Hense A, Toldam-Andersen TB, Braun P (2005) Large-scale climate variability and its effects on mean temperature and flowering time of Prunus and Betula in Denmark. Theor Appl Climatol 82:41–50

Gouveia C, Trigo RM (2008) Influence of climate variability on wheat production in Portugal. In: Soares A, Pereira MJ, Dimitrakopoulos R (eds) geoENV VI – geostatistics for environmental applications. Springer, Berlin, pp 335–345

Gouveia C, Trigo RM, Da Camara CC, Libonati R, Pereira JMC (2008) The North Atlantic Oscillation and European vegetation dynamics. Int J Climatol 28:1835–1847

Grifoni D, Mancini M, Maracchi G, Orlandini S, Zipoli G (2006) Analysis of Italian wine quality using freely available meteorological information. Am J Enol Vitic 57:339–346

Hurrell JW, Kushnir Y, Ottersen G, Visbeck M (eds) (2003) The North Atlantic Oscillation: climatic significance and environmental impact. Geophys Monogr 134:1–35. Copyright 2003 by the American Geophysical Union 10.1029/134GM01

Hurrell JW, Van Loon H (1997) Decadal variations in climate associated with the North Atlantic Oscillation. Clim Change 36:301–326

Jones PD, Jonsson T, Wheeler D (1997) Extension to the North Atlantic Oscillation using early instrumental pressure observations from Gibraltar and South-West Iceland. Int J Climatol 17:1433–1450

Kettlewell PS, Sothern RB, Koukkari WLJ (1999) U.K. wheat quality and economic value are dependent on the North Atlantic Oscillation. J Cereal Sci 29:205–209

Kettlewell PS, Stephenson DB, Atkinson MD, Hollins PD (2003) Summer rainfall and wheat grain quality: relationships with the North Atlantic Oscillation. Weather 58:155–164

Klaveness D, Wielgolaski FE (1996) Plant phenology in Norway – a summary of past and present first flowering dates (FFDs) with emphasis on conditions within three different areas. Phenol Season 1:47–61

Menzel A (2003) Plant phenological anomalies in Germany and their relation to air temperature and NAO. Clim Change 57:243–263

Nordli Ø, Wielgolaski FE, Bakken AK, Hjeltnes SH, Måge F, Sivle A, Skre O (2008) Regional trends for bud burst and flowering of woody plants in Norway as related to climate change. Int J Biometeorol 52:625–639

Post E, Stenseth NC (1999) Climatic variability, plant phenology, and northern ungulates. Ecology 80:1322–1339

Riou C (1994) The effect of climate on grape ripening: application to the zoning of sugar content in the European Community. European Commission, Brussels, 319 pp

Sivle A (2005) Climatic oscillations in the period 1910–2004 (in Norwegian). Geophysical Institute, University of Bergen, Bergen

Vose RS, Menne MJ (2004) A method to determine station density requirements for climate observing networks. J Clim 17:2961–2971

Wanner H, Bronnimann S, Casty C, Gyalistras D, Luterbacher J, Schmutz C, Stephenson DB, Xoplaki E (2001) North Atlantic Oscillation – concepts and studies. Surv Geophys 22:321–381

World Meteorological Organization (1981) Guide to agricultural meteorological practices, 2nd edn. Secretariat of World Meteorological Organization, Geneva

The Impacts of the NAO on the Vegetation Activity in Iberia

Célia Gouveia and Ricardo M. Trigo

Abstract The present study analyses the relation between North Atlantic Oscillation (NAO) and satellite-based measures of vegetation dynamics (as obtained using Normalized Difference Vegetation Index, NDVI), carbon absorption by living plants (obtained by Net Primary Production, NPP estimates) over Iberian Peninsula and wheat yield in Portugal. There is a strong evidence that positive (negative) values of winter NAO induce low (high) vegetation activity in the following spring and summer seasons. Consequently, significant correlations patterns between NAO and NPP were also obtained, identifying a negative impact in spring and summer over southern Iberia. These features are mainly associated with the impact of NAO on winter precipitation, together with the strong dependence of the spring and summer vegetation activity on water availability during the previous winter. The different role played by NAO along the vegetative cycle of wheat yield in Portugal is also assessed and results obtained suggest using spring NDVI together with NAO in April and June to build up a simple model of wheat yield in Portugal. Results reveal to be satisfactory and are expect to be useful to estimate crop production and to perform agricultural monitoring.

Keywords NAO · Vegetation activity · NDVI · NPP · Wheat yield

1 Introduction

Recent global changes in vegetation dynamics have been continuously monitored from space and, additionally, significant connections with changes of surface climatic variables, such as temperature and precipitation, have been established (Myneni et al., 1997; Zhou et al., 2001; Nemani et al., 2003). In particular, the observed increment in temperature during spring and autumn over high latitude regions of the Northern Hemisphere leads to higher photosynthetic activity and larger growing seasons (Groisman et al., 1994; Boagert et al., 2002; Shabanov

C. Gouveia (✉)
Faculty of Sciences, Instituto Dom Luiz (IDL), University of Lisbon, Lisbon, Portugal; EST, Polytechnic Institute of Setubal, Setúbal, Portugal
e-mail: cmgouveia@fc.ul.pt

S.M. Vicente-Serrano, R.M. Trigo (eds.), *Hydrological, Socioeconomic and Ecological Impacts of the North Atlantic Oscillation in the Mediterranean Region*, Advances in Global Change Research 46, DOI 10.1007/978-94-007-1372-7_9,

et al., 2002). However, whereas in spring photosynthesis dominates respiration, the opposite takes place in autumn and therefore it is in spring that an increase in CO_2 sequestration is expected to occur (Piao et al., 2008). Accordingly, in a future warmer world, carbon sequestration capacity by northern hemisphere ecosystems may decrease faster than previously suggested due to a faster warming in autumn than in spring (Sitch et al., 2008). Nevertheless, changes in the seasonality of temperature and precipitation may have distinct impacts, depending on local characteristics. The strong dependence of vegetation dynamics on water availability in the Mediterranean regions has been now widely recognized (e.g. Gouveia et al., 2009; Vicente-Serrano, 2007; Lindner et al., 2010). It has been showed that the lack of precipitation over a certain period, combined with other climatic anomalies, such as high temperature, strong wind and low relative humidity over a particular area, may result in reduced green vegetation cover.

The above mentioned changes in the vegetation annual cycle can be detected, using satellite information, namely from observed changes in the most widely used vegetation-related satellite variable, the Normalised Difference Vegetation Index (NDVI) (Stöckli and Vidale, 2004; Vicente-Serrano and Heredia-Laclaustra, 2004). However, there is a strong need of long term studies of the impact of atmospheric circulation variability on vegetation greenness. In this context, special attention has been devoted to study the links between vegetation dynamics and the North Atlantic Oscillation (NAO). D'Odorico et al. (2002) showed that spring phenology and timing of pollen season in British Isles, Poland, Norway and Sardinia are influenced by NAO, with leaf unfolding occurring later (earlier) for negative (positive) phases of NAO. Stöckli and Vidale (2004) found that spring phenology over Europe, as obtained using AVHRR Pathfinder NDVI data, correlates well with anomalies in winter NAO index, as well as in winter temperature. More recently, Maignan et al. (2008) showed the strong impact of NAO on the vegetation onset over a large fraction of Northern Europe. The pattern of the NAO control on the vegetation dynamics is clearly apparent over a large extension of Europe revealing in general an evident North–South gradient.

On a more regional context, Vicente-Serrano and Heredia-Laclaustra (2004) analysed the NAO control on Iberian vegetation productivity trends, as represented by the annual integral of monthly AVHRR Pathfinder NDVI values. The NAO impact revealed to be stronger in southern Iberia that corresponds to areas of stable or decreasing vegetation productivity. On the other hand the NAO impact on vegetation greenness is weaker in the north of the peninsula, over areas where significant positive productivity trends occur. Gouveia et al. (2008) expanded the analysis on the relationship between vegetation greenness and NAO to the entire Europe. The authors relate the different vegetation behaviours particularly over Iberian Peninsula and Northeastern Europe, with a strong dependence of vegetation dynamics from water availability in Iberia and from winter temperature in Northeastern Europe.

Terrestrial Net Primary Production (NPP) is a measure of the amount of carbon fixed by the living plants and converted into plant biomass. Nemani et al. (2003) showed that changes in climate, related to increasing temperature and solar radiation, have eased several critical constrains to plant growth, leading to an upward

trend in global NPP from 1982 to 1999. However, they have also shown that NPP respond differentially with respect to latitude to major climate events, such as volcanic eruptions and El Niño episodes. Globally, NPP decreases during the three major El Niño events (particularly over tropical regions), while the decrease after Mount Pinatubo eruption was mostly restricted to northern hemisphere high latitudes. Recently, Zhao and Running (2010) showed that large-scale droughts have reduced NPP during the last decade, imposing a drying trend for the entire southern hemisphere, whereas an increased NPP can be found over the majority of the northern hemisphere.

Finally, climate is a key factor for the majority of agricultural systems, due to their dependence on interannual climate variability and in particular, their vulnerability to extreme events such as; droughts, floods and frost or hail occurrences. Important relations between regional distributions of temperature and precipitation and wheat yield for the European countries have been found (Cantelaube et al., 2004). In particular the winter NAO index has been associated with UK wheat crop and with better wheat, rye, oat and citrus yields in the Iberian Peninsula (Atkinson et al., 2005; Gimeno et al., 2002). Although not directly related to NAO, some authors have analysed in detail the links to climate variables that are ultimately related to this mode of variability. Iglesias and Quiroga (2007) evaluated over five sites in Spain the effects of climate variability on the final crop. Similarly drought indices and remote-sensed data were used to predict wheat and barley yields in the Ebro valley (Vicente-Serrano et al., 2006). Rodríguez-Puebla et al. (2007) have derived a model that integrates the effects of abundant precipitation together with dynamic aspects of the air masses during the maturation in order evaluate their effects on winter cereals productivity in Spain.

1.1 Objectives

Here we intend to provide a comprehensive assessment on the impact of NAO on various vegetation related variables, namely NDVI, NPP and crop yields. While some of these analyses were initially performed for other regions in northern Europe we opted here to focus only on the western Mediterranean sector.

1.2 Data

A number of restrictions on access to different types of dataset implied slightly distinct periods of analysis. The NPP monthly estimates were obtained in the frame of a Portuguese funded project entitled CARBERIAN, using the Carnegie Ames Stanford Approach (CASA) terrestrial biogeochemical model (Potter et al., 1993; Friedlingstein et al., 1999) for the period 1982–1999. The CASA simulates plant and soil processes allowing the estimation of both NPP and NEP (Net Ecosystem Productivity). Besides climate variables (precipitation, temperature and radiation)

the model incorporates information from vegetation and land cover types using satellite data, as well as soil texture and soil depth (Seixas et al., 2009).

The NDVI monthly anomalies, with 8 km of spatial resolution, were obtained from the Global Inventory Modeling and Mapping Studies (GIMMS) dataset. This dataset holds 24 values per year (i.e. twice a month) and corresponds to the most complete and longest remote sensing dataset covering the entire Mediterranean region. In the case of the assessment of the NAO impact on vegetation dynamics we have performed the analysis for the period 1982–2002. Finally, the evaluation of the NAO impact on Portuguese crop yields covers the period spanning between 1982 and 2005.

The NAO index used in this study is based on the one developed by the Climatic Research Unit (CRU), which was originally defined, on a monthly basis, as the difference between the normalized surface pressure at Gibraltar, in the southern tip of the Iberian Peninsula and Stykkisholmur, in Iceland (Jones et al., 1997). For each year covering the above mentioned periods, we have derived a late winter NAO index (hereafter referred only as NAO), defined as the average of the monthly values for January, February and March of the corresponding year.

2 Results

2.1 NAO and Vegetation Dynamics

Figure 1 displays the spatial patterns over Iberia of point correlation values of winter NAO and NDVI for spring ($NDVI_{MAM}$), summer ($NDVI_{JJA}$) and autumn ($NDVI_{SON}$) for the period between 1982 and 2002. Results show a negative correlation over the Iberian Peninsula in spring and summer, with some values reaching as low as –0.8. Positive correlations are found in northern and eastern Iberia, as well as in Murcia region in spring (values around 0.7). However in summer the positive

Fig. 1 Point correlation fields of NAO vs. $NDVI_{MAM}$ (*left panel*) and NAO vs. $NDVI_{JJA}$ (*middle panel*) and NAO vs. $NDVI_{SON}$ (*right panel*) for the period from 1982 to 2002. The *colorbar* identifies values of correlation and the *two arrows* indicate the ranges that are significant at 5% level

correlations are restricted to the Murcia region. A weaker pattern of correlations is found during autumn, presenting lower but still significant positive correlation (values around 0.4) spreading over eastern and Northern Iberia and presenting a negative pattern in Southern Iberia. These patterns are in good agreement with those obtained previously in the works of Buermann et al. (2003) and Vicente-Serrano and Heredia-Laclaustra (2004). This behaviour reflects the response of the annual variability of meteorological variables to large-scale atmospheric variability associated to the NAO mode, as well as the different response of vegetation to atmospheric variability, namely the changes induced by temperature and precipitation in the annual cycle of heat and moisture.

It should be stressed that other factors linked to the human influence can disturb the relationship between atmospheric parameters and vegetation, such as the nature and quality of the plant substrate, the over-use of agriculture land and the employment of irrigation (Gouveia et al., 2008). In order to isolate the NAO effect on vegetation dynamics, as obtained by NDVI, from factors related to the human influence, we have compared the NDVI anomalies for two subsets of years associated to extreme NAO indices: (i) higher than usual values of seasonal NAO index (whenever the NAO index is higher than the percentile 75 of its distribution) and (ii) lower than usual (whenever the NAO index is lower than the percentile 75 of its distribution). Figure 2 shows a comparison of seasonal NDVI anomalies for the two subsets of years associated with NAO– (upper panel) and NAO+ (lower panel). In the case of spring NDVI anomalies present well defined pattern over central and western Iberia. The southern anomaly center is particularly intense, positive (negative) anomalies being observed in spring NAO– (NAO+). The pattern is remaining in

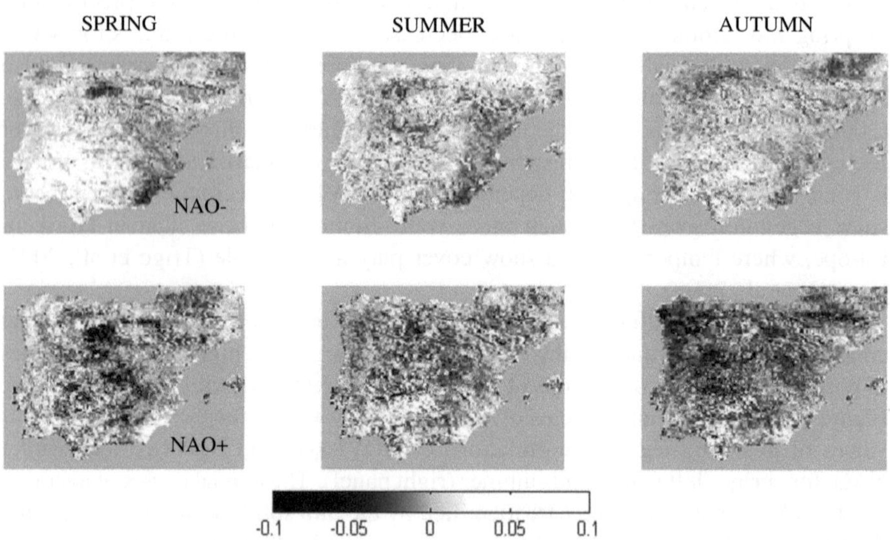

Fig. 2 Seasonal anomalies of NDVI for years of NAO– (*top panel*) and NAO+ (*bottom panel*) for spring (*left*), summer (*middle*) and autumn (*right*)

the following seasons, albeit with lower correlation values. Therefore, it means that for years characterised with positive winter NAO values, the vegetation is in vegetative stress, with lower than usual NDVI values, a feature that is common during drought years, like the recent 2004/2005 drought event (Gouveia et al., 2009).

In fact, Gouveia et al. (2008) have found significant correlations between winter NAO and contemporaneous late winter means of surface temperature and precipitation. For precipitation a well-developed meridional dipolar structure over Europe, delimiting two well-defined zonal bands of positive and negative correlation values along Northern Europe and Mediterranean region respectively. However it is also known that the connection between NAO and temperature over Iberia is not as clear-cut as for precipitation (Castro-Díez et al., 2002; Trigo et al., 2002). Although it is known that the impact of NAO on temperature and precipitation described above is especially prominent in winter (e.g. Vicente-Serrano and Heredia-Laclaustra, 2004; Trigo et al., 2004), such behaviour contrasts with that obtained for vegetation activity (Fig. 1), where the impact of NAO is clearly significant in spring, summer and autumn. Consequently the relationship between late winter temperature and precipitation with vegetation greenness in the following seasons are very important.

Additionally the work of Gouveia et al. (2008) showed that almost two thirds (64%) of the pixels that exhibit the highest (lowest) values of positive (negative) correlations of $NDVI_{MAM}$ and $NDVI_{JJA}$ with NAO correspond to areas of spring crops and about one sixth (17%) to forests and shrublands. The relative proportion of the two types undergoes a significant change in summer with only 29% relative to forest and shrubland, and an increase to 47% associated to cultivated areas. These results reflect the distinct responses of the various land cover types to moisture and heat conditions prevailing during the previous winter.

The Iberian Peninsula presents the same response of vegetation to precipitation in spring and summer, i.e. an increase (decrease) of vegetation greenness for NAO– (NAO+) years. A small dependence of vegetation greenness on temperature is also apparent in the case of spring (Gouveia et al., 2008). Additionally, Gouveia et al. (2008) have emphasized that the impact on precipitation is three times larger over Iberian Peninsula than the corresponding impact over Northern Europe. This feature is consistent with the specific dependence of NDVI on precipitation over Iberia, since vegetation growth is much more water-limited in Iberia than in Northern Europe, where temperature and snow cover play a major role (Trigo et al., 2002; Gouveia et al., 2008).

The remarkable differences in the response of vegetation to moisture and heat conditions, lead us to analyse the NDVI annual cycle for the highest (lowest) values of positive (negative) correlations of $NDVI_{MAM}$ and $NDVI_{JJA}$ with NAO over Iberia. Figure 3 shows the annual cycles of NDVI monthly values for the highest (lowest) values of positive (negative) correlations of $NDVI_{MAM}$ and $NDVI_{JJA}$ with winter NAO, for spring (left panel) and summer (right panel). The annual cycles of averages for the NAO+ (NAO–) subsets are identified by the thin solid (dashed) curves, while the annual cycles of mean NDVI for the entire period (1982–2002) are represented by thick solid lines.

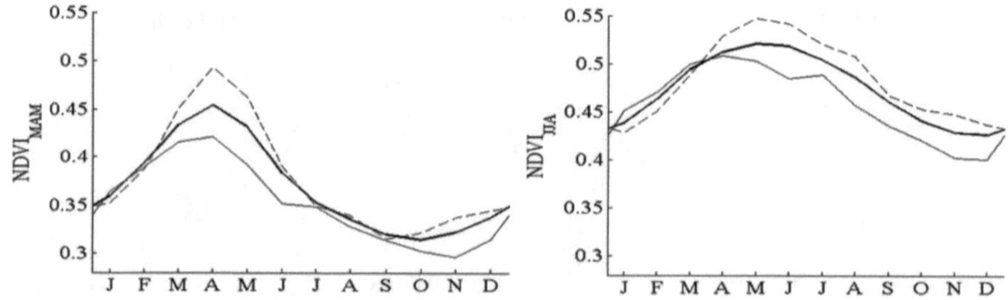

Fig. 3 Annual cycles of monthly values of NDVI for the highest (lowest) values of positive (negative) correlations of $NDVI_{MAM}$ and $NDVI_{JJA}$ with winter NAO, for spring (*left panel*) and summer (*right panel*), over Iberia. The annual cycles of averages for the NAO+ (NAO–) subsets are identified by the *thin solid (dashed) curves* and the annual cycles of mean NDVI for the entire period (1982–2002) are represented by *thick solid lines*

The highest impact of NAO is observed to occur during the periods of the year characterised by more intense vegetation activity (Ji and Peters, 2003), i.e. around April (June) in the case of highest correlated pixels for spring (summer). In the case of spring vegetation, two thirds of the pixels correspond to cultivated areas that are mostly associated to non irrigated crops, adapted to the relatively dry Iberian conditions. Vegetation has a short growth cycle, due to the generally observed high temperatures, starting as soon as water is available. This is a typical situation that can be favoured by NAO–. In the case of summer, the growing period of the most impacted vegetation starts later and therefore the response to precipitation tends to extend late in the year (Gouveia et al., 2008).

2.2 NAO and NPP Estimates

Figure 4 displays the spatial patterns over Iberia of point correlation values of winter NAO (NAO_{JFM}) and NPP for spring (NPP_{MAM}), summer (NPP_{JJA}) and autumn (NPP_{SON}) for the period between 1982 and 1999. This slightly shorter period results from the limited period of available data obtained with the CASA model by our colleagues within the CARBERIAN project that started in the early 2000s. Results show a negative correlation between NAO and NPP over the Iberian Peninsula in spring and summer, with values as low as –0.7 (Fig. 4, top panel). Positive correlations are found in northern and eastern Iberia and Northwestern Portugal. The highest positive correlations (values around 0.6) are found in Pyrenees and in coastal Mediterranean region of Murcia in spring. However in summer the positive correlations are mostly restricted to the Murcia region. A weaker pattern of correlations is found during autumn, presenting lower but still significant positive correlation (values around 0.4) spreading over eastern and Northern Iberia, Murcia region and southern Portugal. Interesting to notice the strong pattern of negative correlations located in the Southern interior of Iberia.

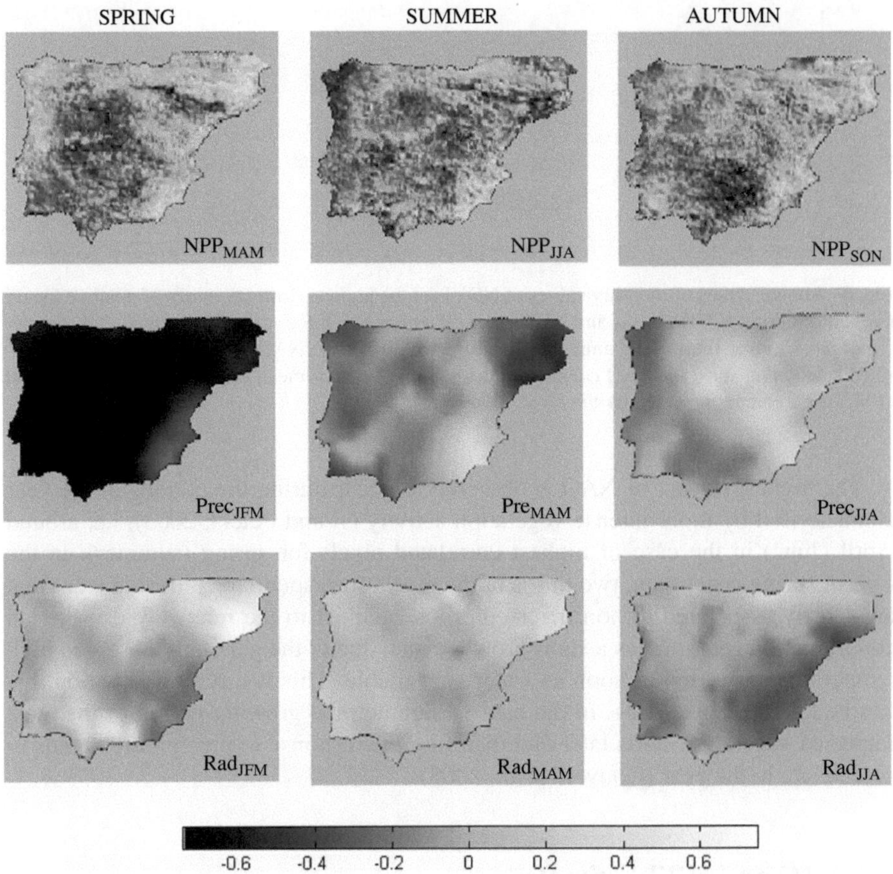

Fig. 4 Correlation fields of NAO vs. NPP, for spring (*left panel*), summer (*middle panel*) and autumn (*right panel*) over the period from 1982 to 1999 (*top panel*) and NAO vs. Precipitation (*middle panel*) and NAO vs. short wave radiation (*bottom panel*), for winter (*left panel*), spring (*middle panel*) and summer (*right panel*) over the period from 1982 to 1999

The impact of NAO in the previous winter, as well as in spring and summer in precipitation and short wave radiation is also provided in Fig. 4 (middle and bottom panels, respectively). As expected the negative pattern of correlations between NAO and NPP in spring is highly related to the corresponding impact of NAO on winter precipitation and short wave radiation. Thus, during the years characterised with a positive NAO mode in winter, lower precipitation and higher short-wave radiation (and fewer clouds) can be observed in the central Iberia, corresponding to lower NPP estimates. These results are in good agreement with the works of Nemani et al. (2003) and Zhao and Running (2010). This behaviour reflects the response of short-rooted grassland and croplands characteristic of the northern mid-latitudes, where water availability is a dominant control for plant growth. Accordingly to the analysis

undertaken by the above mentioned authors, NPP has a significant positive correlation with both total precipitation and Palmer Drought Severity Index (PDSI), which means that higher precipitation values and consequently lowers than usual NAO values, usually lead to higher NPP estimates. This result is compatible with the negative values of correlations between winter NAO and NPP that are prevailing during spring, summer and autumn. The exceptions for this behaviour are the forest areas typical over the Northwestern Iberia and higher altitude mountains in Pyrenees, that are more insensitive to precipitation and temperature and also the Murcia region where the NAO plays a negligible role.

The above mentioned differences in the response of vegetation to moisture and heat conditions prompted the analysis of the evolution (for NAO+ and NAO–) of NPP annual cycles for two sites over Iberia, used to calibrate and validate the CASA Model over Iberia (Seixas et al., 2009). Figure 5 (left panel) shows the annual cycles of monthly NPP for the above selected sites. The two selected sites correspond to two contrasting situations; one site being characterised mainly by negative correlation between NAO and NPP estimates located in the south of Iberia (site 1 in Fig. 5, top panel) while the other is characterised by positive correlations located in the North of Iberia (site 2 in Fig. 5, bottom panel). Moreover, while site 1 presents a maximum of NPP estimates during spring and site 2 corresponds to pixels with a maximum of NPP estimates during summer. Figure 5 (left panel) shows the annual cycles of NPP monthly anomalies for the above selected sites.

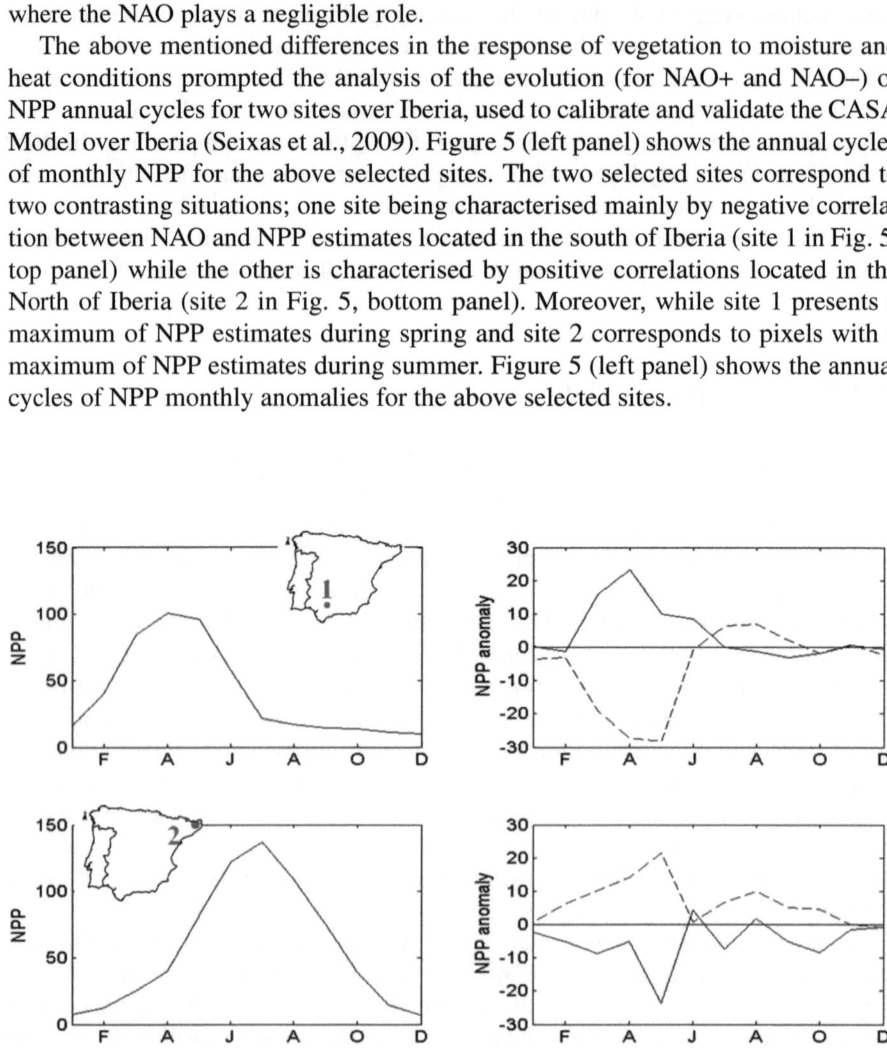

Fig. 5 *Left panel*: Annual cycles of monthly NPP estimates for site 1 located in the south of Iberia (*top panel*) and a site 2 located in the north of Iberia (*bottom panel*). *Right panel*: corresponding annual cycles of monthly anomalies of NPP estimates for the NAO– (*thin solid line*) and NAO+ (*thin dashed line*) subsets over site 1 (*top panel*) and site 2 (*bottom panel*)

The annual cycles of NPP anomalies for the NAO– (NAO+) subsets are identified by the thin solid (dashed) curves. The highest impact of NAO is observed to occur during spring and also in summer in the case of site 2. Differences in topography should be taken into account, as site 1 corresponds to low altitude areas while site 2 is located in a mountainous region. Moreover there are disparities in the main vegetation type characteristic of these sites; while site 1 are mainly grasses, cereal crops, site 2 is dominated by needleleaf forest. During years characterized by positive winter NAO, NPP estimates present low values during spring over site 1. This situation corresponds to years with considerably lower (higher) than usual precipitation (short wave radiation) that may inhibit the vegetation greenness and consequently low NPP estimates. This is the typical situation during drought events (García-Herrera et al., 2007). In the case of mountainous areas (site 2) where the vegetation activity is not dependent of the scarcity of water in part due to the lower temperature, the NPP estimates are relatively high during spring and even during summer although with relatively lower values than in spring. These features are in good agreement with the previously mentioned authors that related the climate variability and the carbon absorption from living plants (Nemani et al., 2003; Zhao and Running, 2010).

2.3 NAO and Wheat Yield

In order to quantify the climate impact on the wheat yield in Portugal it must be stressed that it is considerably smaller than the corresponding wheat yield obtained in the North-western European countries, with cold (but not too wet) winters and relatively wet summers (Gouveia and Trigo, 2008). During the grain filling until the complete grain ripening phase, the Mediterranean conditions may even get worse, due to the short period of time between frost episodes and relatively high temperatures at the end of spring (May/June). Another adverse situation, especially when compared with the same period for North-western European countries, consists in the low photoperiod (number of sun hours) and the high temperatures, which may occur at the end of the maturation phase. During this phase large values of potential evapotranspiration may lead to weak photosynthetic activity, since the plant mostly spends most of energy in the transpiration process in order to offset the warm season effects, thus reducing its capacity to produce dry matter. An excess of evapotranspiration may therefore lead to a decrease of wheat quality (Gouveia and Trigo, 2008).

In order to evaluate the seasonal evolution of the correlation between NAO and wheat yield we have performed a monthly evaluation (Table 1) reflecting the different role played by NAO along the vegetative cycle of wheat. In fact, the contrast between late winter/early spring and late spring/early summer is well apparent, with low correlations between NAO and yield in January to March being replaced by positive correlations for April and June (the last one significant at the 5% level). The role of NAO in June appears to be related to the need of warm temperatures during the wheat maturation phase, and simultaneously the absence of precipitation.

Table 1 Correlation coefficient values between annual wheat yield and monthly NAO (from January to June) for the pixels coded as arable land not irrigated. Bold values highlighted when significant at the 5% level

	Jan	Feb	Mar	Apr	May	Jun
NAO	0.06	0.08	–0.04	0.27	–0.06	**0.53**

The first and simplest assessment on the NAO effective impact on wheat production is shown in Fig. 6 (top panel) that depicts the inter-annual evolution of the NAO during April and June (NAO_{AJ}) and detrended wheat yield, considering the 24 year period between 1982 and 2005, presenting a correlation value of 0.49. With the aim of isolating the effect NAO on annual wheat yield from human-related factors we adopted the methodology described in Section 2.1, comparing the detrended wheat yield for two subsets of years associated to extreme NAO indices: (i) higher than usual values of seasonal NAO index (NAO > percentile 75) and (ii) lower than usual (NAO lower than percentile 25). Figure 6 (lower panel) shows a comparison of annual wheat yield for the three subsets of years associated with NAO+, NAO– and intermediate class. Generally speaking, years presenting positive (negative) anomalies for wheat yield are usually characterized by positive (negative) NAO index values in late spring. Therefore, it is possible to confirm that late spring correspond to an important moment for the wheat vegetative cycle in Portugal and that the NAO index controls, at least partially, what is happening in this moment of the vegetative cycle.

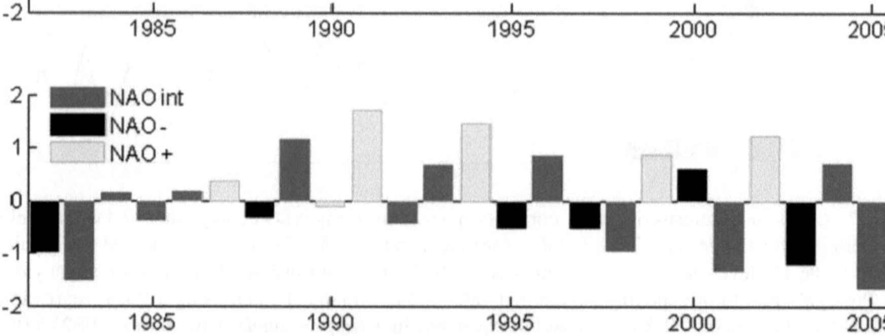

Fig. 6 *Top panel*: Inter-annual variability of spring NAO Index (*dashed line*) and wheat yield over the period 1982–2005. *Bottom panel*: Inter-annual evolution of annual wheat yield for years characterised by NAO_{AJ}+, NAO_{AJ}– and intermediate subsets

In summary, a good wheat yield usually corresponds to years characterized by positive NAO values in spring, particularly in April and June. This behaviour may be viewed as reflecting the integrated impact on radiation, temperature and precipitation fields of the large scale atmospheric circulation patterns associated to the different phases of NAO. The role of NAO in June could be especially related to the necessity of warm temperatures during the maturation phase of wheat.

Figure 7 (left panel) shows the grid point correlations between spring NDVI composites ($NDVI_{MAM}$) and detrended wheat yield, considering the 24 year period between 1982 and 2005. The highest positive and significant correlations (at the 1% level) are found over the southern region of Alentejo (Fig. 7, left panel). The analysis was restricted to the set of pixels that were coded, using Corine Land Cover Map (CLC2000) as non irrigated arable land. The average of spring NDVI for this subset of pixels holds a correlation value of 0.70 (significant at the 1% level). Results obtained above suggest using spring NDVI together with late spring NAO to build up a simple model of wheat yield for Portugal for the period 1982–2005. Spring NDVI is an indicator of the healthiness of wheat during the growing stage which in turn reflects the meteorological conditions in terms of radiation, temperature and

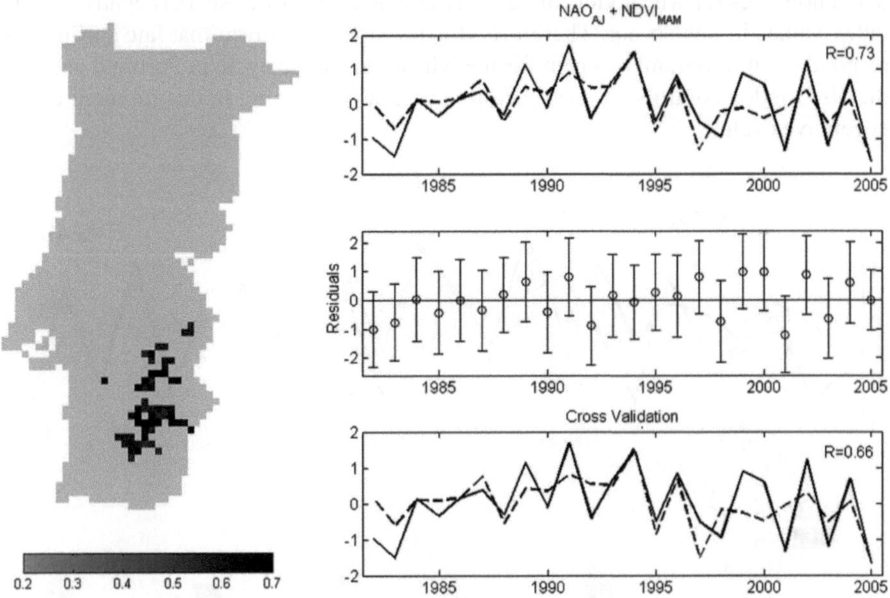

Fig. 7 *Left panel*: Patterns of simple correlation between spring NDVI composites and wheat yield in Portugal, for the period of 1982–2005, for pixels coded as arable land non-irrigated and significant at the 1% level. *Right panel*: Time series (1982–2005) of observed (*dark curve*) wheat yield in Portugal and of corresponding modelled values (*dashed curve*) when using a linear regression model based on spring NDVI and NAO in April and June (*upper panel*). Time series (1982–2005) of residuals and respective 95% level confidence intervals (*central panel*). Time series (1982–2005) of observed (*dark curve*) wheat yield in Portugal and of corresponding modelled values (*dashed curve*) as obtained from the leave-one-out cross-validation procedure

precipitation regimes. On the other hand late spring NAO is an indicator of the large-scale circulation affecting Portugal which in turn is related to regional conditions in terms of radiation, temperature and precipitation that have an important role in the wheat maturation process.

In fact a simple multi-linear regression model was calibrated and validated for the 24-year period of available data, using as predictors the spring $NDVI_{MAM}$ together with NAO_{AJ}. Figure 7 (right-upper panel) presents the time series of observed and modeled wheat yield. The overall agreement is worth being noted, with the two time-series presenting a correlation of 0.73. Figure 7 (right-central panel) shows the time series of residuals, defined as departures of observed values from modeled ones. The 95% confidence intervals relative to these residuals are plotted as error bars and it may be noted that there is not presenting any outlier since its error bar does not cross the zero reference line. Simple calibration approaches, using the entire period of available data, can lead to misleading results, namely over-fitting (Wilks, 2006). Therefore it is advisable to use more robust validation approaches, such as the leave-one-out cross validation (Fig. 7, right-lower panel). The good agreement between the modelled time series by the regression model (upper panel) and the one obtained by the cross validation indicate that the developed model is relatively robust with only a slight decrease (from 0.73 to 0.66) on the correlation between original and modelled time series. In summary, these results are statistically robust but also physically sound, revealing a very satisfactory performance and are expect to be useful to estimate crop production as well as to perform agricultural monitoring.

3 Conclusions

The behaviour of vegetation reflects the different response of surface climate to large-scale atmospheric variability associated to the NAO mode. Over Iberia there is strong evidence that positive (negative) values of winter NAO induce low (high) vegetation activity in the following spring and summer seasons. This feature is mainly associated with the impact of NAO on winter precipitation, together with the strong dependence of spring and summer NDVI on contemporary water availability. It is also related with the strong impact of climate variability in semi-arid areas, namely regarding effects of drought conditions on vegetation activity (Vicente-Serrano and Heredia-Laclaustra, 2004), in particular during the intense spring vegetation growth period.

In Iberia the NAO impact is greater on non-forested vegetation which responds rapidly to spatio-temporal variations in precipitation and soil moisture. During summer, forests and other dense vegetation areas display the highest sensitivity to NAO dynamics, as this type of vegetation shows slower response to precipitation, and therefore the NAO impacts are delayed until late in the year. Consequently, the NAO impact negatively on NPP estimates in southern and central Iberia and positively over the Northern mountainous areas. In fact, in Iberia the NAO impacts negatively the short-rooted grassland and croplands which responds rapidly to spatio-temporal

variations in precipitation and soil moisture. This type of vegetation is very sensitive to drought events.

It should be stressed that our lagged relationships between winter NAO and NDVI values for spring, summer and autumn already represent an added value since they allow formulating, by the end of March, simple outlooks of vegetation greenness for certain land cover types over the Iberian Peninsula that may provide useful information for a wide range of applications, including long-lead wildfire risk assessment and crop forecasts. It is within this context that we have evaluated the impact of NAO on cereal production over a specific region of Iberia with sufficient data. Accordingly, a strong negative correlation (range from 0.6 to 0.8) between NDVI and wheat yield for the 24 year-long period 1982–2005, over the southern part of Portugal (Alentejo) was obtained. A good year for wheat yield is usually characterized by positive NAO values for late spring. To obtain a good wheat yield, the NAO index signal in late spring should be positive, inducing higher values of short wave radiation and lower values of precipitation in the Iberian Peninsula. The low precipitation wheat requirements in this period allow a slow maturation to origin well formed grains and avoid the pests development and potential decrease of the wheat quality as mentioned before.

Acknowledgements This work was supported by the Portuguese Science Foundation (FCT) through project ENAC (Evolution of North Atlantic Climate; the role of Blocking and Storm-tracks in the Past, Present and Future climate of Southern Europe) PTDC/AAC-CLI/103567/2008. The authors would like to thank J. Seixas and N. Carvalhais for the NPP dataset derived in the frame of CARBERIAN project. The wheat yield was provided by Portuguese Statistical Institute (INE). The NDVI dataset was kindly supplied by Global Inventory Modeling and Mapping Studies (GIMMS, http://glcf.umiacs.umd.edu/data/gimms/) project.

References

Atkinson MD, Kettlewell PS, Hollins PD, Stephenson DB, Hardwick NV (2005) Summer climate mediates UK wheat quality response to winter North Atlantic Oscillation. Agri Forest Meteorol 130:27–37

Bogaert J, Zhou L, Tucker CJ, Myneni RB, Ceulemans R (2002) Evidence for a persistent and extensive greening trend in Eurasia inferred from satellite vegetation index data. J Geophys Res 107:4119. doi:10.1029/2001JD00107

Buermann W, Anderson B, Tucker CJ, Dickinson RE, Lucht W, Potter CS, Myneni RB (2003) Interannual covariability in Northern Hemisphere air temperatures and greenness associated with El Niño-southern oscillation and the Arctic Oscillation. J Geophys Res 108:4396. doi:10.1029/2002JD002

Cantelaube P, Terres J-M, Doblas-Reyes FJ (2004) Influence of climate variability on European agriculture – analysis of winter wheat production. Clim Res 27:135–144

Castro-Díez Y, Pozo-Vázquez D, Rodrigo FS, Esteban-Parra MJ (2002) NAO and winter temperature variability in southern Europe. Geophys Res Lett 29:1160. doi:10.1029/2001GL014042

D'Odorico P, Yoo JC, Jaeger S (2002) Changing seasons: an effect of the North Atlantic Oscillation? J Clim 15:435–445

Friedlingstein P, Joel G, Field CB, Fung IY (1999) Toward an allocation scheme for global terrestrial carbon models. Glob Change Biol 5:755–770

García-Herrera R, Paredes D, Trigo RM, Trigo IF, Hernández E, Barriopedro D, Mendes MA (2007) The outstanding 2004/05 drought in the Iberian Peninsula: associated atmospheric circulation. J Hydrometeorol 8:483–498

Gimeno L, Ribera P, Iglesias R, Torre L, Garcia R, Hernández E (2002) Identification of empirical relationships between indices of ENSO and NAO and agricultural yields in Spain. Clim Res 21:165–172

Gouveia C, Trigo RM (2008) Influence of climate variability on wheat production in Portugal. In: Soares A, Pereira MJ, Dimitrakopoulos R (eds) geoENV VI – geostatistics for environmental applications. Springer, Berlin, pp 335–345

Gouveia C, Trigo RM, DaCamara CC (2009) Drought and vegetation stress monitoring in Portugal using satellite data. Nat Hazards Earth Syst Sci 9:185–195

Gouveia C, Trigo RM, DaCamara CC, Libonati R, Pereira JMC (2008) The North Atlantic Oscillation and European vegetation dynamics. Int J Climatol 28:1835–1847

Groisman PY, Karl TR, Knight RW (1994) Observed impact of snow cover on the heat balance and the rise of continental spring temperatures. Science 263:198–200

Iglesias A, Quiroga S (2007) Measuring the risk of climate variability to cereal production at five sites in Spain. Clim Res 34:47–57

Ji L, Peters AJ (2003) Assessing vegetation response to drought in the northern Great Plains using vegetation and drought indices. Rem Sens Environ 87:85–98

Jones PD, Jónsson T, Wheeler D (1997) Extension to the North Atlantic Oscillation using early instrumental pressure observations from Gibraltar and south-west Iceland. Int J Climatol 17:1433–1450

Lindner M, Maroshek M, Netherer S, Kremer A, Barbati A, García-Gonzalo J, Seidl R, Delzon S, Corona P, Kolström M, Lexer M, Marchetti M (2010) Climate change impacts, adaptive capacity and vulnerability of European forest ecosystems. Forest Ecol Manag 259:698–709

Maignan F, Bréon F-M, Bacour C, Demarty J, Poirson A (2008) Interannual vegetation phenology estimates from global AVHRR measurements: comparison with in situ data and applications. Rem Sens Environ 112:496–505

Myneni RB, Keeling CD, Tucker CJ, Asrar G, Nemani RR (1997) Increase plant growth in the northern high latitudes from 1981–1991. Nature 386:698–702

Nemani RR, Keeling CD, Hashimoto H, Jolly WM, Piper SC, Tucker CJ, Myneni RB, Running SW (2003) Climate-driven increases in global terrestrial net primary production from 1982 to 1999. Science 300:1560–1563

Piao S, Ciais P, Friedlingstein P, Peylin P, Reichstein M, Luyssaert S, Margolis H, Fang J, Barr A, Chen A, Grelle A, Hollinger DY, Laurila T, Lindroth A, Richardson AD, Vesala T (2008) Net carbon dioxide losses of northern ecosystems in response to autumn warming. Nature 451: 49–52

Potter CS, Randerson JT, Field CB, Matson PA, Vitousek PM, Mooney HA, Klooster SA (1993) Terrestrial ecosystem production – a process model-based on global satellite and surface data. Global Biogeochem Cycles 7:811–841

Rodríguez-Puebla C, Ayuso SM, Frias MD, Garcia-Casado LA (2007) Effects of climate variation on winter cereal production in Spain. Clim Res 34:223–232

Seixas J, Carvalhais N, Nunes C, Benali A (2009) Comparative analysis of MODIS-FAPAR and MERIS-MGVI datasets: potential impacts on ecosystem modeling. Rem Sens Environ 113:2547–2559

Shabanov NV, Zhou L, Knyazikhin Y, Myneni RB (2002) Analysis of interannual changes in northern vegetation activity observed in AVHRR data during 1981 to 1994. IEEE Trans Geosci Rem Sens 40:115–130

Sitch S, Huntingford C, Gedney N, Levy PE, Lomas M, Piao SL, Betts R, Ciais P, Cox P, Friedlingstein P, Jones CD, Prentice IC, Woodward FI (2008) Evaluation of the terrestrial carbon cycle, future plant geography and climate-carbon cycle feedbacks using five Dynamic Global Vegetation Models (DGVMs). Glob Change Biol 14:2015–2039

Stöckli R, Vidale PL (2004) European plant phenology and climate as seen in a 20-year AVHRR land-surface parameter dataset. Int J Rem Sens 25:3303–3330

Trigo RM, Osborn TJ, Corte-Real JM (2002) The North Atlantic Oscillation influence on Europe: climate impacts and associated physical mechanisms. Clim Res 20:9–17

Trigo RM, Pozo-Vazquez D, Osborn TJ, Castro-Diez Y, Gámis-Fortis S, Esteban-Parra MJ (2004) North Atlantic Oscillation influence on precipitation, river flow and water resources in the Iberian Peninsula. Int J Climatol 24:925–944

Vicente-Serrano SM (2007) Evaluating the Impact of drought using remote sensing in a Mediterranean, semi-arid region. Nat Hazards 40:17–208

Vicente-Serrano SM, Cuadrat-Prats JM, Romo A (2006) Early prediction of crop production using drought indices at different time-scales and remote sensing data: application in the Ebro valley (North–East Spain). Int J Rem Sens 27:511–518

Vicente-Serrano SM, Heredia-Laclaustra A (2004) NAO influence on NDVI trends in the Iberian Peninsula (1982–2000). Int J Rem Sens 25:2871–2879

Wilks DS (2006) Statistical methods in the atmospheric sciences, International Geophysics Series, 2nd edn, vol 59. Academic, New York, NY, 627 pp

Zhao M, Running SW (2010) Drought-induced reduction in global terrestrial net primary production from 2000 through 2009. Science 20:940–943

Zhou L, Tucker CJ, Kaufmann RK, Slayback D, Shabanov NV, Myneni RB (2001) Variations in northern vegetation activity inferred from satellite data of vegetation index during 1981 to 1999. J Geophys Res 106:20069–20083

Direct and Indirect Effects of the North Atlantic Oscillation on Tree Growth and Forest Decline in Northeastern Spain

Jesús Julio Camarero

Abstract NAO effects on tree growth have not been assessed in the Northeastern Iberian Peninsula where: (i) Atlantic (NAO) and Mediterranean influences affect tree growth and (ii) contrasting climatic conditions determine very different forest types. It is hypothesized that positive (negative) values of winter NAO are linked to drier winter conditions and low (high) forest growth on the following spring. To understand the effects of climate and NAO on tree growth, first I quantified the growth responses to monthly and seasonal climatic variables (mean temperature, total precipitation and cloud cover) and NAO indices across a climatic gradient in Northeastern Spain. Dendrochronological methods are used and an extensive tree-ring network (48 sites, 687 trees) based on ten tree species with contrasting habitats and plausibly different growth responses to climate. Then, I focused on the longest tree-ring chronologies to estimate how unstable are the relationships between tree growth and NAO indices. Finally, the resultant findings are applied to understand how NAO has affected climatic variability during the last decades and how this may be related to the recent silver fir (*Abies alba*) decline in the Pyrenees. Climatic variables and NAO indices explained on average 40.1 and 15.9% of the growth variance, respectively. The maximum growth variance (52.1%) explained by NAO indices was found in a *Juniperus thurifera* forest under continental Mediterranean conditions. High NAO values during the previous December were linked to a reduction in precipitation and a low radial growth the following spring, but high February NAO values were also associated with warmer early-spring conditions and enhanced growth. The growth responses to climate and NAO also changed through time. In the case of silver fir, a shift in the positive relationship between growth and NAO indices of the previous winter (November, February) coincided with the triggering of decline in 1986 suggesting an alteration in the growth response to NAO and related climatic variables.

Keywords Atmospheric circulation patterns · Dendrochronology · Forest dieback · *Juniperus thurifera* · Tree ring-width

J.J. Camarero (✉)
ARAID-Instituto Pirenaico de Ecología, Spanish National Research Council (CSIC), Zaragoza, Spain
e-mail: jjcamarero@ipe.csic.es

S.M. Vicente-Serrano, R.M. Trigo (eds.), *Hydrological, Socioeconomic and Ecological Impacts of the North Atlantic Oscillation in the Mediterranean Region*, Advances in Global Change Research 46, DOI 10.1007/978-94-007-1372-7_10,

1 Introduction

Climatic projections foresee strong warming trends and a decrease in rainfall for the Mediterranean region during the twenty-first century (IPCC, 2007). Such tendencies toward drier conditions have been linked to an increasing anticyclonic activity associated with a northward shift of the north Atlantic storm track (Giorgi and Lionello, 2008). Thus, the predicted changes in atmospheric circulation and climatic conditions should affect forest growth in the Mediterranean basin directly through changes in large scale temperature but also indirectly through modifications in atmospheric patterns and proximate local climate factors such as wind patterns and precipitation (Hirschboeck et al., 1996; Girardin and Tardif, 2005).

Large-scale climatic indices may outperform local climatic conditions in explaining variation in ecological processes (Hallett et al., 2004). Such distinct results may be explained by the more spatially and temporally variable associations between local climate data (e.g., monthly temperature means or precipitation sums) and ecological process as compared with large-scale indices which can partially capture these processes giving insights on the underlying mechanisms (Garfin, 1998). However, few studies have assessed the influence of large-scale indices on tree growth in areas subjected to diverse climatic conditions, as is the case of the Iberian Peninsula, where climates range from mild to continental and from humid to semi-arid resulting in diverse forest types subjected to different constraints for tree growth (Nahal, 1981).

In the Iberian Peninsula the climatic variability is under the influence of several large-scale atmospheric circulation patterns including the North Atlantic Oscillation (NAO), which is defined by a north-south dipole of sea level pressures and geopotential heights in the North Atlantic region (Hurrell et al., 2003). The NAO variability is related to changes in the position of the Iceland low-pressure and the Azores high-pressure systems, and to modifications in the direction and strength of westerly winds in southern Europe (Hurrell, 1995). The NAO has a strong effect on winter climate over the Iberian Peninsula Iberia where high (low) NAO values in winter are linked to low (high) precipitation (Hurrell and Van Loon, 1997), being these associations particularly strong in western Iberia (Rodó et al., 1997; Rodríguez-Puebla et al., 2001). The NAO does not exert a relevant influence on winter temperature variability in the western Mediterranean Basin (Sáenz et al., 2001).

Spatially heterogeneous and lagged effects of the previous winter NAO on tree growth of the following year have been described in mesic forests from western Spain (Roig et al., 2009; Rozas et al., 2009) and in drier locations from eastern Spain (Bogino and Bravo, 2008). However, few data exist in transitional areas closer to the Mediterranean Sea and subjected to xeric conditions, where the NAO influence on growth might differ as compared with western mesic areas. In this work, it is evaluated if high NAO values (increased westerlies, storms track towards Southern Europe) during the previous winter are linked to a reduction in tree radial growth through lower precipitation and water availability at the beginning of the growing season the following spring. The objectives of this study are: (i) to quantify the growth responses to monthly and seasonal climatic variables (mean temperature,

total precipitation and cloud cover) and NAO indices across a climatic gradient in Northeastern Spain using dendrochronological methods and an extensive tree-ring network (48 sites); (ii) to estimate how unstable are the relationships between tree growth and NAO indices, and (iii) to understand how NAO has affected climatic variability during the last decades and how this may be related to recent silver fir (*Abies alba*) decline episodes. Such decline events were observed in the Spanish Central Pyrenees since the 1980s and showed a marked geographical pattern characterized by a greater abundance of declining forests in more xeric sites from the western Pyrenees than elsewhere (Camarero et al., 2002).

2 Materials and Methods

2.1 Study Sites

In total 687 trees of ten tree species (*Abies alba* Mill., *Fagus sylvatica* L., *Juniperus oxycedrus* L., *Juniperus thurifera* L., *Pinus halepensis* Mill., *Pinus sylvestris* L., *Pinus uncinata* Ram., *Pistacia terebinthus* L., *Quercus faginea* Lam., *Quercus robur* L.) were sampled in 48 forests located in an area of approximately 150,000 km^2 within Northeastern Spain (Table 1). The studied species belong to diverse families (Anacardiaceae, Cupressaceae, Fagaceae, Pinaceae) and the studied forests occupy sites encompassing wide gradients of elevation (340–1800 m) and climatic types (mean annual temperature: 8.3–13.8°C; total annual precipitation: 317–2350 mm).

In Northeastern Spain, the rainfall decreases from north to south and from the coast towards inland sites resulting in diverse climatic conditions (González-Hidalgo et al., 2009). In the study area, Atlantic (e.g., higher winter and spring precipitation derived from the Atlantic storm track) and Mediterranean (e.g., higher autumn precipitation caused by cyclonic influence from the close Mediterranean Sea) influences increase westwards and eastwards, respectively. Such contrasting climatic conditions determine different forest types including Mediterranean coniferous forests growing under semi-arid continental climates (*J. thurifera*, *P. halepensis*), mesic Mediterranean shrub formations (*P. terebinthus*), sub-Mediterranean transitional oak forests (*Q. faginea*), humid temperate forests (*A. alba*, *Q. robur*), and cold mountain conifer forests (*P. sylvestris*, *P. uncinata*). Some of the studied sites corresponded to relict and isolated tree populations as is the case of *Q. robur* in the Moncayo site or *J. thurifera* in the Retuerta de Pina site. Other forests represent ecological limits near the "xeric" limit of the species distribution such as the *F. sylvatica* and *P. sylvestris* forests in the Monrepos site. Finally, the *P. uncinata* population in the Moncayo site was a plantation outside the main Iberian range of the species in the Pyrenees. Most sites are located on marls and limestones, which generate basic soils.

Table 1 Characteristics of the study sites

Species	Site (code)	Latitude (N)	Longitude	Elevation (m)	P (mm)[a]
Quercus robur	Bértiz (QrBE)	43° 09′	1° 36′ W	405	1677
Abies alba	Irati (AaIR)	42° 59′	1° 03′ W	1020	780
Quercus faginea	Luesia (QfLU)	42° 22′	1° 01′ W	950	684
Pinus sylvestris	Luesia (PsLU)	42° 22′	1° 01′ W	920	684
Quercus robur	Moncayo (QrMO)	41° 48′	1° 51′ W	1400	1020
Pinus uncinata	Moncayo (PuMO)	41° 47′	1° 49′ W	1850	1350
Fagus sylvatica	Monrepos (FsMR)	42° 19′	0° 25′ W	1290	1200
Pinus sylvestris	Monrepos (PsMR)	42° 19′	0° 25′ W	1300	1200
Abies alba	Declining sites – Pyrenees (AadPY)[b]	42° 44′	0° 45′ W	1196	1150
Abies alba	Non declining sites – Pyrenees (AanPY)	42° 35′	0° 23′ W	1390	1800
Juniperus oxycedrus	Agüero (JoAG)	42° 18′	0° 47′ W	750	635
Pistacia terebinthus	Mediano (PtME)	42° 19′	0° 14′ E	488	732
Pinus halepensis	Peñaflor (PhPE)	41° 47′	0° 43′ W	345	340
Juniperus thurifera	Peñaflor (JtPE)	41° 47′	0° 43′ W	340	340
Juniperus thurifera	Retuerta de Pina (JtRE)	41° 27′	0° 16′ W	350	395
Juniperus thurifera	Olmedilla (JtOL)	40° 19′	0° 44′ W	1400	700
Juniperus thurifera	Ademuz (JtAD)	40° 04′	1° 15′ W	1518	750

[a]Annual precipitation based on local climatic data and corrected taking into account the site elevation.
[b]Mean values of the referred characteristics were obtained for declining and non-declining sites.

2.2 Field Sampling and Dendrochronological Methods

In each site 8–88 dominant trees were randomly selected and their size (Dbh, diameter at 1.3 m) was measured (Table 2). At least two radial cores per tree were sampled at 1.3 m using a Pressler increment borer. Cores were prepared following standard dendrochronological methods (Fritts, 2001). They were mounted and sanded until tree-rings were clearly visible under a binocular microscope. All samples were visually cross-dated and tree-ring widths were measured to a precision of 0.001 mm using a LINTAB measuring device (Rinntech, Heidelberg, Germany). Cross-dating was evaluated using the program COFECHA (Holmes, 1983).

Each individual series was standardized using a spline function with a 50% frequency response of 32 years to keep high-frequency variability. Standardization involved transforming the measured value into a dimensionless index by dividing it by the expected values given by the spline function. Autoregressive modelling was carried out on each series to remove temporal autocorrelation (Table 2). Then the indexed residual series were averaged using a biweight robust mean to obtain site chronologies for each species using the program ARSTAN (Cook, 1985). In the case of *A. alba* the chronologies were averaged in two groups of sites with common growth trends and similar decline symptoms corresponding to declining (8 sites) and non-declining (25 sites) stands. Declining sites were regarded as those with more than 25% trees with crown defoliation greater than 50% (Camarero et al., 2002). In the case of the raw width series the mean and standard deviation (SD) and the first-order autocorrelation (AR1), which measures the year-to-year persistence in growth, were calculated. The statistical quality of the residual chronologies was evaluated using several dendrochronological statistics (Briffa and Jones, 1990): the mean correlation (r) of individual series with the mean site chronology; the mean sensitivity (MSx) which quantifies the relative change in width among consecutive years; and the Expressed Population Signal (EPS) of residual series which indicates to what extent the sample size is representative of a theoretical infinite population. The common interval of each site and species was selected for EPS values above the 0.85 threshold which is widely used in dendrochronological studies (Wigley et al., 1984).

2.3 NAO and Climatic Data

Monthly and season values of the NAO index were calculated based on the differences between time series of sea-level pressures recorded at Gibraltar and Reykjavik (Jones et al., 1997). NAO data are available in the Climate Research Unit (CRU) homepage (http://www.cru.uea.ac.uk/cru/data/nao.htm). The monthly climatic data used (mean temperature, total precipitation, cloud cover) were also produced by CRU, they are freely available at http://badc.nerc.ac.uk/data/cru/ and correspond to the TS 2.1 and 3.0 gridded (0.5° resolution) datasets (Mitchell and Jones, 2005). The missing data of cloud cover were replaced by the mean values of each monthly variable. In addition to the local climatic datasets compiled at 0.5° resolution, a

Table 2 Dendrochronological statistics of the growth series

Site	N° trees (cores)	Dbh (cm)	RW (mm)	*r*	MSx	AR1	Common interval[a]	Climate R^2_{adj} (%)	NAO R^2_{adj} (%)
QrBE	15 (30)	45.0	2.11 ± 0.76	0.54	0.23	0.74	1924–2008	16.3	3.0
AaIR	14 (29)	32.0	1.83 ± 0.89	0.54	0.17	0.87	1944–2003	31.8	5.1
QfLU	88 (176)	13.4	1.43 ± 0.60	0.61	0.27	0.68	1955–2002	14.1	6.1
PsLU	8 (15)	36.0	2.21 ± 0.83	0.64	0.33	0.77	1946–2004	27.6	10.1
QrMO	12 (24)	23.4	1.45 ± 0.55	0.43	0.26	0.75	1931–2000	71.5	14.4
PuMO	15 (30)	20.5	2.41 ± 0.95	0.58	0.15	0.70	1960–2008	18.2	10.8
FsMR	11 (22)	26.3	1.34 ± 0.09	0.61	0.28	0.71	1920–2009	66.5	3.5
PsMR	11 (23)	40.6	1.50 ± 0.11	0.65	0.29	0.79	1915–2009	41.5	13.0
AadPY	66 (140)	43.3	1.77 ± 0.43	0.58	0.24	0.79	1886–2000	66.0	12.5
AanPY	322 (660)	55.3	2.36 ± 0.67	0.49	0.20	0.82	1889–2000	53.0	10.9
JoAG	8 (21)	12.0	1.07 ± 0.53	0.50	0.27	0.59	1958–2009	38.7	14.1
PtME	24 (41)	8.8	1.06 ± 0.51	0.53	0.40	0.49	1959–2006	70.9	9.9
PhPE	15 (29)	17.2	1.12 ± 0.73	0.73	0.37	0.66	1900–2006	45.0	33.8
JtPE	10 (20)	16.0	1.47 ± 0.78	0.60	0.34	0.74	1956–2006	37.8	25.5
JtRE	15 (25)	17.6	1.00 ± 0.59	0.72	0.35	0.58	1955–2004	51.8	52.1
JtOL	13 (25)	26.8	0.97 ± 0.47	0.63	0.32	0.65	1949–2005	20.0	9.7
JtAD	40 (80)	15.0	1.10 ± 0.37	0.56	0.30	0.42	1960–2004	11.2	36.0

Abbreviations: RW, mean (± SD) tree-ring width for the common interval; r, mean correlation with the mean master site chronology; MSx, mean sensitivity of the residual chronologies; AR1, first-order autocorrelation; Climate R^2_{adj}, adjusted percentage of the variance of growth explained by monthly climatic variables (mean temperature, total precipitation); NAO R^2_{adj}, adjusted percentage of the variance of growth explained by monthly NAO indices.

[a]The common interval is the period regarded as statistically well replicated with EPS > 0.85, where the EPS is the Expressed Population Signal (a measure of the tree-to-tree common growth variance).

regional climatic dataset was obtained considering for the area contained within coordinates 40° 00′–43° 30′ N and 1° 30′ W–0° 30′ E.

2.4 Statistical Analyses

The spatio-temporal of ring-width residual chronologies was described using a Principal Component Analysis (PCA) calculated from their correlation matrix for the common period 1960–1999. The broken stick test was performed to determine the significance of the first three principal components (Holmes, 1992). To quantify the influence of the monthly and seasonal climatic data (mean temperature, total precipitation) and NAO indices on growth, bootstrapped correlation analyses were carried out for the common period 1960–1999 using the program Dendroclim2002 (Biondi and Waikul, 2004). Spatial patterns of climate-growth relationships were explored using the KNMI Climate Explorer (http://climexp.knmi.nl/). The significances of the correlations were based on 999 bootstrapped estimates. The long-term serial correlation in the NAO monthly or seasonal series (winter, December–February; spring, March–May; summer, June–August; autumn, September–November) were removed assuming linear trends. Furthermore, correlation analyses considering all NAO data ($n = 40$) or only positive ($n = 16-27$) or negative ($n = 13-24$) NAO indices were performed to emphasize the effects of different NAO phases on growth. The relationships between climate or NAO data (monthly and seasonal values) and growth (residual indices) were analyzed from the previous September up to the current September. In the case of seasonal data, I focused on the relationships between previous-winter and current-spring NAO indices and growth data based on previous results in nearby areas (Rozas et al., 2009). To estimate the percentage of variance (R^2) and its adjusted value (R^2_{adj}) in growth data explained by climatic (temperature, precipitation) and NAO data, multiple regressions were performed based on the forward selection of significant ($P < 0.05$) monthly predictors. Cloud cover was not used in the climatic model because most dendroclimatic studies are based on temperature and precipitation data (Fritts, 2001). To detect if NAO-growth associations were unstable through time, moving correlations for 30-year periods were performed between NAO monthly and seasonal indices and growth data of the longest chronologies which covered most of the past twentieth century (PsMR, PhPE and the silver fir series AadPY, AanPY).

3 Results

3.1 Growth and Dendrochronological Data

Tree size changed considerably across the study area with mean diameter (Dbh) ranging between 8.8 (*P. terebinthus*) and 55.3 cm (*A. alba*) (Table 1). The mean ring widths varied also notably among sites and species with minimum values of 1.0 mm

(open *J. thurifera* stands under semi-arid and continental conditions) and maximum values of 3.4 mm (young *P. uncinata* mountain forests) (Table 2). High growth values were also observed in mesic temperate forests (*A. alba*, *P. sylvestris*, *Q. robur*), whereas tree rings were usually narrow in sites with a pronounced drought period under Mediterranean conditions dominated by *P. terebinthus*, *Juniperus* species and *P. halepensis*. The maximum correlations between tree series and the mean site chronology were observed in semi-arid sites dominated by *J. thurifera* and *P. halepensis* followed by transitional *P. sylvestris* forests, whereas low correlations were observed in more mesic sites both in temperate (*Q. robur*, *A. alba*) and in Mediterranean conditions (*J. oxycedrus*, *P. terebinthus*). Mesic temperate forest showed the lowest mean sensitivity values, while *J. thurifera* and *P. halepensis* semi-arid sites showed the highest mean relative change in width among consecutive years. Conversely, these last forests and Mediterranean sites (*P. terebinthus*) showed the lowest first-order autocorrelation values, whereas *A. alba* and *P. sylvestris* trees showed the highest year-to-year persistence in growth. Finally, the maximum interspecific correlations between growth series were observed for the Mediterranean sites PtME and PhPE ($r = 0.56, P < 0.001$), separated by 112 km, and for the temperate or transitional sites AadPY and PsMR ($r = 0.53, P < 0.001$), which were located 59 km apart. However, negative associations were detected between Mediterranean continental *J. thurifera* sites and mesic *Q. robur* forests separated by about 200 km (QrMO-JtAD, $r = -0.17, P = 0.40$; QrBE-JtRE, $r = -0.16, P = 0.30$).

Forests with trees of greater diameter showed a higher mean ring-width ($r = 0.68, P = 0.003$) and first-order autocorrelation of tree-ring series ($r = 0.71$, $P = 0.001$) but a lower mean sensitivity ($r = -0.48, P = 0.04$) (Table 2). The percentage of growth variance explained by climatic variables was not significantly related to tree size or to any dendrochronological statistic. However, the percentage of growth variance explained by NAO indices was negatively related to tree-ring width ($r = -0.47, P = 0.05$) and first-order autocorrelation ($r = -0.50, P = 0.04$) but positively associated with the mean sensitivity ($r = 0.48, P = 0.05$).

The first three principal components of the PCA explained, 40.5, 10.9 and 7.1% of the total variance of growth data for all sites (48 series), respectively, being significant ($P < 0.05$) the first two components (*results not presented*). Considering the mean growth series for the two groups of *A. alba* forests based on their decline symptoms and the rest of sites (17 series), the first (PC1) and second (PC2) components were again significant and explained, respectively, 34 and 13% of the total variance (Fig. 1). The PC1 arranged sites according to their main climatic conditions related to precipitation differences since forests located in the more xeric sites (Mediterranean continental climate) showed high scores along this component whereas forests in more mesic habitats (temperate and mountain forests) had the lowest scores. Mountain conifer forests and *Quercus* forests, both mesic and transitional sites, showed high scores in the PC2.

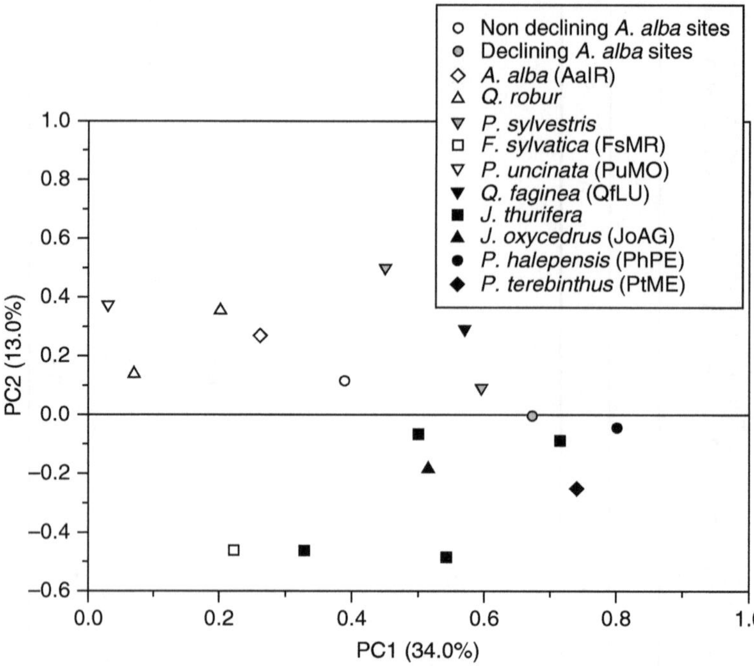

Fig. 1 Scatter diagram of growth chronologies based on the scores of the first two principal components of a PCA considering the mean series of non-declining and declining *A. alba* sites and the rest of study sites (17 series). Codes are as in Table 1

3.2 Relationships Between Growth and Climatic Variables

Tree growth in most study sites was enhanced by warmer April temperatures, wetter and cloudier conditions in June during the year of tree-ring growth (Table 3). In addition, higher February temperatures and more rainfall and cloud cover in May and wet conditions in August of the current year also favored wood formation. However, a previous warm September with low values of cloud cover was usually related to decreased growth. In mesic or transitional sub-Mediterranean sites, a wet January was associated to lower growth the following spring, but in Mediterranean sites tree growth was enhanced by wet conditions in January and the previous December.

The spatial patterns of climate-growth relationships changed according to the climatic variable. For instance, the positive effect of winter precipitation (previous December) on tree growth was stronger in the southern than in the northern part of the study area in Northeastern Spain (Fig. 2). However, the positive effects of late spring and early summer precipitation (current June) on growth did not show such divergence and it was observed near the core of the study area.

Table 3 Significant ($P < 0.05$) correlations (Pearson coefficients) obtained between monthly climatic data (T, mean temperature; P, total precipitation; CC, cloud cover) and growth for the study sites including the first and second principal components (PC1, PC2). Months abbreviated by lower- and uppercase letters correspond to the years before (*t*–1) and during (*t*) tree-ring formation. Codes are as in Table 1

Year *t*–1					Year *t*								
T	s	o	n	d	J	F	M	A	M	J	J	A	S
QrBE								0.32					
AaIR	0.29		0.30					0.42					
QfLU								0.29					
PsLU	–0.39					0.35		0.34					0.32
QrMO			0.29				–0.41						
PuMO								0.29					
FsMR							–0.28						
PsMR	–0.42							0.38	0.29				
AadPY	–0.49					0.34		0.38					
AanPY	–0.50		0.40			0.39		0.40					
JoAG								0.29					
PtME		0.36											
PhPE	–0.33			0.32	0.31	0.29		0.29		–0.29			
JtPE													
JtRE					0.31	0.38		0.42			–0.45		
JtOL				0.29		0.33							
JtAD							0.29	0.30					
PC1	–0.36					0.45		0.46					
PC2	–0.28									0.32			0.29

Table 3 (continued)

Year t–1					Year t								
P	s	o	n	d	J	F	M	A	M	J	J	A	S
QrBE					–0.30								
AalR	–0.29											0.31	
QfLU									0.40	0.33			
PsLU									0.46				
QrMO						0.38					0.39		
PuMO												0.33	
FsMR			–0.29		–0.29			0.42		0.36			
PsMR									0.32	0.37		0.29	
AadPY	0.30				–0.33					0.56			
AanPY	0.31				–0.32					0.39			
JoAG			0.38							0.44		0.33	
PtME					0.33			0.45	0.33	0.59			
PhPE				0.29	0.56					0.48			
JtPE					0.40				0.33				
JtRE				0.34	0.40				0.35	0.59			
JtOL				0.39	0.30				0.33			0.30	
JtAD												0.38	
PC1				0.38	0.42				0.47	0.61	0.47		
PC2													

Table 3 (continued)

Year *t*–1					Year *t*								
CC	s	o	n	d	J	F	M	A	M	J	J	A	S
QrBE												0.29	
AaIR	0.31												–0.35
QfLU									0.41	0.31			
PsLU									0.29				
QrMO	0.31		–0.34				0.29						
PuMO								–0.33		–0.30			
FsMR								0.29					
PsMR	0.42												
AadPY	0.42	0.39								0.30			–0.30
AanPY	0.48	0.33			–0.34								
JoAG			0.31							0.44			
PtME								0.30	0.38	0.49			
PhPE				0.51	0.52	0.30			0.34	0.39			
JtPE					0.33				0.46				
JtRE				0.36						0.30	0.30		
JtOL										0.33			
JtAD						–0.32				0.37			
PC1				0.34					0.37	0.46			
PC2						0.29							

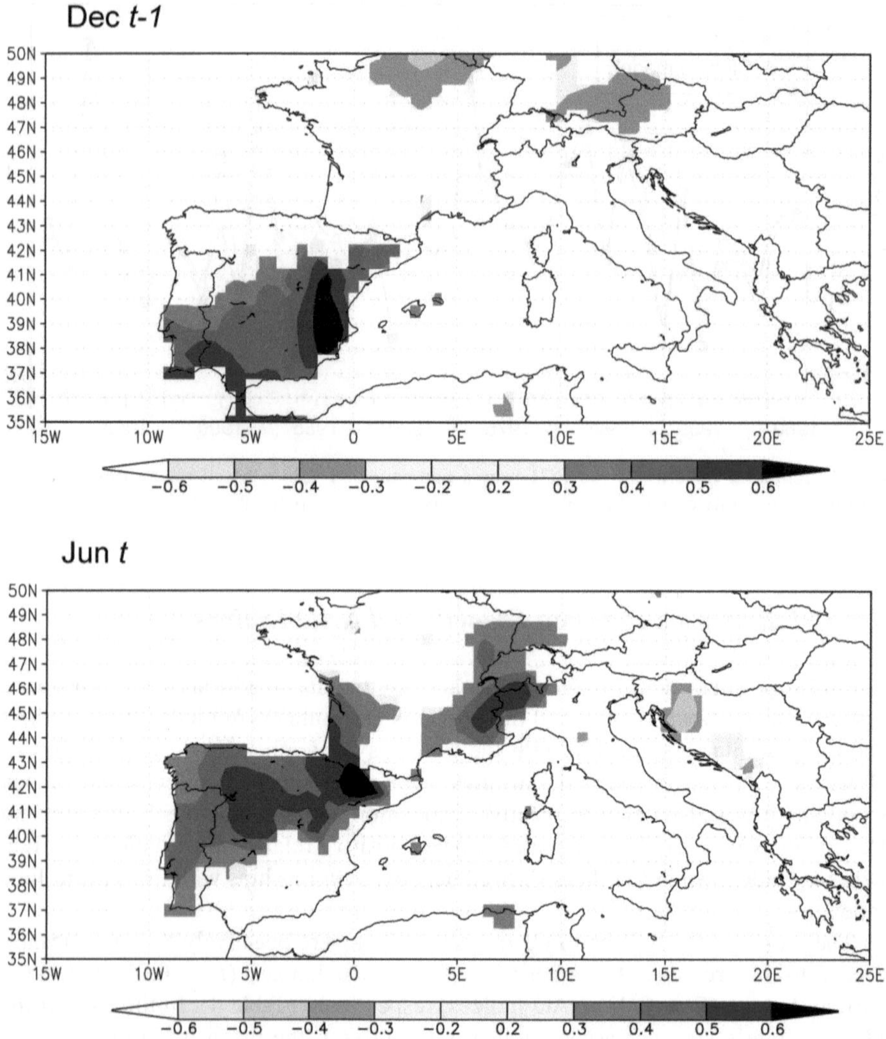

Fig. 2 Spatial patterns of some of the main growth-climate relationships in Northeastern Spain. The maps show correlations between scores of the first principal component (PC1) of growth chronologies and precipitation data (0.5° resolution) for the previous (Dec *t*–1) and the current June (Jun *t*)

The best climatic model for the PC1 scores, capturing the main growth trends of the study sites during the period 1960–1999, explained 78.3% of their variance ($R^2_{adj} = 0.75, F = 24.6, P < 0.001$) and included the following variables: April and late spring-early summer (May, June, July) precipitation, previous October and current April temperature (Fig. 3).

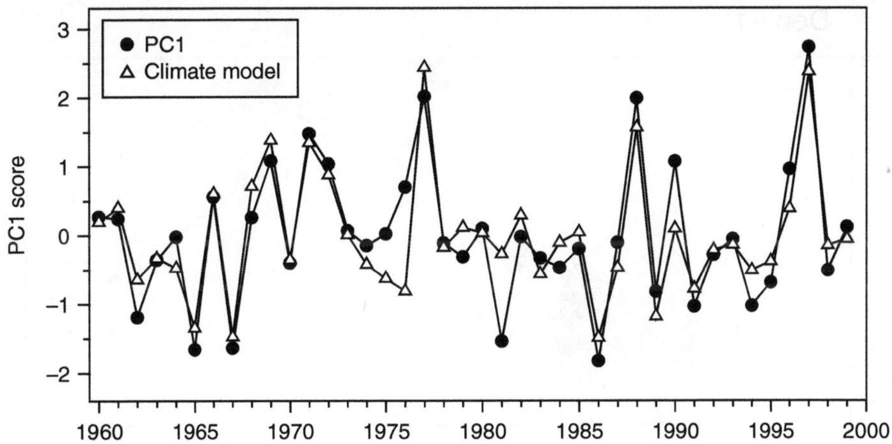

Fig. 3 Temporal patterns of the PC1 scores and modeled values based on a climatic model considering monthly climate variables (mean temperature, total precipitation)

3.3 Relationships Between Growth and NAO Indices

Several xeric sites under Mediterranean and continental conditions showed negative associations between growth and the previous December NAO index, while others showed a similar relationship with the previous September NAO index (Table 4). However, in mesic or transitional sites such as mountain conifer forest or oak woodland it was also detected positive association with the previous November and December NAO indices. Some sites showed positive relationships with the current February index, whereas others showed negative relationships with the current June index.

The scores of the PC1 and PC2 of growth data were negatively and significantly related to the previous December ($P = 0.003$) and January ($P = 0.05$) and to the current August ($P = 0.02$) NAO indices, respectively (Table 4). In the case of the PC1, a slightly stronger association was found with the mean NAO index of the previous December and the current January ($r = -0.47, P = 0.002$). On the other hand, the PC2 was positively but not significantly related to the mean NAO index of the previous November and December ($r = 0.20, P = 0.07$).

The analyses based on positive NAO indices mostly reflected local effects since no significant association was found with the scores of the PC1 (Table 5). Most significant correlations corresponded to previous December and current February NAO indices. Considering only negative NAO indices, several sites and the PC1 showed positive associations with the values of the current February index and negative relationships with the values of the previous December index, whereas the PC2 showed negative associations with the NAO indices of the previous September and the current August and a positive association with the NAO index of the current March (Table 6).

Table 4 Significant ($P < 0.05$) correlations obtained between monthly and seasonal NAO indices and growth for the study sites including the first and second principal components (PC1, PC2). Values in bold are highly significant ($P = 0.001$) and months or seasons without significant values are not presented. Codes are as in Tables 1 and 3

Year *t*–1					Year *t*						
Site	Summer	sep	nov	dec	Spring	Jan	Feb	Mar	Apr	Jun	Aug
QrBE			0.22	0.21							
AaIR										–0.30	
PsLU			0.23								
QrMO								–0.41			
PuMO	0.33		**0.45**								
FsMR							–0.29		–0.25	–0.21	
PsMR		–0.30	0.23	–0.28							
AadPY			0.20								
AanPY			0.27				0.28				
JoAG				–0.28				0.32		–0.31	
PtME				–0.32		–0.31					
PhPE				**–0.49**							
JtPE				–0.29	–0.36				–0.33		
JtRE				**–0.49**							
JtOL							0.35			–0.30	0.34
JtAD							**0.54**			–0.35	0.40
PC1				–0.46		–0.31				–0.28	
PC2		–0.28									–0.36

Table 5 Significant ($P < 0.05$) correlations obtained between monthly and seasonal positive NAO indices and growth for the study sites. Values in bold are highly significant ($P = 0.001$). Codes are as in Tables 1 and 4

Year *t*–1				Year *t*				
Site	sep	dec	Winter	Spring	Fall	Feb	Mar	Apr
QrBE					0.28			
AaIR								0.35
QfLU							–0.32	
PuMO			0.29					
FsMR			–0.22			–0.39		
PsMR	–0.34							
AadPY		**0.45**		–0.22				
JoAG						0.50		
PhPE		0.26	0.20					
JtRE		**–0.49**						
JtOL	–0.38		0.28			0.43		
JtAD			0.35			**0.53**		

Table 6 Significant ($P < 0.05$) correlations obtained between monthly and seasonal negative NAO indices and growth for the study sites. Values in bold are highly significant ($P = 0.001$). Codes are as in Tables 1 and 4

	Year *t*–1					Year *t*						
Site	Summer	Fall	sep	oct	dec	Spring	Summer	Feb	Mar	May	Jun	Aug
QrBE		–0.26		–0.36								
AaIR								0.69				
QfLU					–0.42							
QrMO								0.51				
PsMR			–0.28		–0.32							
AadPY								0.30		–0.27		
AanPY						–0.25				**–0.37**		
JoAG							–0.42					
PhPE					–0.33			0.33				
JtPE						–0.39	–0.31	0.50		–0.33		
JtRE	0.38				**–0.52**			0.44				
JtOL					–0.44			0.39				
JtAD					–0.43			0.53				
PC1				0.45	**–0.58**		–0.35	0.48			–0.39	
PC2			–0.45						0.62			–0.56

The negative associations between the PC1 and PC2 scores and the previous December and the current August NAO indices were more noticeable in the last 2 decades of the past century than before (Fig. 4). In fact, the moving correlations indicated that the negative associations between growth and the previous December NAO index in PhPE and PsMR were unstable and became significant since the 1960s (Fig. 5).

In contrast with the climatic model of the PC1 scores which explained 78% of growth variance (Fig. 3), the best model of the PC1 based on the NAO index explained only 27.9% of their variance ($F = 8.5, P = 0.001$) and contained as predictors the averaged index of the previous December and the current January and the current May index. Actually, only in one study site (JtRE) the NAO indices were able to account for a similar variance of growth as the monthly climatic variables did (Fig. 6). In this *J. thurifera* forest subjected to semiarid and continental Mediterranean conditions, the climatic model explained 51.8% of the growth variance ($F = 11.7, P < 0.001$), whereas the NAO model captured 52.1% of the growth variance ($F = 11.9, P < 0.001$). The climatic model included the following variables (effects between parentheses): June precipitation (+), February, April (+) and July temperature (−) (Table 3; Fig. 6). The NAO indices which were selected in the model corresponded to the previous December (−) and to current February (+) and September (+) (Tables 4, 5 and 6; Fig. 6).

3.4 NAO and Silver Fir Decline

The instability in the NAO-growth relationships was also observed in the case of silver fir forests which showed positive associations with monthly November and

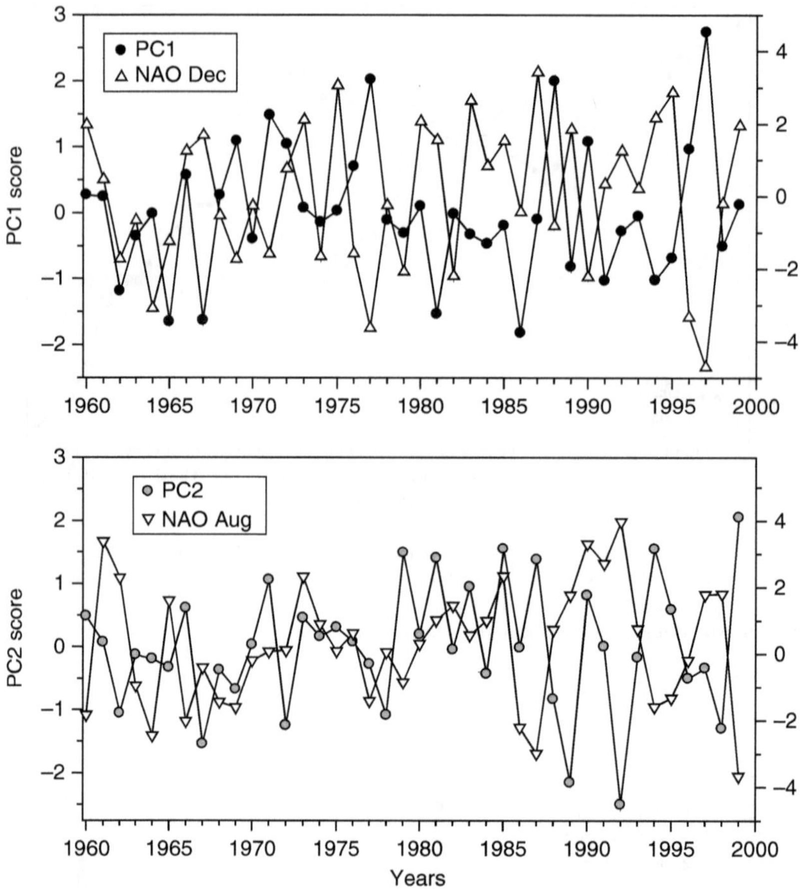

Fig. 4 Temporal patterns of the scores of the first two principal components (PC1, PC2) of growth chronologies and related December and August NAO indices for the years before (t–1) and during (t) tree-ring formation, respectively

February indices of winter before tree-ring growth (Table 4). In some of these forests a decline episode characterised by massive defoliation and mortality was observed in the 1980s after a severe growth reduction in 1986 (Camarero et al., 2002), when the minimum value of the November–February NAO index during the twentieth century was also observed (Fig. 7). The analyses based on moving correlations detected a shift towards positive and significant associations between such index and silver fir growth since the growth reduction in 1986, i.e. the responsiveness of silver fir growth to NAO winter indices increased in the last 2 decades of the past century. Furthermore, this rise in the NAO-growth correlation was greater in non-declining than in declining forests (Fig. 7).

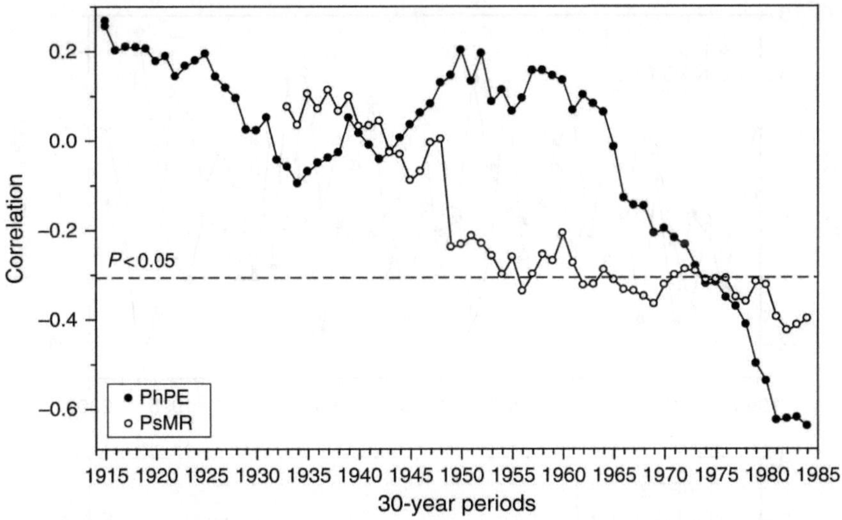

Fig. 5 Moving correlations (30-year periods) between the NAO index of the previous (year *t*–1) December and growth of two selected sites (PhPE, PsMR; see sites codes in Table 1). The years in the *x* axis indicate the mid-point of each 30-year period. The *dashed line* indicates the significance ($P < 0.05$) threshold for negative correlations lower than −0.3

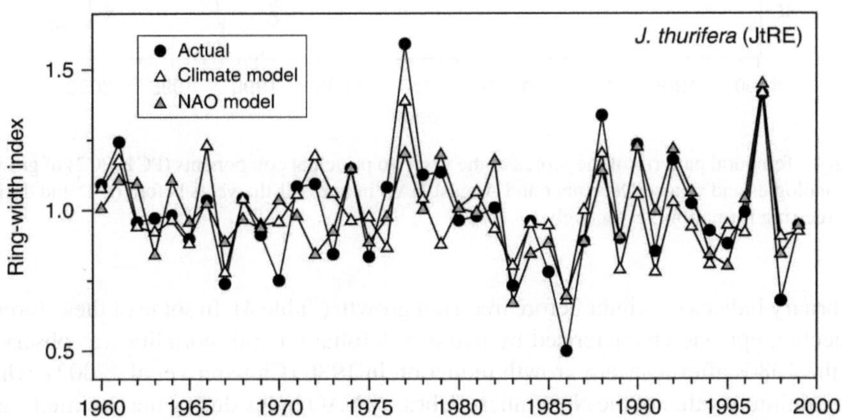

Fig. 6 Temporal patterns of the growth indices of *J. thurifera* in the Retuerta de Pina site (JtRE) as compared with models of tree growth based on climatic (monthly mean temperature and total precipitation) or NAO (indices) predictors

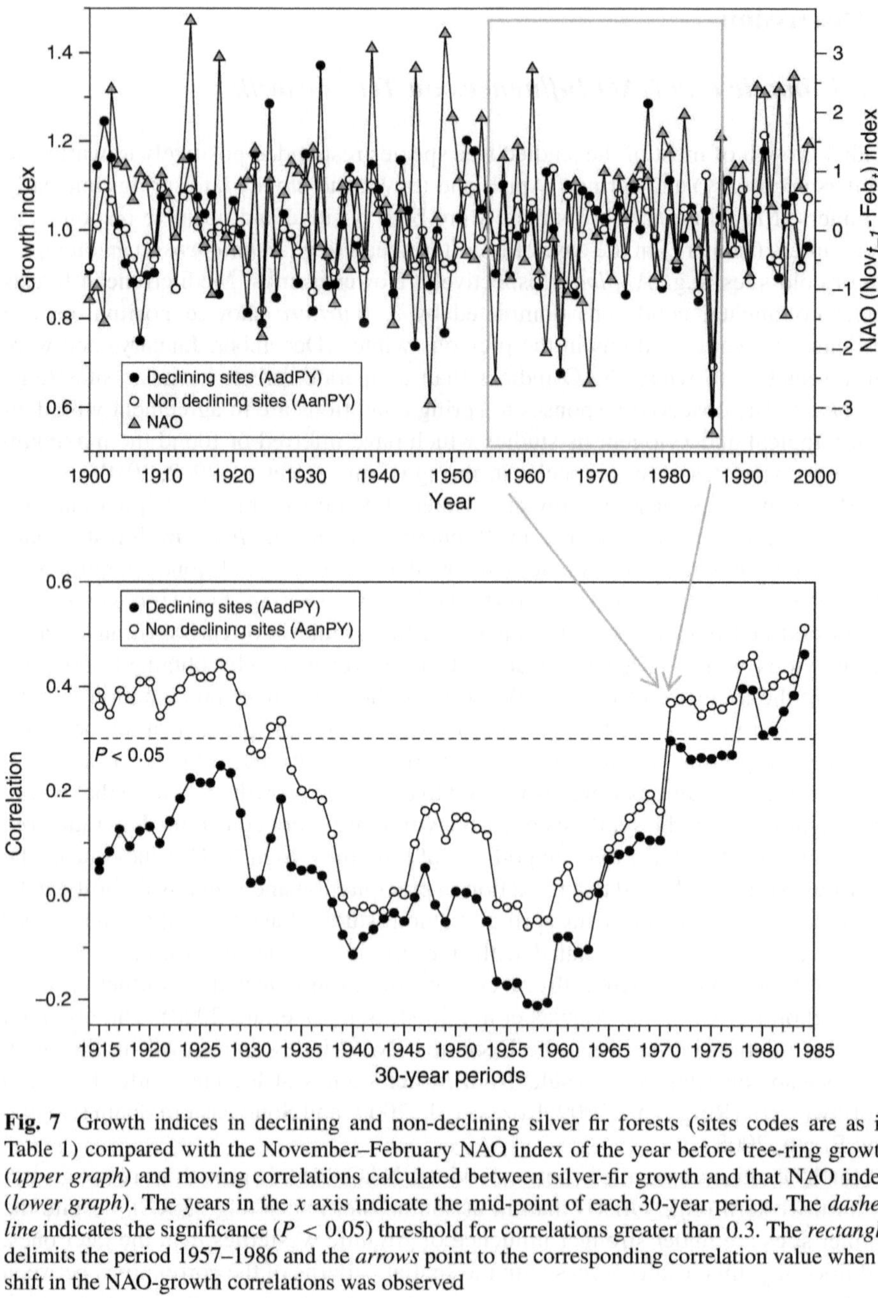

Fig. 7 Growth indices in declining and non-declining silver fir forests (sites codes are as in Table 1) compared with the November–February NAO index of the year before tree-ring growth (*upper graph*) and moving correlations calculated between silver-fir growth and that NAO index (*lower graph*). The years in the *x* axis indicate the mid-point of each 30-year period. The *dashed line* indicates the significance ($P < 0.05$) threshold for correlations greater than 0.3. The *rectangle* delimits the period 1957–1986 and the *arrows* point to the corresponding correlation value when a shift in the NAO-growth correlations was observed

4 Discussion

4.1 Climatic and NAO Influences on Tree Growth

Radial growth of most of the studied tree species responded positively to warm late winters (February) and springs (April) and to wet and cloudy late springs and early summers (May, June), whereas there were both positive and negative responses to wet winters (January) in the case of Mediterranean (e.g., *P. halepensis*) and temperate or cold sites (e.g., *A. alba*), respectively. However, most Mediterranean forests under continental conditions dominated by *J. thurifera* showed positive growth responses to wet conditions in the previous winter (December, January) and were more sensitive to winter NAO indices than temperate forests in mesic sites (e.g., *Q. robur*). The observed responses to spring conditions are in agreement with dendroecological and xylogenesis studies which have inferred or found the maximum radial-growth rates of these species in spring (Camarero et al., 2002, 2010).

The positive responses to previous-winter climatic variables (precipitation, temperature, cloud cover) detected in *P. halepensis* and *J. thurifera* forests under continental Mediterranean climates indicate that warm, wet and cloudy conditions in December and January enhance growth the following spring. In evergreen conifers, earlywood and tree-ring growth depend on the synthesis of carbohydrates during the previous winter (Kagawa et al., 2006). However, results obtained here suggest that the moisture reserves in the soil supplied by winter precipitation may be more relevant to enhance tree growth of these species in spring than temperature-mediated responses which were less important than rainfall effects. In addition, cloudy winter conditions were also linked to enhanced growth indicating that winter photosynthesis and the synthesis of carbohydrate were not constrained by radiation.

In agreement with the main hypothesis of this study, high NAO values during the previous winter are linked to a reduction in precipitation and water availability and a low radial growth the following spring. Previous studies have found that winter NAO indices are negatively associated with tree growth in Southern Europe (Piovesan and Schirone, 2000), whereas the reverse response was detected in Northern Europe (Linderholm et al., 2003; Macias et al., 2004; Schultz et al., 2008). The observed negative relationship between the December NAO index and growth in the study area is also consistent with results from earlier studies of Iberian conifer forests in Northwestern (Roig et al., 2009; Rozas et al., 2009) and Southeastern Spain (Bogino and Bravo, 2008).

Such negative associations between winter NAO and tree growth were not greater in northwestern (e.g., QfLU) than in at southeastern sites (e.g., JtRE), despite the former sites are under stronger influence of Atlantic westerlies than the later ones. Additionally, this results agrees with the spatial patterns of the correlations between the first principal component of growth and previous December precipitation which were higher in southeastern than in northwestern sites. Although the NAO mainly affects the precipitation patterns in southwestern Iberia, the southwest flows are reactivated when they reach the Pyrenees, explaining why tree growth of forests growing in the Middle Ebro Basin (e.g., JtRE) in eastern sites with continental

Mediterranean climates are also greatly affected by NAO patterns (Vicente-Serrano and Lopez-Moreno, 2006). In this *J. thurifera* site, both climatic and NAO-based models were able to capture about half of the growth variance which suggests that long chronologies of this species in semiarid and continental sites may provide a valuable proxy for reconstructions of winter NAO indices. The response of growth to NAO in this and other sites was not related o elevation as suggested Bogino and Bravo (2008) in the case of *Pinus pinaster*. Furthermore, this suggestion agrees with the negative association between the percentage of growth variance explained by NAO indices and tree-ring width, which indicate that series from sites with low growth and high inter-annual variability may reflect well NAO effects on the winter precipitation in the high-frequency domain. It is also noticeable that the average explained variance by climatic variables was much higher (40.1%) than that explained by NAO indices (15.9%), but the NAO-models explained a much higher percentage of growth if only *J. thurifera* and *P. halepensis* forests from semiarid and continental sites were considered (36.8%). Contrastingly, tree growth in mesic and transitional forests was less related to NAO variability as indicated the low growth variance captured by NAO indices (8.6%) (see similar values in Bogino and Bravo, 2008; Rozas et al., 2009). This result was observed even if these sites were located in Western areas near the Atlantic Ocean, probably because of wet conditions, and a low growth responsiveness of dominant tree species to winter precipitation. For instance, sites QrBE (*Q. robur*) and JtRE (*J. thurifera*) are located at linear distances of 32 and 252 km from the Atlantic Ocean, respectively, but NAO indices explained 3.0 and 52.1% of their recent growth variability, respectively.

Most dendroecological studies have focused on the effect of winter NAO on climate and tree growth in the low-frequency domain (Bogino and Bravo, 2008; Roig et al., 2009) because the NAO mainly affects long-term winter and spring climatic variability in Western Europe (Hurrell, 1995). However, several authors have also noted that summer precipitation at high-frequency timescales is related to tree growth in southern Europe where rainfall in the dry season may alleviate drought stress (Piovesan and Schirone, 2000). In this study, it was observed that tree growth was inversely related to current NAO summer indices, a season when growth is mainly enhanced by wet (June, August) and cloudy (June) conditions in xeric and transitional sites. The spatial pattern of the positive effect of June precipitation on growth was centered in Northeastern Spain and resembled related NAO-precipitation associations suggesting that the NAO was involved in this summer effect on growth, but further studies are required to test these assumptions. These results contrast with the positive effect of late-spring and early-summer NAO indices and temperature on tree growth in Fennoscandia (Macias et al., 2004), and even in Northwest Spain (Rozas et al., 2009). In Fennoscandia the positive effect of NAO on growth of Scots pine was mostly observed for negative NAO indices series indicating that these spring NAO values corresponded to climatic conditions (late snowmelt and beginning of the growing season) constraining tree growth whereas positive NAO values did not greatly affect growth (Macias et al., 2004). In this study, a similar effect was observed for February NAO values which were positively related to growth in

temperate (*A. alba*, *Q. robur*) and continental Mediterranean sites (*J. thurifera*, *P. halepensis*). Thus, a warmer early-spring linked to high NAO February values in negative NAO series may be related to an earlier growth resumption in some of these sites where growth was also enhanced by February and April temperatures.

4.2 NAO Variability and Silver Fir Decline

Silver fir growth showed positive associations with monthly November and February NAO indices of the previous winter but such association steeply rose in the mid 1980s when silver fir growth fell abruptly and decline started in xeric sites. Since the 1980s, southwestern Europe has been under the influence of extended periods of warm winters, which were associated with a persistent positive phase of the NAO index for that season (Hurrell, 1995). These results and the growth-climate relationships found for silver fir suggest that high NAO winter indices were possibly linked to warmer late-winter conditions and an enhanced growth through an earlier cambial reactivation of the silver fir xylogenesis, particularly in non-declining sites. However, silver firs produced an extremely narrow ring in 1986 probably because of the previous September 1985 which was very warm and dry (Camarero et al., 2002). Further studies are required to understand if the NAO or other circulation patterns were involved in the climatic determination of such silver fir decline.

One of the main shortcomings of the present study results from the short length of some of the dendrochronological series used. To assess the value of dendrochronological records as a proxy of NAO and as sources of climatic information contained in the NAO index, further analyses are required to investigate the NAO signals in long tree-ring chronologies of different species, particularly in sensitive sites with species whose growth is responsive to NAO variability. These longer dendrochronological records should be analyzed focusing on the temporal instability of NAO-growth relationships on decadal and multidecadal scales (5–15 years) which are the scales of the highest influence of the NAO on climate variability (Hurrell et al., 2003).

5 Conclusions

The proper assessment of the relationships between NAO and tree growth requires an approach based on the analysis of multi-species networks across contrasting geographical and climatic gradients. Regional monthly climatic variables were shown to be usually better predictors of tree growth than large-scale NAO indices. In northeastern Spain, tree growth was enhanced by warm and wet conditions in spring during the year of tree-ring formation. High NAO values during the previous December were linked to a reduction in precipitation and a low radial growth the following spring, but high February NAO values were also associated with warmer early-spring conditions and enhanced growth. Forests growing under xeric

conditions with semiarid and continental climates such as Mediterranean *Juniperus thurifera* sites showed a higher sensitivity in their recent growth patterns to NAO variability than forests growing under more mesic conditions. In fact, only in one *J. thurifera* forest under continental Mediterranean conditions the growth variance explained by NAO indices was high and similar to that explained by climatic data. Further research efforts to monitor the effects of NAO on forest growth and decline should build long *J. thurifera* tree-ring chronologies in continental Mediterranean xeric sites which have proved to be very sensitive to high-frequency changes in winter precipitation and NAO indices of the previous December.

Acknowledgements This work has been supported by the research projects RTA01-071-C3-1 and CGL2008-04847-C02-01/BOS financed by the Spanish Ministry of Science. I thank the support of ARAID and several colleagues for providing samples and driving me to many impressive forests (E. Arrechea, J.M. Gil, R. Hernández, G. Montserrat-Martí, C. Lastanao, N.A. Laskurain, V. Pérez, M.A. Ros, A. Solla).

References

Biondi F, Waikul K (2004) DENDROCLIM2002: a C++ program for statistical calibration of climate signals in tree-ring chronologies. Comput Geosci 30:303–311

Bogino S, Bravo F (2008) SOI and NAO impacts on *Pinus pinaster* Ait. growth in Spanish forests. In: Elferts D, Brumelis G, Gärtner H, Helle G, Schleser G (eds) TRACE – Tree rings in archaeology, climatology and ecology, vol. 6. GFZ Potsdam, Scientific technical report STR 08/05, Potsdam, pp 21–26

Briffa KR, Jones PD (1990) Basic chronology statistics and assessment. In: Cook ER, Kairiuktsis LA(eds) Methods of dendrochronology. Kluwer Academic, Dordrecht, pp 137–152

Camarero JJ, Olano JM, Parras A (2010) Plastic bimodal xylogenesis in conifers from continental Mediterranean climates. New Phytol 185:471–480

Camarero JJ, Padró A, Martín-Bernal E, Gil-Pelegrín E (2002) Aproximación dendroecológica al decaimiento del abeto (*Abies alba* Mill.) en el Pirineo aragonés. Montes 70:26–33

Cook ER (1985) A time series approach to tree-ring standardization. PhD Dissertation, University of Arizona, Tucson, AZ

Fritts H (2001) Tree rings and climate. Academic Press, London

Garfin GM (1998) Relationships between winter atmospheric circulation patterns and extreme tree growth anomalies in the Sierra Nevada. Int J Climatol 18:725–740

Giorgi F, Lionello P (2008) Climate change projections for the Mediterranean region. Glob Planet Change 63:90–104

Girardin MP, Tardif J (2005) Sensitivity of tree growth to the atmospheric vertical profile in the boreal plains of Manitoba, Canada. Can J For Res 35:48–64

González-Hidalgo JC, López-Bustins JA, Štepánek P, Martín-Vide J, de Luis M (2009) Monthly precipitation trends on the Mediterranean fringe of the Iberian Peninsula during the second half of the 20th century (1951–2000). Int J Climatol 29:1415–1429

Hallett TB, Coulson T, Pilkington JG, Clutton-Brock TH, Pemberton JM, Grenfell BT (2004) Why large-scale climate indices seem to predict ecological processes better than local weather. Nature 430:71–75

Hirschboeck KK, Ni F, Wood ML, Woodhouse C (1996) Synoptic dendroclimatology: overview and outlook. In: Dean JS, Meko DM, Swetnam TW (eds) Tree rings, environment, and humanity. Radiocarbon, Tuscon, pp 205–223

Holmes RL (1983) Computer-assisted quality control in tree-ring dating and measurement. Tree-Ring Bull 43:69–78

Holmes RL (1992) Dendrochronology program library. Version 1992-1. Laboratory of Tree-Ring Research, University of Arizona, Tucson, AZ
Hurrell J (1995) Decadal trends in North Atlantic Oscillation and relationship to regional temperature and precipitation. Science 269:676–679
Hurrell J, Kushnir Y, Ottersen G, Visbeck M (2003) An overview of the North Atlantic Oscillation. The North Atlantic Oscillation: climate significance and environmental impact. Geophys Monogr, vol. 134, Amer Geophys Union, pp 1–36
Hurrell J, Van Loon H (1997) Decadal variations in climate associated with the North Atlantic Oscillation. Clim Change 36:301–326
IPCC (2007) Climate change 2007: the physical science basis. Contribution of Working Group I to the 4th assessment report of the Intergovernmental panel on climate change. Cambridge University Press, Cambridge, UK
Jones PD, Jónsson T, Wheeler D (1997) Extension to the North Atlantic Oscillation using early instrumental pressure observations from Gibraltar and south-west Iceland. Int J Climatol 17:1433–1450
Kagawa A, Sugimoto A, Maximov T (2006) $^{13}CO_2$ pulse-labelling of photoassimilates reveals carbon allocation within and between tree rings. Plant Cell Environ 29:1571–1584
Linderholm HW, Solberg BO, Lindholm M (2003) Tree-ring records from central Fennoscandia: the relationship between tree growth and climate along a west-east transect. The Holocene 13:887–895
Macias M, Timonen M, Kirchhefer AJ, Lindholm M, Eronen M, Gutiérrez E (2004) Growth variability of Scots Pine (*Pinus sylvestris*) along a west-east gradient across Northern Fennoscandia: a dendroclimatic approach. Arct Antarct Alp Res 36:565–574
Mitchell TD, Jones PD (2005) An improved method of constructing a database of monthly climate observations and associated high-resolution grids. Int J Climatol 25:693–712
Nahal I (1981) The Mediterranean climate from a biological viewpoint. In: di Castri F, Goodall FW, Specht RL (eds) Mediterranean-type shrublands. Elsevier, Amsterdam, the Netherlands, pp 63–86
Piovesan G, Schirone B (2000) Winter North Atlantic Oscillation effects on the tree rings of the Italian beech (*Fagus sylvatica* L.). Int J Biometeorol 44:121–127
Rodó X, Baert E, Comin F (1997) Variations in seasonal rainfall in Southern Europe during the present century: relationships with the North Atlantic Oscillation and the El Niño-Southern Oscillation. Clim Dyn 13:275–284
Rodríguez-Puebla C, Encinas AH, Sáenz J (2001) Winter precipitation over the Iberian Peninsula and its relationship to circulation indices. Hydrol Earth Syst Sci 5:233–244
Roig FA, Barriopedro D, Herrera RG, Patón Domínguez D, Monge S (2009) North Atlantic Oscillation signatures in western Iberian tree-rings. Geogr Ann 91:141–157
Rozas V, Lama S, García-González I (2009) Differential tree-growth responses to local and large-scale climatic variation in two *Pinus* and two *Quercus* species in northwest Spain. Écoscience 3:299–310
Sáenz J, Rodríguez-Puebla C, Fernández J, Zubillaga J (2001) Interpretation of interannual winter temperature variations over southwestern Europe. J Geophys Res 106:20641–20652
Schultz J, Neuwirth B, Friedrichs DA, Löffler J, Winiger M (2008) Growth responses to NAO along a Central European West-East transect. In: Elferts D, Brumelis G, Gärtner H, Helle G, Schleser G (eds) TRACE – Tree rings in archaeology, climatology and ecology, vol. 6. GFZ Potsdam, Scientific technical report STR 08/05, Potsdam, pp 17–24
Vicente-Serrano SM, Lopez-Moreno JI (2006) The influence of atmospheric circulation at different spatial scales on winter drought variability through a semi-arid climatic gradient in northeast Spain. Int J Climatol 26:1427–1453
Wigley TML, Briffa KR, Jones PD (1984) On the average value of correlated time series, with applications in dendroclimatology and hydrometeorology. J Clim Appl Meteorol 23:201–203

Ecological Impacts of the North Atlantic Oscillation (NAO) in Mediterranean Ecosystems

Oscar Gordo, Carles Barriocanal, and David Robson

Abstract Large-scale climate indices have received much attention in recent years in ecology-climate research due to the advantages they have over typically used local weather variables, such as temperature or rainfall. In the Mediterranean, the North Atlantic Oscillation (NAO) is a major forcing of climate patterns, especially precipitation. More than 60 studies to date have demonstrated the effects of the NAO on both terrestrial and aquatic Mediterranean ecosystems. In terrestrial ecosystems, the NAO affects the phenology and growth of plants and crop yields. It also affects the condition and diet of mammals and disease-related mortality in amphibians. The effects of the NAO are probably better known in marine ecosystems, where the impact on the hydrodynamics of the water column and currents is felt in the dynamics of populations from plankton to fishes in both pelagic and benthonic environments. Additionally, birds are an especially well studied taxon when it comes to effects of the NAO. The NAO has been shown to affect the population dynamics of water birds by impacting the availability and extent of their habitat and by influencing dispersal decisions of individuals. The NAO plays an essential role in the migration of birds throughout the Mediterranean basin, and it is probably a reason for the observed advance of arrival dates during the spring in Europe. In spite of the notable number of studies carried out to date, we are far from knowing accurately the ecological impacts of the NAO on Mediterranean ecosystems. More efforts are needed to understand regional differences in the NAO effects within the Mediterranean basin and how they compare with more northern latitudes of Europe.

Keywords Birds · Climate · Ecosystem · Phenology · Review

1 Why Using Large Scale Climate Indices in Ecology?

We are all aware that environmental conditions, defined as the set of abiotic variables to which organisms are subjected, shape the distribution, abundance, behavior

---

O. Gordo (✉)
Departamento de Zoología y Antropología Física, Universidad Complutense de Madrid, Madrid, Spain
e-mail: ogordo@bio.ucm.es

S.M. Vicente-Serrano, R.M. Trigo (eds.), *Hydrological, Socioeconomic and Ecological Impacts of the North Atlantic Oscillation in the Mediterranean Region*, Advances in Global Change Research 46, DOI 10.1007/978-94-007-1372-7_11,

and ultimately the evolution of organisms (Begon et al., 2006). Among environmental drivers, climate is recognized as one of the most important abiotic factors. Therefore, one should not be surprised that organisms are responding to the ongoing climate change, even though observed changes in Earth's climate have so far been relatively small (e.g., the global mean annual temperature has increased only by about 0.7°C during the last century (Solomon et al., 2007)). Even though many other factors operate on a long-term temporal scale (e.g., density-dependent processes, land-use transformations, etc.) and may prevent us from establishing unambiguously a causal link between climate change and some responses of organisms (Sala et al., 2000; Forchhammer and Post, 2004), evidence of climate change impacts on organisms is abundant. The impacts are widespread, ranging from the equator to the poles in both terrestrial and aquatic ecosystems and suggesting a climate change imprint on the biosphere at a global scale (Parmesan and Yohe, 2003; Root et al., 2003). Climate-induced responses range from the plasticity of many traits to adapt to weather variability to persistent alterations in ecosystem functioning, community composition and functional interactions among species. Therefore, climate change signals are already apparent on all biological scales (Parmesan, 2006; Rosenzweig et al., 2008).

Given the current global climate change scenario, there is an urgent need to understand better how organisms are affected by climate. This understanding is essential to the accurate prediction of the biological consequences of the ongoing alterations of climate and from this to delineate the necessary actions to ensure conservation of natural populations, communities and ecosystems. One of the main challenges in the study of the effects of climate on organisms is to find the key climate parameter controlling a certain biological phenomenon and its temporal and spatial scales (Hallett et al., 2004; Stenseth and Mysterud, 2005; Gordo, 2007). Temperature is probably the most frequently used climate variable in ecological research. In fact, many biological processes are temperature-dependent (e.g., metabolism, development, activity, etc.), and thus, one may make sound predictions about its effect. In addition, temperature is the climate variable with the most homogeneous and coherent pattern of change across the globe during the last century (i.e., global warming; Solomon et al., 2007). However, temperature is not the only important variable. Other variables, such as precipitation (rainfall and snow), solar radiation, wind, etc., are also essential for individual survival and ecosystem functioning and are, in some cases, as in arid regions, more important than temperature. Therefore, the key climate variable may differ according to the species or the biological process under study. Moreover, the effect of a certain climate variable may vary in space, being more influential for some populations than for others (Forchhammer et al., 2002a, b; Stenseth et al., 2004; Doi and Takahashi, 2008). Such complex responses of organisms to climate are just a reflection of the fact that weather cannot be described simply by temperature or rainfall. Weather is complex to quantify because it results from a collection of physical features of the atmosphere at a certain moment along with their interactions. In addition, weather does not affect organisms in isolation. Climate fluctuations affect all organisms in a certain ecosystem, and it is essential to understand how climate influences organisms and their interactions

with other individuals. In summary, the influence of climate on the individual seems too complex to be described simply by one or a few climate variables.

Large-scale climate indices may help one to deal with these complexities resulting from the heterogeneous influence of climate across taxa and populations. It is obvious that organisms adapt to climate at a local scale by responding to temperature, amount of rainfall, intensity of winds, and so on. However, climate is a paradigm of spatial autocorrelation because sites that are close together resemble each other in their patterns of temporal variability of climate. This feature arises from the fact that local weather is controlled by atmospheric circulatory patterns, which work on a large scale. Climate indices are able to summarize in a single value the weather conditions for a large area. This synthetic nature has been demonstrated to be of great value and use in ecology-climate research, and the application of climatic indices has become increasingly frequent during the last decade (Fig. 1).

Climatologists have defined over a dozen large-scale climate indices that govern most of the observable interannual variability of climate at the planetary scale. Among them, the El Niño-Southern Oscillation (ENSO) and the North Atlantic Oscillation (NAO) are probably the most commonly used by ecologists and are those for which the ecological consequences are best known (Stenseth et al., 2002, 2003; Forchhammer and Post, 2004). As broadly described in other parts of this book, the NAO is a large-scale fluctuation in atmospheric pressure difference along a north-south gradient between the subtropical North Atlantic region (centered on the Azores) and the subpolar region (centered on Iceland). The differences in atmospheric pressure at sea level between the subtropical and the subpolar centers affect the speed and direction of the westerly surface winds across the North Atlantic from North America to Europe (40°N–60°N). As a consequence, the numbers and paths of storms and their associated weather are affected (Hurrell, 1995; Hurrell and van Loon, 1997). Furthermore, there are effects on ocean surface temperature and currents, which in turn have direct effects on heat transport in the North Atlantic and the sea ice coverage in the Arctic region (Ottersen et al., 2001; Stenseth et al., 2003). Although present throughout the year, the greatest interannual and decadal variability is seen during the boreal winter, when the atmosphere is most dynamically active

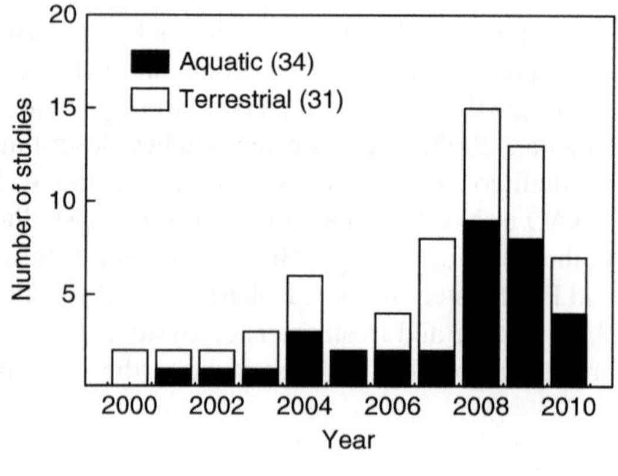

Fig. 1 Number of published studies using the North Atlantic Oscillation as an explanatory variable of any biological phenomena in terrestrial or aquatic ecosystems of the Mediterranean region Source: Isi Web of Knowledge (access date 1-10-2010)

(Hurrell, 1995). The NAO is the most robust mode of recurrent atmospheric behavior in the North Atlantic and has been suggested as the most probable cause for the remarkable changes in the climate over the middle and high latitudes of the Northern Hemisphere during recent decades (Hurrell and van Loon, 1997; Hoerling et al., 2001; Visbeck et al., 2001).

Positive values of the NAO are associated with more frequent and stronger winter storms crossing the Atlantic Ocean on a more northerly track, leading to warmer and wetter conditions in northern Europe and in the eastern coastal areas of the United States (Ottersen et al., 2001; Visbeck et al., 2001; Mysterud et al., 2003; Stenseth et al., 2003; Forchhammer and Post, 2004). In contrast, during negative phases of the NAO, the westerlies are weakened and become less frequent, and the warm and wet winter weather remains over North America, leaving northern Europe cold and dry. The most important feature arising from a comparison between the positive and the negative phases of the NAO is that this large-scale atmospheric circulatory pattern has a geographically marked differential impact on local weather. Correlations between the NAO and other climate variables, such as temperature and rainfall, vary regionally (Hurrell and van Loon, 1997; Ottersen et al., 2001; Visbeck et al., 2001; López-Moreno and Vicente-Serrano, 2008), and thus, the value and interpretation of the NAO as a proxy of climate fluctuations is variable. A paradigmatic example of regional variability is found between northern and southern Europe. During a positive phase of the NAO, the northward track of storms leads to drought anomalies in the Mediterranean basin, while northern latitudes enjoy a wetter climate than average. Unfortunately, the heterogeneous effects of the NAO at the continental scale remain poorly studied in ecology (but see Sanz, 2002, 2003; Jonzén et al., 2006; Gouveia et al., 2008). Moreover, the overwhelming majority of studies about the ecological impact of the NAO have been focussed on central and northern Europe (Ottersen et al., 2001; Mysterud et al., 2003; Forchhammer and Post, 2004), where the intensity and sign of its effect are quite different from other areas.

2 Impacts of the NAO in Mediterranean Ecosystems

Although local climate is also driven by the NAO in the Mediterranean basin, especially precipitation patterns (Lamb and Peppler, 1987; Rodó et al., 1997; Quadrelli et al., 2001; Trigo et al., 2004; Rodrigo and Trigo, 2007; Vicente-Serrano and Cuadrat, 2007), there are few studies describing ecological impacts of the NAO in Mediterranean latitudes. A search in the Web of Science using the next chain "(NAO or North Atlantic Oscillation) and Mediterranean" provided only 65 studies in the fields of Ecology, Biodiversity and Conservation, Zoology, Forestry, Marine and Freshwater Biology, Fisheries, and Agriculture (Fig. 1). Terrestrial and aquatic (both marine and freshwater) ecosystems have received similar attention (31 vs. 34 articles, respectively). Most of the studies have been published in the last few years

in journals such as *Climate Research*, *Fisheries Oceanography*, *Global Change Biology*, *International Journal of Biometeorology*, *Journal of Animal Ecology*, *Journal of Marine Systems*, or *Progress in Oceanography*. Therefore, the study of ecological impacts of the NAO in Mediterranean ecosystems is recent and has occupied many pages in some of the top journals in ecology.

In terrestrial ecosystems, it has been demonstrated that positive NAO values are related to increased growth and advancement of spring phenological events, such as flowering, leaf unfolding or pollen release, of trees (Piovesan and Schirone, 2000; Avolio et al., 2008; Campelo et al., 2009; Rozas et al., 2009; Gordo and Sanz, 2010; Orlandi et al., 2010). Nonetheless, proposed mechanisms for the NAO effect in growth and phenology were different. In the case of plant growth, the main driver is soil moisture, which is dependent mainly on winter rainfalls, which are in turn strongly dependent on the NAO (Piovesan and Schirone, 2000; Campelo et al., 2009; Rozas et al., 2009). However, the effect of the NAO on spring plant phenology would be mediated by its effect on spring temperatures (Gouveia et al., 2008; Gordo and Sanz, 2010). From a more applied point of view, the NAO has been found to affect the yields of wheat and citruses in Spain (Gimeno et al., 2002; Rodríguez-Puebla et al., 2007). Interestingly, wine production and quality were not affected by the NAO in this country (Rodó and Comín, 2000), but they were affected in Portugal (Esteves and Orgaz, 2001). Such differences in crop effects may be due to differences in the regional influence of the NAO on local weather. Therefore, the NAO influence could vary even among areas close together within the Mediterranean basin.

Evidence of NAO effects in terrestrial animals is almost entirely limited to studies with birds. We still lack in depth knowledge on the NAO effect on invertebrate, amphibian, reptile or mammal distributions, abundances and behavior. In a recent study, Martínez-Jauregui et al. (2009) demonstrated that the NAO did not have an effect on seasonal variation of weight in females of red deer (*Cervus elaphus*) in a locality in central Spain. Interestingly, the NAO did play an important role in understanding changes in body weight of females of the same species in populations in Scotland and Norway. Geographical differences in the NAO effect are seen once more, suggesting that one needs to exercise caution in generalizing the NAO effect over broad areas because its effect shows marked regional differences, especially between the Mediterranean and more northern latitudes in Europe. However, the NAO has been an important predictor of the observed changes during the last 3 decades in the diet of the nearly extinct population of Iberian brown bears (*Ursus arctos*; Rodríguez et al., 2007). By altering the phenology and productivity of some plant species (Gordo and Sanz, 2010), the NAO may play an important role in the conservation of this iconic species. The NAO has also been demonstrated to be important to understanding amphibian declines in mountainous lakes of central Spain (Bosch et al., 2007). In this particular case, the NAO has driven increases in local temperatures, which have promoted the spread of a fungus disease that causes mass mortality among amphibians.

The NAO effects in marine ecosystems of the Mediterranean Sea are probably known with greater certainty than in terrestrial ecosystems. The NAO affects the water column hydrodynamics, which in turn affect vertical gradients of mixing, nutrient flow and irradiance, by its control of weather conditions (e.g., temperature, winds, or storms). Knowledge of water column hydrodynamics is essential to an understanding of plankton functioning, abundance, and diversity, so it is not surprising that plankton community dynamics are closely related to the NAO fluctuations in the Mediterranean Sea as well as in other areas (Katara et al., 2008; Kamburska and Fonda-Umani, 2009; Gladan et al., 2010; Pérez et al., 2010). For instance, in the northwestern Mediterranean, dominant zooplankton species vary under positive and negative phases of the NAO because each phase favors different species by imposing different hydrodynamic features (Molinero et al., 2005, 2008; de Puelles et al., 2007; de Puelles and Molinero, 2008). However, the NAO can also show regional differences in its effects in the sea. For instance, *Centropages typicus*, the commonest calanoid copepod in Mediterranean waters, showed a response to the NAO in its abundance in populations in the Gulf of Naples that was opposite to its response in the Bay of Villefranche (Mazzochi et al., 2007).

The NAO effect on hydrodynamics, and consequently on plankton, may have a cascade effect on higher trophic levels. Several studies have found significant associations between the NAO and fish and cephalopod abundances in the Mediterranean (Grbec et al., 2002; Relini et al., 2006, 2010; Bridges et al., 2009; Pérez et al., 2010; but see Ravier and Fromentin, 2004). In the northwestern region of the Mediterranean Sea, the collapse of the sardine *Sardina pilchardus* stocks in the 1990s has been related to the dominant positive phase of the NAO during recent decades (Guisande et al., 2004). Similarly, the collapse of the anchovy *Engraulis encrasicolus* population in 1987 in the Adriatic Sea has been attributed to low recruitment due to the extreme positive value of the NAO during the autumn of 1986 (Santojanni et al., 2006). Along with the cascade effect through hydrodynamics and plankton, fish abundances could also be indirectly affected by water discharge of Mediterranean rivers, which is closely related to the NAO due to precipitation patterns, because the rivers supply nutrients to the sea (Lloret et al., 2001; Cartes et al., 2009).

The NAO has also been related to the strength of some currents within the Mediterranean basin. This relationship has direct effects on the horizontal transport of nutrients from some regions to others of the sea, enhancing productivity and, as consequence, the richness of marine communities. For instance, population dynamics of the hake (*Merluccius merluccius*), an economically important fishery in the Mediterranean, have been linked to the NAO through its effect on currents in the Ligurian and Tyrrhenian Seas (Abella et al., 2008; Massutí et al., 2008).

It is not only the abundance and dynamics of marine organisms that are affected by the NAO. The NAO has also been suggested as the cause of the increase of thermophilic species in the Adriatic Sea as a result of its control over sea temperature (Dulcic et al., 2004; Grubelic et al., 2004).

The NAO fingerprint has also been detected in benthonic communities. For instance, periodical surveys of the benthic macrofauna (especially polychetes) in

the Bay of Banyuls-sur-Mer (France) since the 1960s have shown notable changes that have been partially attributed to NAO fluctuations (Lebrune et al., 2007). This fingerprint can reach even those benthic communities living at depths greater than 1000 m. Off the Catalan coast, it was found that changes in the abundance and composition of deep megabenthos were related to the contrasting values of the NAO between the surveys carried out at the end of the 1980s and the end of the 2000s (Cartes et al., 2009). Similarly, landings of the shrimp *Aristeus antennatus*, which lives at depths between 600 and 800 m, were higher following positive values of the winter NAO (Maynou, 2008a, b). In this case, there was a time lag in the response to the NAO, supporting the idea that the NAO effect may be carried by a population over a number of years following a particular NAO situation (Ottersen et al., 2001).

3 Effects of the NAO on Bird Populations

The NAO has a direct effect on demographic parameters, such as survival, dispersal or fecundity, of many animal species (Ottersen et al., 2001; Mysterud et al., 2003; Forchhammer and Post, 2004). Therefore, one expects to find effects of the NAO on abundances and dynamics of bird populations, and they have indeed been found (Sæther et al., 2004). For instance, a long-term study in a Norwegian population of dippers (*Cinclus cinclus*) demonstrated that year-to-year variations in the population size were positively related to the NAO index, i.e., the population increased under positive NAO values (Sæther et al., 2000). Positive NAO values were related to milder winters, which reduced the duration of the frozen surface of rivers where these individuals forage. These improved conditions for foraging enhanced survival, especially of the younger individuals, which recruited in larger numbers the next spring, leading to an increase in the breeding population.

In the Mediterranean region, the NAO effects on dynamics and demography of bird populations have been studied in a wide variety of species, such as water birds (Almaraz and Amat, 2004a, b; Figuerola, 2007; Bechet and Johnson, 2008; Jiguet et al., 2008), seabirds (Jenouvrier et al., 2009), raptors (Rodríguez and Bustamante, 2003; Costantini et al., 2010), and songbirds (Sanz, 2002, 2003; Grosbois et al., 2006; Rubolini et al., 2007; Balbontín et al., 2009). Such diversity of species demonstrates that NAO signals are apparent in the Mediterranean avifauna as a whole and supports the notion that the NAO effect can be found across all habitats of a large area like the Mediterranean.

Several studies have demonstrated effects of the NAO on water birds, a fact that is not surprising if we take into account the strong dependence of these species on lakes, ponds and marshlands. The occurrence and extent of wetlands in the Mediterranean region are strongly dependent on precipitation patterns, which in turn are driven by the NAO (see other chapters in this book). Almaraz and Amat (2004a, b) showed for the first time that large-scale indices, such as the NAO and the ENSO, were related to population dynamics of the white-headed duck (*Oxyura leucocephala*), a threatened water bird species. Low values of the winter NAO were

linked with increased rainfall, which led to a large expansion of the range of the species during the following breeding season as a result of an increase in wetland surface in Spain (Almaraz and Amat, 2004a). Furthermore, El Niño events (i.e., high ENSO values) were linked with increased rainfall during the summer in the Mediterranean, which reduced breeding densities and increased markedly the recruitment in the next year.

In similar studies, some breeding populations of waders from the Mediterranean showed clear fluctuations in their numbers related to the NAO. In the Doñana National Park (southwestern Spain), the black-winged stilt (*Himantopus himantopus*) breeds in numbers from a dozen pairs up to more than 14,000 pairs (which represents about 30% of the European population). Such extreme variability was related to positive and negative phases of the NAO, respectively, which affect the extent of the Doñana marshlands and consequently the phylopatry and dispersal behavior of this species (Figuerola, 2007). The NAO also drives the hydrological conditions of another major wetland area of the Mediterranean, the Camargue (southern France). Interestingly, the NAO signal is even detected in those populations that use strongly human-managed habitats, such as the commercial salt pans, for breeding, as in the case of the greater flamingo (*Phoenicopterus roseus*; Bechet and Johnson, 2008). Even though the date of the start of salt production determines the potential number of breeding pairs of the species each year, the observed breeding population is also dependent on the water levels in the Camargue lagoons, which are under the NAO influence. A larger extent of marshlands increases the food availability and thus the opportunities for breeding.

Kestrels are small falcons that inhabit open landscapes in Europe and Asia. During the twentieth century, the lesser kestrel (*Falco naumanni*) has shown marked declines, which have led to a threatened status in most European countries and even to extinction in some. Rodríguez and Bustamante (2003) investigated whether recent climatic changes were related to declines in populations from southern Spain. They found significant influences of the winter NAO and annual precipitation patterns on kestrel demography. However, rainfall outperformed the predictive ability of the NAO in demographic models because rainfall probably captured better the environmental variability at the local scale, i.e., at the scale of the breeding colonies of the species. Nevertheless, the rainfall in southern Spain and the NAO showed strong correlations, and thus the NAO influence on bird demography could be shown for this terrestrial species via precipitation amount and seasonal distribution. However, in spite of the drier conditions recorded during recent decades in the kestrel colonies as a result of the prevailing positive phase of the NAO, the authors concluded that the effect was not sufficient to explain the population declines and that causes must be sought in other environmental factors. In Italy, reproductive decisions and reproductive success in a population of kestrels (*Falco tinnunculus*) were also closely related to rainfall, temperature and the NAO index during both the spring and the previous winter (Costantini et al., 2010). Contrary to predictions, this population laid smaller clutches after mild and dry winters (i.e., positive winter NAO index). The authors hypothesized as possible causes for this result a greater survival of low-quality individuals during the winter and/or a disruption in prey phenology.

Nevertheless, the mechanism is unclear, and this fact demonstrated the complexities of the NAO influence on animal population dynamics. The NAO effect can be mediated by conditions in previous seasons, and the balance between current and past weather determines environmental conditions and thus the observable reproductive success.

In a similar way, the NAO effect may be different among different populations (e.g., Sanz, 2002), and this fact may lead to divergent patterns of response to climate change. For instance, barn swallows (*Hirundo rustica*) from Spain are more likely to survive during their first year of life when the winter NAO is positive, while Danish swallows show increased mortality during positive phases of the winter NAO (Balbontín et al., 2009). Unlike the black-winged stilt recruitment (Figuerola, 2007), the recruitment probability of the barn swallows to their natal areas was not affected by the NAO. In these studied populations of swallows, dispersal was explained by another large-scale climate index, the Southern Oscillation Index (SOI), which also showed an opposite effect between the two populations (Balbontín et al., 2009).

4 Effects of the NAO on Bird Migration

Every year millions of individuals move between breeding and wintering areas in one of the most fascinating events of nature: bird migration. Bird migration has long been known to be influenced by weather (Newton, 2008). In fact, the arrival of some migratory species, such as swallows or cuckoos (*Cuculus canorus*), has been traditionally used as an unequivocal signal of spring onset. This sensitivity of bird migratory phenology to weather has led to a renewed interest in this phenomenon in light of the ongoing climate change. If bird migration is sensitive to climate fluctuations, we can use it as a bioindicator of climate change impacts. A plethora of studies has demonstrated shifts in the timing of migration for many species along with significant relationships between climate from departure, passage and arrival areas and bird phenology (see reviews in Gordo, 2007; Lehikoinen and Sparks, 2010). Although climate change seems to be the cause for the alteration of bird migration timing during recent decades, the precise ecological and evolutionary mechanisms by which birds are changing their migratory schedule in response to the new environmental conditions remain to be elucidated.

The arrival date of a migrant to a certain site depends on two factors: (1) the onset date of migration and (2) the speed of progression across the areas between the start and end points of its journey (Gordo, 2007). Experiments with captive birds have demonstrated that the onset of migration is controlled by inflexible endogenous rhythms with a genetic basis (Berthold, 2001; Newton, 2008). However, a growing number of correlative studies have recently suggested some plasticity in the onset of migration; i.e., birds may adjust their departure date according to the prevailing weather or to the ecological conditions found at their departure areas (Saino et al., 2004; Gordo et al., 2005; Saino et al., 2007; Gordo and Sanz, 2008). In addition, many other studies have demonstrated that temperature, precipitation, wind, or vegetation phenology in the passage areas affect bird progression (Gordo, 2007). Overall,

the evidence suggests that climate may influence bird migration both directly and indirectly at several times throughout the year and at several sites. Such complexities of climate effects are not surprising if one takes into account the fact that migrants spend much time moving across vast areas. In this situation, large-scale climate indices become ideal explanatory variables for bird migratory phenology because they are capable of summarizing these complexities of local weather in a single variable that is easy to interpret and obtain (see Section 1).

The NAO index was employed for the first time by Forchhammer et al. (2002a) to explain migratory phenology of some common migrants in Norway. Since then, more than 30 studies have used this variable to explain interannual variability in arrival dates of migrants (Gordo, 2007; Hubálek and Čapek, 2008; Donnelly et al., 2009; Balbontín et al., 2009; Tøttrup et al., 2010). As a norm, most of the time-series of arrival dates show a negative relationship with the NAO; i.e., birds arrive early when the NAO (especially the winter NAO index) is positive (Gienapp et al., 2007). Positive NAO values imply milder winters and thus an advance of the spring (e.g., earlier flowering and leafing of plants, Gordo and Sanz, 2010) and an enhancement of weather conditions during migratory flights (e.g., more tail winds, Hüppop and Hüppop, 2003; Sinelschikova et al., 2007). Therefore, the proposed mechanism for the effect of positive NAO values on migration is the improved environmental conditions for migratory progression across Europe in terms of both food availability and the weather.

However, this picture may be less clear than it seems because most of the studies have been conducted in northern and western Europe, where the NAO effect is stronger and bird populations are subjected along their entire migratory route to the NAO influence. If we focus on the particular case of Mediterranean populations, there are few studied cases, and those studies that do exist usually suggest no effect of the NAO. In Spain, the arrival dates of five common migrants, the white stork (*Ciconia ciconia*), the cuckoo, the swift (*Apus apus*), the barn swallow or the nightingale (*Luscinia megarhynchos*), were not related to the NAO, either for earlier individuals or for later ones (Gordo and Sanz, 2008). In a locality of central Italy, the arrival dates of the first swifts, swallows and nightingales were not related to either the winter NAO (December–March) or the spring NAO (March–May) (Rubolini et al., 2007). Here, only one species, the house martin (*Delichon urbicum*), advanced significantly their arrival in response to positive winter NAO. In a long-term monitoring study of the arrival dates of nightingales in Croatia (Kralj and Dolenec, 2008), the winter NAO did not show any effect. Although the number of studies of Mediterranean bird populations is still small, the evidence suggests that the NAO has little influence on their arrival dates. This fact supports the idea that the NAO effect would be felt mainly from the Mediterranean latitudes northwards and thus, that only those populations breeding in the northernmost regions of the continent, and consequently with extensive migratory routes over Europe, would be significantly influenced by the NAO (Forchhammer et al., 2002a; Hüppop and Hüppop, 2003). Moreover, some studies have demonstrated that the negative

relationship with the winter NAO is stronger in short-distance migrants (i.e., those species overwintering in the Mediterranean basin) than in long-distance ones (i.e., those species overwintering to the south of the Sahara) (Rainio et al., 2006; Tøttrup et al., 2010). If it is considered that positive NAO values are related to drought in the Mediterranean and thus to harsh ecological conditions, it follows that short-distance migrants must benefit from a positive NAO *en route* but not during their wintering stay in the Mediterranean. Unfortunately, the environmental factors operating on the dynamics of wintering populations in the Mediterranean are poorly known, as is the effect of the NAO.

Finally, the Mediterranean basin is an obligatory passage area for all long-distance migrants coming from Africa and moving into Europe for breeding (Newton, 2008). In spite of this key role in the European-African migratory system, only two studies have analyzed the NAO effect on passing migratory bird populations. The *Istituto Nazionale per la Fauna Selvatica* from Italy has been promoting since 1988 the monitoring scheme *Progetto Piccole Isole*. This is a network of ringing sites located in islands and coastal sites, mainly in the central and western parts of the Mediterranean Sea (Italy, Spain, France and Malta), that uses standardized protocols for traping birds (Spina et al., 1993). One of the pioneering and most active ringing stations is located on the Italian island of Capri. There, the passage dates of nine trans-Saharan species were positively correlated to the winter NAO both in early and in late individuals, in opposition to the relationship found between the NAO and passage dates for the same species in ringing sites in Scandinavia (Jonzén et al., 2006). While a positive NAO would enhance migration across Europe (see above), it would impair ecological conditions in northern Africa by droughts in the Mediterranean basin and rainfall deficits during the monsoon season in West Africa (Lamb and Peppler, 1987; Hurrell, 1995; McHugh and Rogers, 2001; Oba et al., 2001; Stige et al., 2006). Therefore, the contrasting effect of the NAO in passage dates of the same species between Mediterranean and Scandinavian latitudes reinforces the idea of the heterogeneous effect of this variable among regions and thus their populations. Populations in Scandinavia have traveled for weeks across Europe and have benefited from enhanced conditions for migratory progression due to positive phases of the NAO. However, in Italy, just after the birds leave Africa, the positive phase of the NAO has the opposite consequences in the environment, and this is reflected in a delayed passage date. However, this is just one study in a single locality. A recent study with 11 migratory species on 3 Spanish islands demonstrated that the NAO did not have any effect on passage dates (Robson and Barriocanal, 2011). Moreover, it was the worst explanatory variable of the timing of migration compared to other predictors, such as temperature, vegetation productivity or the SOI. This extreme contrast in the results of different studies highlights the urgent need to develop more studies in the Mediterranean region to elucidate the true effects of the NAO on migratory bird phenology and the potential regional variability in the effects in relation to each migratory route across the Mediterranean.

5 Conclusions

There is already a notable number of studies that have assessed the effect of NAO fluctuations on a wide variety of species and biological phenomena in the Mediterranean region (Fig. 1). However, after a comprehensive review of the available evidence, we conclude that the impacts of the NAO on Mediterranean ecosystems are still poorly understood, especially on terrestrial ecosystems. We have identified three major non-excluding possible causes for the present state of the art:

(i) There is a considerable heterogeneity in the species and biological phenomena studied, and consequently, there are only one or a few populations studied in each case and usually from a single site. This fact hinders seriously our capacity to make generalizations because the NAO effect on local weather (the relevant scale for organisms) varies among regions (even within the Mediterranean), and we must be very cautious about extrapolating results from one site to others. In addition, many studies, especially in aquatic ecosystems, just report significant associations between the NAO and biological phenomena without providing a mechanistic interpretation of these results. We are firmly convinced that without a biological interpretation, results are of limited use.

(ii) Most of the studies have assessed only the effect of the winter NAO index (December–March), probably imitating the pioneering studies carried out in northern latitudes. Indeed, this phase shows the greater variability (Hurrell, 1995), but the potential influence of the NAO in organisms through climate patterns of other seasons has been completely neglected (but see Gordo and Sanz, 2010). The NAO effect on climate over the whole annual cycle is well-known by the climatologists (e.g., this book), but this is another point where considerably additional progress needs to be made. The climatologic literature is extensive, but unfortunately it is difficult to understand for most ecologists. In addition, the method of studying and defining climate patterns by climatologists is not necessarily the most relevant for organisms (e.g., see Doi et al., 2008). It seems obvious that there is an urgent need to establish multidisciplinary collaborations between climatologists and ecologists to deal accurately with the complex interactions between climate and biosphere.

(iii) Despite the NAO influence on the variability of Mediterranean climate, the NAO may indeed have little effect on Mediterranean ecosystems, especially on terrestrial ecosystems, as the remarkable number of studies reporting non-significant results may demonstrate (e.g., Ravier and Fromentin, 2004; Grosbois et al., 2006; Gordo and Sanz, 2008; Jiguet et al., 2008; Kralj and Dolenec, 2008; Jenouvrier et al., 2009; Robson and Barriocanal, 2011). Moreover, some studies have shown that the NAO effect is overridden by local weather (e.g., Rodríguez and Bustamante, 2003; Gordo and Sanz, 2010). The extremely variable nature of the Mediterranean climate among years as well as the complex balances between temperature and rainfall could be the cause for the poor correlations with the NAO. However, a larger sample of studies would be needed to distinguish unambiguously between a real absence of an effect and a simple lack of evidence.

In view of the current climate change, we need to improve our knowledge about climate effects on organisms. The NAO, as a major driver of climate in the Mediterranean region, may become a keystone to the understanding of past, present and future impacts of climate change on Mediterranean ecosystems. However, the mechanisms by which the NAO influences organisms in Mediterranean ecosystems are different and to some extent more complex than they are in more northern latitudes. These facts demand of the Mediterranean ecologists more original thought and more effort to start long-term monitoring programs of biodiversity and to enhance and maintain those already in progress.

Acknowledgements Oscar Gordo was supported by a Juan de la Cierva contract (ref. JCI-2009-05274).

References

Abella A, Fiorentino F, Mannini A, Relini LO (2008) Exploring relationships between recruitment of European hake (*Merluccius merluccius* L. 1758) and environmental factors in the Ligurian Sea and the Strait of Sicily (Central Mediterranean). J Mar Syst 71:279–293

Almaraz P, Amat JA (2004a) Complex structural effects of two hemispheric climatic oscillators on the regional spatio-temporal expansion of a threatened bird. Ecol Lett 7:547–556

Almaraz P, Amat JA (2004b) Multi-annual spatial and numeric dynamics of the white-headed duck *Oxyura leucocephala* in southern Europe: seasonality, density dependence and climatic variability. J Anim Ecol 73:1013–1023

Avolio E, Pasqualoni L, Federico S, Fornaciari M, Bonofiglio T, Orlandi F, Bellecci C, Romano B (2008) Correlation between large-scale atmospheric fields and the olive pollen season in Central Italy. Int J Biometeorol 52:787–796

Balbontín J, Møller AP, Hermosell IG, Marzal E, Reviriego M, de Lope F (2009) Divergent patterns of impact of environmental conditions on life history traits in two populations of a long-distance migratory bird. Oecologia 159:859–872

Bechet A, Johnson AR (2008) Anthropogenic and environmental determinants of Greater Flamingo *Phoenicopterus roseus* breeding numbers and productivity in the Camargue (Rhone delta, southern France). Ibis 150:69–79

Begon M, Townsend CR, Harper JL (2006) Ecology: from individuals to ecosystems, 4th edn. Blackwell Publishing, Oxford

Berthold P (2001) Bird migration: a general survey. Oxford University Press, Oxford

Bosch J, Carrascal LM, Durán L, Walker S, Fisher MC (2007) Climate change and outbreaks of amphibian chytridiomycosis in a montane area of Central Spain: is there a link? Proc Roy Soc Lond B 274:253–260

Bridges CR, Krohn O, Deflorio M, de Metrio G (2009) Possible NAO and SST influences on the eastern bluefin tuna stock – the in-exfish approach. Collect Vol Sci Papers ICCAT 63: 138–152

Campelo F, Nabais C, García-González I, Cherubini P, Gutiérrez E, Freitas H (2009) Dendrochronology of *Quercus ilex* L. and its potential use for climate reconstruction in the Mediterranean region. Can J For Res 39:2486–2493

Cartes JE, Maynou F, Fanelli E, Papiol V, Lloris D (2009) Long-term changes in the composition and diversity of deep-slope megabenthos and trophic webs off Catalonia (western Mediterranean): Are trends related to climatic oscillations? Prog Oceanogr 82:32–46

Costantini D, Carello L, Dell'Omo G (2010) Patterns of covariation among weather conditions, winter North Atlantic Oscillation index and reproductive traits in Mediterranean kestrels. J Zoology 280:177–184

de Puelles MLF, Alemany F, Jansa J (2007) Zooplankton time-series in the Balearic Sea (western Mediterranean): Variability during the decade 1994–2003. Prog Oceanogr 74:329–354

de Puelles MLF, Molinero JC (2008) Decadal changes in hydrographic and ecological time-series in the Balearic Sea (western Mediterranean), identifying links between climate and zooplankton. ICES J Mar Sci 65:311–317

Doi H, Gordo O, Katano I (2008) Heterogeneous intra-annual climatic changes drive different phenological responses at two trophic levels. Clim Res 36:181–190

Doi H, Takahashi M (2008) Latitudinal patterns in the phenological responses of leaf colouring and leaf fall to climate change in Japan. Glob Ecol Biogeogr 17:556–561

Donnelly A, Cooney T, Jennings E, Buscardo E, Jones M (2009) Response of birds to climatic variability: evidence from the western fringe of Europe. Int J Biometeorol 53:211–220

Dulcic J, Grbec B, Lipej L, Paklar GB, Supic N, Smircic A (2004) The effect of the hemispheric climatic oscillations on the Adriatic ichthyofauna. Fresenius Environ Bull 13:293–298

Esteves MA, Orgaz MDM (2001) The influence of climatic variability on the quality of wine. Int J Biometeorol 45:13–21

Figuerola J (2007) Climate and dispersal: Black-winged stilts disperse further in dry springs. PLoS ONE 2:e539

Forchhammer MC, Post E (2004) Using large-scale climate indices in climate change ecology studies. Popul Ecol 46:1–12

Forchhammer MC, Post E, Stenseth NC (2002a) North Atlantic Oscillation timing of long- and short-distance migration. J Anim Ecol 71:1002–1014

Forchhammer MC, Post E, Stenseth NC, Boertmann DM (2002b) Long-term responses in arctic ungulate dynamics to changes in climatic and trophic processes. Popul Ecol 44:113–120

Gienapp P, Leimu R, Merilä J (2007) Responses to climate change in avian migration time – microevolution or phenotypic plasticity? Clim Res 35:25–35

Gimeno L, Ribera P, Iglesias R, de la Torre L, García R, Hernández E (2002) Identification of empirical relationships between indices of ENSO and NAO and agricultural yields in Spain. Clim Res 21:165–172

Gladan ZN, Marasovic I, Grbec B, Skejic S, Buzancic M, Kuspilic G, Matijevic S, Matic F (2010) Inter-decadal variability in phytoplankton community in the Middle Adriatic (Katela Bay) in relation to the North Atlantic Oscillation. Estuaries Coasts 33:376–383

Gordo O (2007) Why are bird migration dates shifting? A review of weather and climate effects on avian migratory phenology. Clim Res 35:37–58

Gordo O, Brotons L, Ferrer X, Comas P (2005) Do changes in climate patterns in wintering areas affect the timing of the spring arrival of trans-Saharan migrant birds? Glob Change Biol 11: 12–21

Gordo O, Sanz JJ (2008) The relative importance of conditions in wintering and passage areas on spring arrival dates: the case of long-distance Iberian migrants. J Ornithol 149:199–210

Gordo O, Sanz JJ (2010) Impact of climate change on plant phenology in Mediterranean ecosystems. Glob Change Biol 16:1082–1106

Gouveia C, Trigo RM, DaCamara CC, Libonati R, Pereira JMC (2008) The North Atlantic Oscillation and European vegetation dynamics. Int J Climatol 28:1835–1847

Grbec B, Dulcic J, Morovic M (2002) Long-term changes in landings of small pelagic fish in the eastern Adriatic – possible influence of climate oscillations over the Northern Hemisphere. Clim Res 20:241–252

Grosbois V, Henry PY, Blondel J, Perret P, Lebreton JD, Thomas DW, Lambrechts MM (2006) Climate impacts on Mediterranean blue tit survival: an investigation across seasons and spatial scales. Glob Change Biol 12:2235–2249

Grubelic I, Antolic B, Despalatovic M, Grbec B, Paklar GB (2004) Effect of climatic fluctuations on the distribution of warm-water coral *Astroides calycularis* in the Adriatic Sea: new records and review. J Mar Biol Assoc UK 85:599–602

Guisande C, Vergara AR, Riveiro I, Cabanas JM (2004) Climate change and abundance of the Atlantic-Iberian sardine (*Sardina pilchardus*). Fish Oceanogr 13:91–101

Hallett TB, Coulson T, Pilkington JG, Clutton-Brock TH, Pemberton JM, Grenfell BT (2004) Why large-scale climate indices seem to predict ecological processes better than local weather. Nature 430:71–75

Hoerling MP, Hurrell JW, Xu T (2001) Tropical origins for recent North Atlantic climate change. Science 292:90–92

Hubálek Z, Čapek M (2008) Migration distance and the effect of North Atlantic Oscillation on the spring arrival of birds in Central Europe. Folia Zoologica 57:212–220

Hüppop O, Hüppop K (2003) North Atlantic Oscillation and timing of spring migration in birds. Proc Roy Soc Lond B 270:233–240

Hurrell JW (1995) Decadal trends in the North Atlantic Oscillation: regional temperatures and precipitation. Science 269:676–679

Hurrell JW, Van Loon H (1997) Decadal variations in climate associated with North Atlantic Oscillation. Clim Change 36:301–326

Jenouvrier S, Thibault JC, Viallefont A, Vidal P, Ristow D, Mougin JL, Brichetti P, Borg JJ, Bretagnolle V (2009) Global climate patterns explain range-wide synchronicity in survival of a migratory seabird. Glob Change Biol 15:268–279

Jiguet F, Doxa A, Robert A (2008) The origin of out-of-range pelicans in Europe: wild bird dispersal or zoo escapes? Ibis 150:606–618

Jonzén N, Lindén A, Ergon T, Knudsen E, Vik JO, Rubolini D, Piacentini. D, Brinch C, Spina F, Karlsson L, Stervander M, Andersson A, Waldenström J, Lehikoinen A, Edvardsen E, Solvang R, Stenseth NC (2006) Rapid advance of spring arrival dates in long-distance migratory birds. Science 312:1959–1961

Kamburska L, Fonda-Umani S (2009) From seasonal to decadal inter-annual variability of mesozooplankton biomass in the Northern Adriatic Sea (Gulf of Trieste). J Mar Syst 78: 490–504

Katara I, Illian J, Pierce GJ, Scott B, Wang JJ (2008) Atmospheric forcing on chlorophyll concentration in the Mediterranean. Hydrobiologia 612:33–48

Kralj J, Dolenec Z (2008) First arrival dates of the Nightingale (*Luscinia megarhynchos*) to Central Croatia in the early 20th century and at the turn of the 21st century. Cent Eur J Biol 3: 295–298

Lamb PJ, Peppler RA (1987) North Atlantic Oscillation: concept and application. Bull Am Meteorol Soc 68:1218–1225

Lebrune C, Gremare A, Guizien K, Amouroux JM (2007) Long-term comparison of soft bottom macrobenthos in the Bay of Banyuls-sur-Mer (north-western Mediterranean Sea): a reappraisal. J Sea Res 58:125–143

Lehikoinen E, Sparks TH (2010) Bird migration. In: Møller AP, Fiedler W, Berthold P (eds) Effects of climate change on birds. Oxford University Press, Oxford, pp 89–112

Lloret J, Lleonart J, Solé I, Fromentin JM (2001) Fluctuations of landings and environmental conditions in the north-western Mediterranean Sea. Fish Oceanogr 10:33–50

López-Moreno JI, Vicente-Serrano SM (2008) Positive and negative phases of the winter-time North Atlantic Oscillation and drought occurrence over Europe: a multitemporal-scale approach. J Clim 21:1220–1243

Martínez-Jauregui M, San Miguel-Ayanz A, Mysterud A, Rodríguez-Vigal C, Clutton-Brock TH, Langvatn R, Coulson T (2009) Are local weather, NDVI and NAO consistent determinants of red deer weight across three contrasting European countries? Glob Change Biol 15: 1727–1738

Massutí E, Monserrat S, Oliver P, Moranta J, López-Jurado JL, Marcos M, Hidalgo M, Guijarro B, Carbonell A, Pereda P (2008) The influence of oceanographic scenarios on the population dynamics of demersal resources in the western Mediterranean: hypothesis for hake and red shrimp off Balearic Islands. J Mar Syst 71:421–438

Maynou F (2008a) Environmental causes of the fluctuations of red shrimp (*Aristeus antennatus*) landings in the Catalan Sea. J Mar Syst 71:294–302

Maynou F (2008b) Influence of the North Atlantic Oscillation on Mediterranean deep-sea shrimp landings. Clim Res 36:253–257

Mazzochi MG, Christou ED, Di Capua I, de Puelles MLF, Fonda-Umani S, Molinero JC, Nival P, Siokou-Frangou I (2007) Temporal variability of *Centropages typicus* in the Mediterranean Sea over seasonal-to-decadal scales. Prog Oceanogr 72:214–232

McHugh MJ, Rogers JC (2001) North Atlantic Oscillation influence on precipitation variability around the Southeast African convergence zone. J Clim 14:3631–3642

Molinero JC, Ibañez F, Souissi S, Buecher E, Dallot S, Nival P (2008) Climate control on the long-term anomalous changes of zooplankton communities in the northwestern Mediterranean. Glob Change Biol 14:11–26

Molinero JC, Ibañez F, Souissi S, Chifflet M, Nival P (2005) Phenological changes in the north-western Mediterranean copepods *Centropages typicus* and *Temora stylifera* linked to climate forcing. Oecologia 145:640–649

Mysterud A, Stenseth NC, Yoccoz NG, Ottersen G, Langvatn R (2003) The response of terrestrial ecosystems to climate variability associated with the North Atlantic Oscillation. In: Hurrell JW, Kushnir Y, Ottersen G, Visbeck MH (eds) The North Atlantic Oscillation: climatic significance and environmental impact. American Geophysical Union, Washington, pp 235–262

Newton I (2008) The migration ecology of birds. Academic Press, London

Oba G, Post E, Stenseth NC (2001) Sub-Saharan desertification and productivity are linked to hemispheric climate variability. Glob Change Biol 7:241–246

Orlandi F, García-Mozo H, Galán C, Romano B, Díaz de la Guardia C, Ruiz L, del Mar Trigo M, Domínguez-Vilches E, Fornaciari M (2010) Olive flowering trends in a large Mediterranean area (Italy and Spain). Int J Biometeorol 54:151–163

Ottersen G, Planque B, Belgrano A, Post E, Reid PC, Stenseth NC (2001) Ecological effects of the North Atlantic Oscillation. Oecologia 128:1–14

Parmesan C (2006) Ecological and evolutionary responses to recent climate change. Annu Rev Ecol Evol Syst 37:637–669

Parmesan C, Yohe G (2003) A globally coherent fingerprint of climate change impacts across natural systems. Nature 421:37–42

Piovesan G, Schirone B (2000) Winter North Atlantic Oscillation effects on the tree rings of the Italian beech (*Fagus sylvatica* L.). Int J Biometeorol 44:121–127

Pérez FF, Padín XA, Pazos Y, Gilcoto M, Cabanas M, Pardo PC, Doval MD, Farina-Busto L (2010) Plankton response to weakening of the Iberian coastal upwelling. Glob Change Biol 16:1258–1267

Quadrelli R, Pavan V, Molteni F (2001) Wintertime variability of Mediterranean precipitation and its links with large-scale circulation anomalies. Clim Dyn 17:457–466

Rainio K, Laaksonen T, Ahola M, Vähätalo AV, Lehikoinen E (2006) Climatic responses in spring migration of boreal and arctic birds in relation to wintering area and taxonomy. J Avian Biol 37:507–515

Ravier C, Fromentin JM (2004) Are the long-term fluctuations in Atlantic bluefin tuna (*Thunnus thynnus*) population related to environmental changes? Fish Oceanogr 13:145–160

Relini LO, Mannini A, Fiorentino F, Palandri G, Relini G (2006) Biology and fishery of *Eledone cirrhosa* in the Ligurian Sea. Fish Res 78:72–88

Relini LO, Palandri G, Garibaldi F, Relini M, Cima C, Lanteri L (2010) Large pelagic fish, sword-fish, bluefin and small tunas, in the Ligurian Sea: biological characteristics and fishery trends. Chem Ecol 26:341–357

Robson D, Barriocanal C (2011) Ecological conditions in wintering and passage areas as determinants of timing of spring migration in trans-Saharan migratory birds. J Anim Ecol 80:320–331

Rodó X, Baert E, Comín FA (1997) Variations in seasonal rainfall in Southern Europe during the present century: relationships with the North Atlantic Oscillation and the El Niño-Southern Oscillation. Clim Dyn 13:275–284

Rodó X, Comín FA (2000) Links between large-scale anomalies, rainfall and wine quality in the Iberian Peninsula during the last three decades. Glob Change Biol 6: 267–273

Rodrigo FS, Trigo RM (2007) Trends in daily rainfall in the Iberian Peninsula from 1951 to 2002. Int J Climatol 27:513–529

Rodríguez C, Bustamante J (2003) The effect of weather on lesser kestrel breeding success: can climate change explain historical population declines? J Anim Ecol 72:793–810

Rodríguez C, Naves J, Fernández-Gil A, Obeso JR, Delibes M (2007) Long-term trends in food habits of a relict brown bear population in northern Spain: the influence of climate and local factors. Environ Conserv 34:36–44

Rodríguez-Puebla C, Ayuso SM, Frías MD, García-Casado LA (2007) Effects of climate variation on winter cereal production in Spain. Clim Res 34:223–232

Root TL, Price JT, Hall KR, Schneider SH, Rosenzweig C, Pounds JA (2003) Fingerprints of global warming on wild animals and plants. Nature 421:57–60

Rosenzweig C, Karoly D, Vicarelli M, Neofotis P, Wu Q, Casassa G, Menzel A, Root TL, Estrella N, Seguin B, Tryjanowski P, Liu C, Rawlins S, Imeson A (2008) Attributing physical and biological impacts to anthropogenic climate change. Nature 435:353–357

Rozas V, Lamas S, García-González I (2009) Differential tree-growth responses to local and large-scale climatic variation in two *Pinus* and two *Quercus* species in northwest Spain. Ecoscience 16:299–310

Rubolini D, Ambrosini R, Caffi M, Brichetti P, Armiraglio S, Saino N (2007) Long-term trends in first arrival and first egg laying dates of some migrant and resident bird species in northern Italy. Int J Biometeorol 51:553–563

Saino N, Rubolini D, Jonzén N, Ergon T, Montemaggiori A, Stenseth NC, Spina F (2007) Temperature and rainfall anomalies in Africa predict timing of spring migration in trans-Saharan migratory birds. Clim Res 35:123–134

Saino N, Szep T, Romano M, Rubolini D, Spina F, Møller AP (2004) Ecological conditions during winter predict arrival date at the breeding quarters in a trans-Saharan migratory bird. Ecol Lett 7:21–25

Sala OE, Chapin FS, Armesto JJ, Berlow E, Bloomfield J, Dirzo R, Huber-Sanwald E, Huenneke LF, Jackson RB, Kinzig A, Leemans R, Lodge DM, Mooney HA, Oesterheld M, LeRoy Poff N, Sykes MT, Walker BH, Walker M, Wall DH (2000) Global biodiversity scenarios for the year 2100. Science 287:1770–1774

Santojanni A, Arneri E, Bernardini V, Cingolani N, Di Marco M, Russo A (2006) Effects of environmental variables on recruitment of anchovy in the Adriatic Sea. Clim Res 31:181–193

Sanz JJ (2002) Climate change and breeding parameters of great and blue tits throughout the western Palaeartic. Glob Change Biol 8:408–422

Sanz JJ (2003) Large-scale effect of climate change on breeding parameters of pied flycatchers in Western Europe. Ecography 26:45–50

Sæther BE, Sutherland WJ, Engen S (2004) Climate influences on avian population dynamics. Adv Ecol Res 35:185–209

Sæther BE, Tufto J, Engen S, Jerstad K, Rostad OW, Skåtan JE (2000) Population dynamical consequences of climate change for a small temperate songbird. Science 287:855–856

Sinelschikova A, Kosarev V, Panov I, Baushev AN (2007) The influence of wind conditions in Europe on the advance in timing of the spring migration of the song thrush (*Turdus philomelos*) in the south-east Baltic region. Int J Biometeorol 51:431–440

Solomon S, Qin D, Manning M, Chen Z, Marquis M, Averyt KB, Tignor M, Miller HL (eds) (2007) Climate change 2007: the physical science basis. Contribution of Working Group I to the 4th assessment report of the Intergovernmental panel on climate change. Cambridge University Press, Cambridge

Spina F, Massi A, Montemaggiori A, Baccetti N (1993) Spring migration across central Mediterranean: general results from the 'Progetto Piccole Isole'. Die Vogelwarte 37:1–94

Stenseth NC, Ehrich D, Rueness EK, Lingjærde OC, Chan KS, Boutin S, O'Donoghue M, Robinson DA, Viljugrein H, Jakobsen KS (2004) The effect of climatic forcing on population synchrony and genetic structuring of the Canadian lynx. Proc Natl Acad Sci USA 101:6056–6061

Stenseth NC, Mysterud A (2005) Weather packages: finding the right scale and composition of climate in ecology. J Anim Ecol 74:1195–1198

Stenseth NC, Mysterud A, Ottersen G, Hurrell JW, Chan KS, Lima M (2002) Ecological effects of climate fluctuations. Science 297:1292–1296

Stenseth NC, Ottersen G, Hurrell JW, Mysterud A, Lima M, Chan KS, Yoccoz NG, Ådlandsvik B (2003) Studying climate effects on ecology through the use of climate indices: the North Atlantic Oscillation, El Niño Southern Oscillation and beyond. Proc Roy Soc Lond B 270:2087–2096

Stige LC, Stave J, Chan KS, Ciannelli L, Pettorelli N, Glantz M, Herren HR, Stenseth NC (2006) The effect of climate variation on agro-pastoral production in Africa. Proc Natl Acad Sci USA 103:3049–3053

Tøttrup AP, Rainio K, Coppack T, Lehikoinen E, Rahbek C, Thorup K (2010) Local temperature fine-tunes the timing of spring migration in birds. Integr Comp Biol 50:293–304

Trigo RM, Pozo-Vázquez D, Osborn TJ, Castro-Díez Y, Gámiz-Fortis S, Esteban-Parra MJ (2004) North Atlantic oscillation influence on precipitation, river flow and water resources in the Iberian Peninsula. Int J Climatol 24:925–944

Vicente-Serrano SM, Cuadrat JM (2007) North Atlantic Oscillation control of droughts in north-east Spain: evaluation since 1600 A.D. Clim Change 85:357–379

Visbeck MH, Hurrell JW, Polvani L, Cullen HM (2001) The North Atlantic Oscillation: past, present, and future. Proc Natl Acad Sci USA 98:12876–12877

Impacts of the NAO on Atmospheric Pollution in the Mediterranean Basin

Uri Dayan

Abstract There are several inherent problems in attributing pollution concentrations to changes in large-scale atmospheric circulation: (1) the year to year variability being modulated by both, changes in circulation and changes in upwind emissions, (2) the shorter life-time of some pollutants precluding a meaningful relationship with changes in circulation, and (3) the both-ways interaction between trace gases, aerosols and climate. Simulations of transport of anthropogenic CO for high and low phases of the NAO are presented followed by an observational-based study relating the ozone seasonal variability across North Atlantic and the Western Mediterranean to the NAO. Both phases of the NAO controlling dust transport to the Mediterranean are described: the positive phase during summer over the western region and the negative one regulating dust transport over the Eastern Mediterranean in winter. Low NAO indices have been related to a higher cyclonic activity over the western basin. On the contrary, other studies found no correlation between annual wet deposition of African dust-related elements and the NAO. Although, at present, there is no consensus on the process responsible for the observed low-frequency variations in the NAO, most of them predict the continuation of the positive NAO state until the mid twenty-first century implying a predominantly zonal circulation over the western Mediterranean. This imply a reduced import of European trace gases, an enhancement of long range transport of air pollutants from North American sources and conditions in favor of mobilization and transport of North African dust mainly to the western part of this fragile basin.

Keywords Air Pollution · Atmospheric circulation · Environment · Mediterranean Basin · NAO

U. Dayan (✉)
Department of Geography, The Hebrew University of Jerusalem, Jerusalem, Israel
e-mail: msudayan@mscc.huji.ac.il

S.M. Vicente-Serrano, R.M. Trigo (eds.), *Hydrological, Socioeconomic and Ecological Impacts of the North Atlantic Oscillation in the Mediterranean Region*, Advances in Global Change Research 46, DOI 10.1007/978-94-007-1372-7_12,

1 Introduction: Problems in Attributing Pollution Concentrations to Changes in Large-Scale Atmospheric Circulation

The variations of atmospheric pollution concentrations and their relationships to large-scale atmospheric circulation systems have been studied for a while (e.g., Davis and Kalkstein, 1990; Dayan and Lamb, 2008; Dayan et al., 2008; Kalderon-Asael et al., 2009). Secondary pollutants such as ozone are primarily a result of photochemical reactions among anthropogenic precursor emissions and as such, are governed by air mass constituents such as temperature and humidity. Accordingly, variation in trace gases concentrations are modulated by exposure to differing air masses as controlled by changes in atmospheric circulation. Problems rising in attributing pollution concentrations to changes in large-scale atmospheric circulation stem from the fact that changes in upwind emissions and removal processes are not necessarily synchronous with variations in circulation. Some efforts were done, mainly through, coupled climate-chemistry models to treat and analyze simultaneously the changes in general circulation and atmospheric chemistry (Hein et al., 2001; Dastoor and Larocque, 2004).

A second major drawback in trying to relate changes in pollutant concentration to variation in circulation patterns is their different life time and distribution. For example, aerosols are most relevant for short spatial and temporal scales as compared to long-lived greenhouse gases such as CH_4 and CO_2 (Andrea, 2001) for which, various meteorological and emission effects dominate large parts of the globe (Voulgarakis et al., 2009). Radiative forcing of aerosols are of higher spatial variability than Green House Gases (GHG) forcings due to the relatively short aerosol life time of the order of a week (Kostler et al., 2010). Consequently, such short life time of the pollutants preclude a meaningful relationship with changes in large scale circulations.

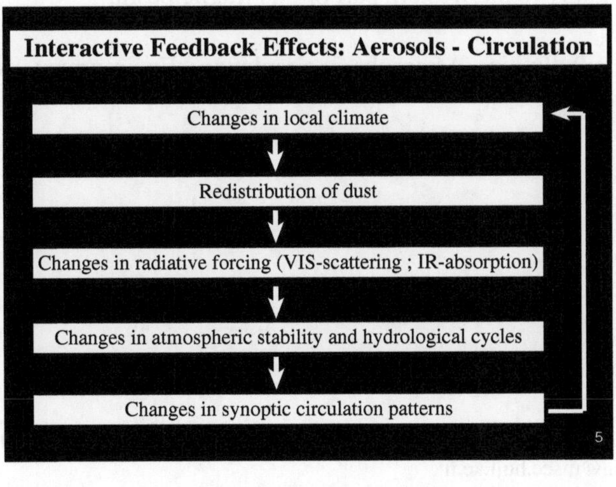

Fig. 1 Flow diagram of the interactive feedback effects: aerosols – circulation

Beside these above mentioned problems, at times, both ways interactions exist between trace gases, aerosols and atmospheric circulation patterns. Tegen et al. (1996) described the interactive feedback effects between mineralogical aerosols and atmospheric circulation emphasizing that: "...both sources and removal of dust are dependent on local climate, changes in radiative forcing by dust may induce changes in regional circulation patterns and therefore to further changes in dust sources and sinks...". Such interactive feedback effects between aerosols and circulation is shown in Fig. 1.

2 Fundamental Approaches Relating Circulation to Environment - Examples

Due to the strong potential for highly interactive feedback effects between typical environmental phenomena related to several modes of the circulation, a methodology consisting of both interactions is needed. This methodology will take into account both approaches: on how the circulation affects the environment (i.e., climatically related variables such as air pollutants) and how the environment affects the circulation. Here, we adopted Yarnal (1993) ideas who addressed these both approaches followed by some examples on studies relating circulation to the environment.

Yarnal explained that in the "Circulation to Environment" classification you first classify the circulation into distinct several patterns and then relate it to the environment, whereas in the second method you classify the circulation data along environment-based criteria (Fig. 2).

The first example demonstrating the impact of circulation on air trace gases is a recent study by Theoharatos et al. (2010) investigating the relationship between synoptic-scale circulation patterns, induced thermal stresses and its possible association to air pollution over Athens, Greece. Here, the authors have shown that poor air quality conditions for SO_2 and O_3 were related to a northward displacement of the Subtropical Jet Stream and large-scale subsidence.

Liu et al. (2009) explained the "Middle East ozone maximum", a summertime enhancement in ozone observed by the Tropospheric Emission Spectrometer satellite instrument as caused by strong descent in the middle and upper troposphere featuring both Arabian and Saharan anticyclones (Fig. 3).

Another study showing the impact of circulation on aerosols was done by Moulin et al. (1998) who used 11-year of daily Meteosat data to develop a satellite-derived climatology of African dust transport. They found that the dust transport patterns are related to cyclogenesis over North Africa.

The last example, this time, depicting the impact of air quality on climate is Lelieveld et al. (2002) study which indicates the influence of elevated European sulfate emissions on cooling of the Mediterranean sea surface temperatures during the mid-1970s. This cooling led to a reduction in evaporation and consequently, less precipitation over the basin.

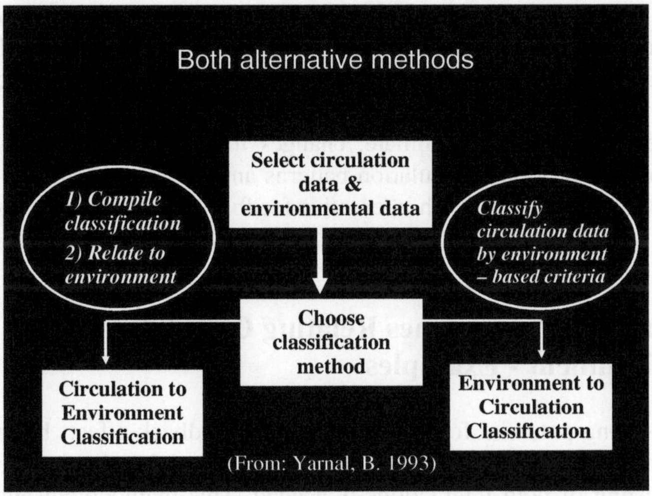

Fig. 2 Both alternative methods to relate environmental phenomena to circulation (adapted from Yarnal, 1993)

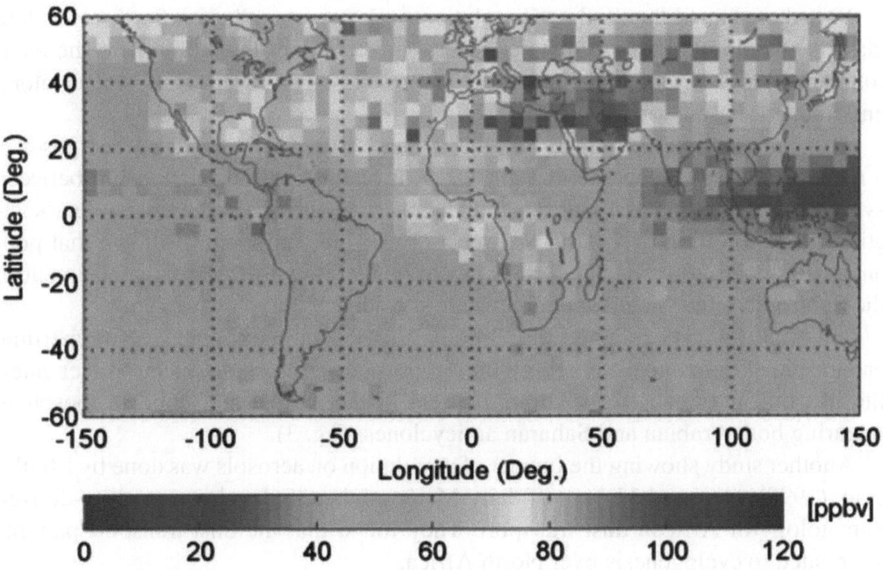

Fig. 3 Monthly mean ozone mixing ratio in July 2005 at 464 hPa from the Tropospheric Emission Spectrometer (from Liu et al., 2009)

3 Circulation to Environment: The Role of the NAO in Controlling the Pathways of Pollution to the Mediterranean

The North Atlantic Oscillation (NAO) index is determined by the intensity and position of the Icelandic low and the Bermuda Azores high. Several studies have identified the NAO as the dominant mode of variability of the surface atmospheric circulation in the Atlantic (Barnston and Livezey, 1987; Kushnir and Wallace, 1989; Hurrell, 1995). The NAO affects not only climatic conditions, but also, climatic-derived conditions such as air pollutant transport and dispersion. Consequently, this is also the oscillation influencing pollution transport over the Northern Hemisphere at large, and more specifically over Western Europe and the western Mediterranean (Eckhardt et al., 2003). Further on, several studies results will be presented to show the strong influence of the NAO on the strengths and locations of the main pollutant transport pathways over the Mediterranean. These changes in these pathways are mainly changes in the mean circulation patterns over Western Europe and the western Mediterranean Basin associated with the NAO and believed to be the main factor for their interannual variations.

Eckhardt et al. (2003) used the Flexpart Lagrangian particle dispersion model to simulate the transport of CO over Europe and the Mediterranean Basin. CO was used as tracer of anthropogenic pollution due to its long life time. The simulation was performed for the 20% lowest and highest NAO for December through February of 1979–1993. They found out that during positive NAO phases, reduced CO outflow from Central Europe to the Mediterranean were obtained due to stronger westerly winds flowing over Europe and most of the pollution was transported northward. These strong westerlies characterizing positive phases of the NAO serve as conveyor belt to transport O_3 formed over the Eastern US to Europe and the Western Mediterranean. This conclusion was reached by Crielson et al. (2003) who used the tropospheric O_3 residual technique which utilizes coincident observations of Total Ozone Mapping Spectrometer (TOMS) and stratospheric O_3 profiles from the Solar Backscatter Ultraviolet instruments to examine the spatial distribution of O_3 and its positive relationship with the NAO (Fig. 4).

NAO controls not only trace gases but particulate as well. Moulin et al. (1997) used Meteosat data from 1983 to 1994 to develop a satellite climatology of African dust transport to the Mediterranean. They found that the interannual variations in dust transport over the Western Mediterranean is well correlated ($R=0.66$, $p=0.027$) with the NAO index during summertime. This positive correlation is explained by the frequent passage of cyclones sweeping the western basin during positive phases of the NAO and mobilizing dust from North Africa.

Contrary to the NAO positive phase impact on the dust transport during summer over the western basin, Dayan et al. (2008) presented evidence on the role of the negative phase of the NAO in controlling dust transport during winter to the South Eastern Mediterranean (SEM). In their study, they examined the atmospheric circulation governing dust over the SEM for the "high dust season" (i.e., October–May)

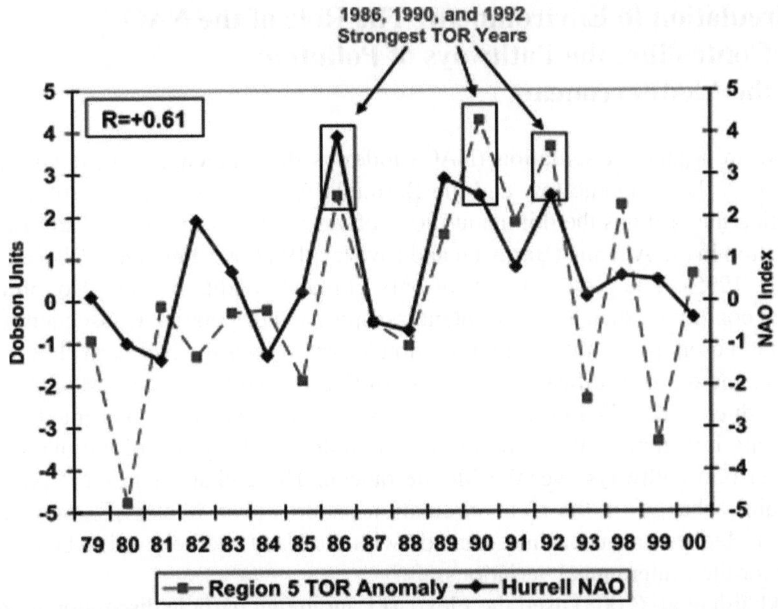

Fig. 4 Springtime depiction showing the interannual variability and positive relationship between springtime tropospheric ozone residual anomaly for Region 5: Eastern North Atlantic and Western Europe and springtime Hurrell NAO index from 1979–2000 (from Crielson et al., 2003)

for 37 years (1968–2004) and found a significant correlation ($R = -0.66$) between the interannual variability of accumulated dust over Beer-Sheba, located in southern Israel which represents the SEM and the NAO modulating the cyclonic activity over the Mediterranean (Fig. 5).

These results indicate that the synoptic system that produces the majority of the dust over the SEM is the Cyprus Low, a cold cyclone, contributing over 2/3 of both, the total yearly dust and rain fields. Namely, the same synoptic system controls both modes of deposition, i.e., wet and dry, and therefore have a similar relationship with the NAO over the SEM.

In another study, Avila and Roda (2002) could not find any correlation between the NAO and wet deposition of African dust over a rural western Mediterranean site in north-east Spain. They suggested that contrary to the Eastern Mediterranean, the two variables controlling wet deposition over the Western Mediterranean vary in an opposite direction with respect to the NAO, i.e., precipitation inversely and dust updraft directly, therefore, canceling each other effects.

In summary we can say that both phases of the NAO have a major role in controlling the transport of particulate matter (mineralogical dust) over the Mediterranean Basin. However, this control is seasonal dependent and differs substantially between the two Mediterranean halves, with the positive phase of the NAO affecting the Western Mediterranean during summertime, whereas, its negative phase produces a larger impact over the Eastern Mediterranean and during winter.

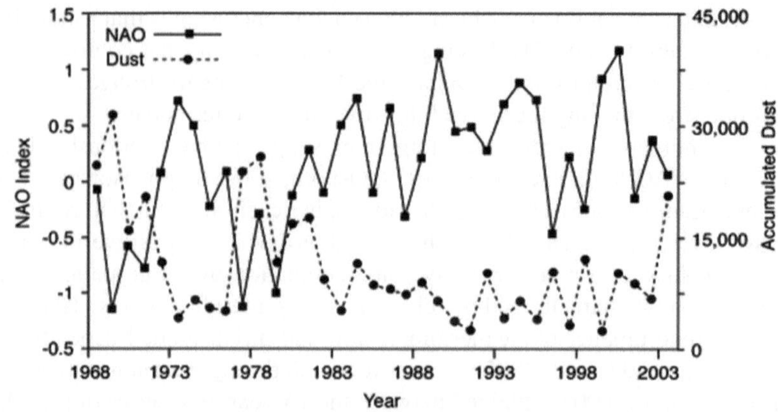

Fig. 5 Annual variation of accumulated dust over southern Israel representing the SEM for October–May (*dashed*) and the NAOI averaged over December–March (*solid*) (from Dayan et al., 2008)

4 Environment to Circulation: Influence of GHG Induced Climate Change on the NAO

In the last section under the "Circulation to Environment" concept, we dealt with the role of the NAO in controlling the export and import pathways of pollution to the Mediterranean Basin based on observations. Here, we will present some studies results, based mainly, on atmospheric model simulations to demonstrate the influence of the increase GHG concentrations on the NAO variability as a sign of global climate change.

Paeth et al. (1999) examined the NAO variability with an observational data set and several model data sets including GHG-induced external forcings. They have shown that the IPCC (BAU) scenario results predict the continuation of the recent positive NAO phase until mid of the twenty-first century. Furthermore, they found statistical evidence that radiative forcing due to increasing GHG concentrations has an impact on the simulated variability of the NAO on time-scale exceeding 60 years. Never the less, such evidence could not be discerned for the real NAO due to its strong interannual variability. The implication of a persisting positive phase of the NAO for the next 30–50 years is a predominantly zonal circulation over Western Europe and less pollution from European sources transported to the Mediterranean Basin.

Osborn (2004) analyzed the simulated response of the winter NAO to GHG forcing using seven coupled climate models. Even though, not all models showed agreement between NAO pattern and regional scale changes expressed as sea level pressure trends for winter atmospheric circulation, a clear correspondence was found for two coupled ocean-atmosphere models, the CCSR (Emori et al., 1999) and the ECHAM-4 (Bacher et al., 1998). All models showed a mean increase of 0.25 hPa/K of global mean temperature increase caused by enhanced GHG forcing

over the Mediterranean Basin. Osborn (2004) main conclusion is that a combination of internal variability and GHG forcing as external variability, has contributed to the NAO increase observed from the 1960s to the 1990s leading to strongest zonal wind over Europe strengthening meridional flow over the Mediterranean.

Based on numerous papers indicating an existing positive relationship between GHG and the NAO (e.g., Paeth et al., 1999; Osborn, 2004; Stephenson et al., 2006), Dayan and Lamb (2005) analyzed the atmospheric circulation that controls wet deposition over Europe and the northern Mediterranean. A positive phase of the NAO is indicative of a predominantly zonal circulation over Europe as compared to a negative phase characterized by relatively strong meridional flows (Paeth et al., 1999). The predominance of the meridional flow and sulfate loaded-deposition years found by Dayan and Lamb (2005) is consistent with the significant negative correlation ($R = -0.43$, $p < 0.005$) obtained between the 16-year time series of the 925 hPa meridional wind and the NAO. A further confirmation of this phenomenon is the significant correlation ($R = -0.39$, $p < 0.005$) found between the meridional wind and sulfate deposition over Europe. The overall downward trend in sulfate deposition observed for the last 16 years over Europe and the northern Mediterranean Basin, the concurrent increases in average temperature (Klein Tank et al., 2002) and the increased frequency of extreme warm-weather events for the summer (Domonkos et al., 2003) suggest, along with the decline in SO_2 emissions, that this period was characterized by predominantly zonal circulation, as indicated by the trend towards greater prevalence of positive NAO. Consequently, Dayan and Lamb (2005) reached the conclusion that negative NAOI years correspond to predominant meridional flow leading to more sulfate deposition over Europe as compared to positive NAO years characterized by a zonal flow and less pollution (Fig. 6). Their conclusion corroborates Duncan and Bey (2004) who emphasized the strong influence of the NAO on the strengths and locations of the export pollution (e.g., O_3) pathways from Europe to the Mediterranean and northern Africa.

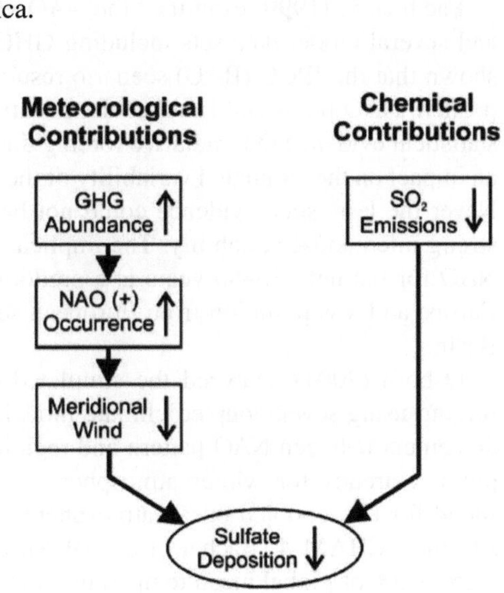

Fig. 6 Block diagram of the meteorological and chemical contributions to changes in sulfate deposition over southern Europe (from Dayan and Lamb, 2005)

5 Future Phases of the NAO and Possible Implications on Air Pollution over the Western and Eastern Mediterranean Basin

Most recent studies point at the difficulty to identify a preferred timescale of variability for the NAO as well as the large amount of its intra-seasonal variability as main reasons for the lack of consensus on the process responsible for its present observed low-frequency variations (Hurrell and Deser, 2010) and, obviously, our inability to predict its trend and variability in the future. Moreover, although this oscillation is natural, it seems to respond to external forcing in addition to its inherent internal variability.

Kuzmina et al. (2005), based on results obtained from 12 coupled climate models, found that the forced model runs resulted in stronger NAO as compared to the control runs. This indicates a possible intensification of the NAO in the future with further increases in GHG concentrations in the atmosphere.

In contrary to Kuzmina et al. (2005) study, Selten et al. (2004) suggest that the strengthening of the westerlies over the North Atlantic is not caused by GHG but rather to a climate variation governed by increased precipitation over the tropical Indian Ocean.

To summarize this section, we can say, with much prudence, that presently there are two opposed scientific views as regarded to the NAO trend and variability, several researchers consider that external climate forcings, such as volcanic eruptions or increased GHG, are the main drivers for a predominance of positive phases of the NAO while other groups consider that physical driven processes internal to the climate system are responsible in driving NAO variations.

Until the scientific community understands better the causes and effects of the observed NAO trend and variability, we should restrict our conclusions on future NAO's implication on air pollution over the Mediterranean Basin by presenting the anticipated impacts of both prevailing modes (positive and negative phases) over this basin as experienced so far. Based on our present limited knowledge, the future NAO's implications are as follow:

A prevailing positive NAO over the Western Mediterranean is likely to:

(1) Reduce the import of European trace gases;
(2) Enhance the long range transport of pollutants from upwind eastward sources, e.g., North America;
(3) Enhance the mobilization and transport of dust from North African sources to the Mediterranean.

A prevailing negative NAO over the Eastern Mediterranean is expected to enhance cyclonic conditions favorizing the passage of more deep cyclones mobilizing and transporting dust from North African sources to this part of the basin.

References

Andrea MO (2001) The dark side of aerosols. Nature 409:671–672

Avila A, Roda F (2002) Assessing decadal changes in rainwater alkalinity at a rural Mediterranean site in the Montseny Mountains (NE Spain). Atmos Environ 36:2881–2890

Bacher S, Oberhuber JM, Roeckner E (1998) ENSO dynamics and seasonal cycle in the tropical Pacific as simulated by the ECHAM4/OPYC3 coupled general circulation model. Clim Dyn 14:1659–1672

Barnston AG, Livezey RE (1987) Classification, seasonality and persistence of low frequency atmospheric circulation patterns. Mon Wea Rev 115:1083–1125

Crielson JK, Fishman J, Wozniak AE (2003) Intercontinental transport of troposheric ozone: a study of its seasonal variability across the North Atlantic utilizing tropospheric ozone residuals and its relationship to the North Atlantic Oscillation. Atmos Chem Phys 3:2053–2066

Dastoor AP, Larocque Y (2004) Global circulation of atmospheric mercury: a modeling study. Atmos Environ 38:147–161

Davis RE, Kalkstein LS (1990) Using a spatial synoptic classification to assess changes in atmospheric pollution concentrations. Phys Geog 11:320–342

Dayan U, Lamb D (2005) Global and synoptic-scale weather patterns controlling wet atmospheric deposition over Central Europe. Atmos Environ 39:521–533

Dayan U, Lamb D (2008) Influence of atmospheric circulation on the variability of wet sulfate deposition. Int J Climatol 28:1315–1324

Dayan U, Ziv B, Shoob T, Enzel Y (2008) Suspended dust over Southeastern Mediterranean and its relation to atmospheric circulations. Int J Climatol 28:915–924

Domonkos P, Kysely J, Piotrowicz P, Petrovic P, Likso T (2003) Variability of extreme temperature events in South-Central Europe during the 20th century and its relationship with large-scale circulation. Int J Climatol 23:987–1010

Duncan BN, Bey I (2004) A modeling study of the export pathways of pollution from Europe: seasonal and interannual variations (1987–1997). J Geophys Res 109:D08301. doi:10.1029/2003JD004079

Eckhardt D, Stohl A, Bierle S, Spichtinger N, James P, Forster C, Junker C, Wagner T, Platt U, Jennings SG (2003) The North Atlantic Oscillation controls air pollution transport to the Arctic. Atmos Chem Phys 3:1769–1778

Emori S, Kozawa T, Abe-Ouchi A, Numaguti A, Kimoto M, Nakajima T (1999) Couple ocean-atmosphere model experiments of future climate change with an explicit representation of sulfate aerosol scattering. J Meteorol Soc Jpn 77:1299–1307

Hein R, Dameris M, Schnadt C, Land C, Grewe V, Kohler I, Ponater M, Sausen R, Steil BB, Landgraf J, Bruhl C (2001) Results of an interactively coupled atmospheric chemistry – general circulation model: comparisons with observations. Annales Geophysicae 19:435–457

Hurrell JW (1995) Decadal trends in the North Atlantic Oscillation: regional temperatures and precipitation. Science 269:676–679

Hurrell JW, Deser C (2010) North Atlantic climate variability: the role of the North Atlantic Oscillation. J Mar Syst 79:231–244

Kalderon-Asael B, Erel Y, Sandler A, Dayan U (2009) Mineralogical and chemical characterization of suspended atmospheric particles over the Eastern Mediterranean based on synoptic-scale circulation patterns. Atmos Environ 43:3963–3970

Klein Tank AMG, Wijngaard JB, Konen JP, Bohm R, Demaree G, Gocheva A, Mileta M, Petrovic P (2002) Daily dataset of the 20th century surface air temperature and precipitation series for the European climate assessment. Int J Climatol 22:1441–1453

Kostler S, Deutener F, Feichter J, Raes F, Lohman U, Roeckner E, Fischer-Bruns I (2010) A GCM study of future climate response to aerosol pollution reductions. Clim Dyn 34:1177–1194

Kushnir Y, Wallace JM (1989) Low frequency variability in the Northern Hemisphere winter – geographical distribution, structure and time-scale dependence. J Atmos Sci 46:3122–3142

Kuzmina SI, Bengtsson L, Jahannenssen OM, Drange H, Bobylev LP, Miles MW (2005) The North Atlantic Oscillation and greenhouse-gas forcing. Geophys Res Lett 32:L04703. doi:10.1029/2004 GL021064

Lelieveld J, Berresheim H, Borrmann S et al (2002) Global air pollution crossroads over the Mediterranean. Science 298:794–799

Liu JJ, Dylan B, Jones A, Worden JR, Noone D, Parington M, Kar J (2009) Analysis of the summertime buildup of tropospheric O_3 abundances over the Mid-East and North Africa as observed by the Tropospheric Emission Spectrometer instrument. J Geophy Res 114:D05304. doi:10.1029/2008JD010993

Moulin C, Lambert CE, Dayan U, Masson V, Ramonet M, Bousquet P, Dulac F (1998) Satellite climatology of African dust transport in the Mediterranean atmosphere. J Geophys Res 103:13137–13144

Moulin C, Lambert CE, Dulac F, Dayan U (1997) Control of atmospheric export of dust from North Africa by the North Atlantic Oscillation. Nature 387:691–694

Osborn TJ (2004) Simulating the winter North Atlantic Oscillation: the role of internal variability and greenhouse gas forcing. Clim Dyn 22:605–623

Paeth H, Hense A, Glowienka-Hense R, Voss R, Cubasch U (1999) The North Atlantic Oscillation as an indicator for greenhouse-gas induced regional climate change. Clim Dyn 15:953–960

Selten FM, Branstator GW, Dijkstra HA, Kliphius M (2004) Tropical origins for recent and future Northern Hemisphere climate change. Geophys Res Lett 31:L21205. doi:10.1029/2004GL020739

Stephenson DB, Pavan V, Collins M, Junge MM, Quadrelli R (2006) North Atlantic Oscillation response to transient greenhouse gas forcing and the impact on European winter climate: a CMIP2 multi-model assessment. Clim Dyn 27:401–420

Tegen I, Lacis AA, Fung I (1996) The influence on climate forcing of mineral aerosols from disturbed soils. Nature 380:419–422

Theoharatos G, Pantavou K, Mavrakis A, Spanou A, Katavoutas G, Efstathiou P, Mpekas P, Assimakopoulos D (2010) Heat waves observed in 2007 in Athens, Greece: synoptic conditions, bioclimatological assessment, air quality levels and health effects. Environ Res 110:152–161

Voulgarakis A, Savage NH, Braesicke P, Wild O, Carver GD, Pyle JA (2009) Interannual variability of tropospheric composition: the influence of changes in emission, meteorology and clouds. Atmos Chem Phys Discuss 9:14023–14057

Yarnal B (1993) Synoptic climatology in environmental analysis. Bellhaven Press, London

Kuzmina SI, Bengtsson L, Johannessen OM, Drange H, Bobylev LP, Miles MW (2005) The North Atlantic Oscillation and greenhouse-gas forcing. Geophys Res Lett 32:L04703. doi:10.1029/2004GL021064
Lelieveld J, Berresheim H, Borrmann S et al (2002) Global air pollution crossroads over the Mediterranean. Science 298:794–799
Liu JJ, Jones DBA, Worden JR, Noone D, Parrington M, Kar J (2009) Analysis of the summertime buildup of tropospheric ozone abundances over the Middle East and North Africa as observed by the Tropospheric Emission Spectrometer instrument. J Geophys Res 114:D05304. doi:10.1029/2008JD010993
Moulin C, Lambert CE, Dayan U, Masson V, Ramonet M, Bousquet P, Dulac F (1998) Satellite climatology of African dust transport in the Mediterranean atmosphere. J Geophys Res 103:13137–13144
Moulin C, Lambert CE, Dulac F, Dayan U (1997) Control of atmospheric export of dust from North Africa by the North Atlantic Oscillation. Nature 387:691–694
Osborn TJ (2004) Simulating the winter North Atlantic Oscillation: the roles of internal variability and greenhouse gas forcing. Clim Dyn 22:605–623
Paeth H, Hense A, Glowienka-Hense R, Voss R, Cubasch U (1999) The North Atlantic Oscillation as an indicator for greenhouse-gas induced regional climate change. Clim Dyn 15:953–960
Selten FM, Branstator GW, Dijkstra HA, Kliphuis M (2004) Tropical origins for recent and future Northern Hemisphere climate change. Geophys Res Lett 31:L21205. doi:10.1029/2004GL020739
Stephenson DB, Pavan V, Collins M, Junge MM, Quadrelli R (2006) North Atlantic Oscillation response to transient greenhouse gas forcing and the impact on European winter climate: a CMIP2 multi-model assessment. Clim Dyn 27:401–420
Tegen I, Lacis AA, Fung I (1996) The influence on climate forcing of mineral aerosols from disturbed soils. Nature 380:419–422
Theoharatos G, Pantavou K, Mavrakis A, Spanou A, Katavoutas G, Efstathiou P, Mpekas P, Asimakopoulos D (2010) Heat waves observed in 2007 in Athens, Greece: synoptic conditions, bioclimatological assessment, air quality levels and health effects. Environ Res 110:152–161
Voulgarakis A, Savage NH, Braesicke P, Wild O, Carver GD, Pyle JA (2009) Interannual variability of tropospheric composition: the influence of changes in emission, meteorology and clouds. Atmos Chem Phys Discuss 9:14023–14057
Yarnal B (1993) Synoptic climatology in environmental analysis. Belhaven Press, London

Evaluation of the Relationship Between the NAO and Rainfall Erosivity in NE Spain During the Period 1955–2006

Marta Angulo-Martínez and Santiago Beguería

Abstract Rainfall erosivity is the ability of precipitation to erode soil. Microerosion processes due to the impact of raindrops on the soil – rainsplash – represent an important mechanism of detachment of soil particles, which might be removed thereafter by surface runoff. The annual and interannual patterns of rainfall erosivity are controlled by the variability of rainfall intensity. This study analyses the interannual variability of daily rainfall erosivity due to precipitation in the NE of Spain during the period 1955–2006, and its connection with the North Atlantic Oscillation (NAO). It was found that the erosive power of rainfall is stronger during the negative phase of NAO and weaker during positive NAO conditions, as are the rainfall amounts. Daily rainfall erosivity series were adjusted to a Generalized Pareto probability distribution for positive and negative NAO phases for assessing the effect of NAO on extreme events. Results showed higher values expected for a given return period in most of the area under negative NAO conditions. These findings would be useful in the implementation of soil conservation strategies.

Keywords Rainfall erosivity · EI · North Atlantic Oscillation · R factor · NE Spain

1 Introduction

The concept of rainfall erosivity refers to the ability of any rainfall event to erode soil. It links together the dynamic properties of rainfall as a consequence of rainfall-generating processes and their impact on the soil surface, and involves two main mechanisms: (i) *rainsplash* – the detachment of soil particles due to the kinetic energy of the raindrops –, and (ii) *runoff erosion* – motion of soil particles by shear stress exerted by the surface runoff. Rainfall erosivity is responsible of changes in the soil properties as: crusting, disruption of aggregates, removal of nutrient-rich parcels of soil, etc. Rainfall erosivity represents one of the main mechanisms of degradation in semiarid landscapes where vegetation is scarce. Its influence depends on the characteristics of soil, topography and land use, as well as on other features of

M. Angulo-Martínez (✉)
Estación Experimental de Aula Dei, Spanish National Research Council (CSIC), Zaragoza, Spain
e-mail: mangulo@eead.csic.es

S.M. Vicente-Serrano, R.M. Trigo (eds.), *Hydrological, Socioeconomic and Ecological Impacts of the North Atlantic Oscillation in the Mediterranean Region*, Advances in Global Change Research 46, DOI 10.1007/978-94-007-1372-7_13,

the rainfall regime (D'Odorico et al., 2001). Rainfall erosivity estimates are needed for the estimation of soil erosion amounts through space and time. In the context of climate change the effect of altered rainfall characteristics combined with other aspects as biomass and changes in the soil moisture are some major concerns of soil conservation studies.

Empirical studies have demonstrated an exponential relationship between rainfall intensity and rainfall erosivity (Brown and Foster, 1987; Coutinho and Tomás, 1995; Van Dijk et al., 2002). This relationship allows introducing some important characteristics of the rainfall events as: (i) the raindrop size distribution and their associated kinetic energy, and (ii) the rainfall intensity at detailed time intervals.

Among the several indices developed to estimate rainfall erosivity the most extensively used is the USLE/RUSLE *R* factor, which is calculated from the EI_{30} index (Wischmeier, 1959; Wischmeier and Smith, 1978; Brown and Foster, 1987; Renard et al., 1997). At many sites worldwide the *R* factor has been shown to be highly correlated with soil loss (Van der Knijff et al., 2000; Diodato, 2004; Shi et al., 2004; Hoyos et al., 2005; Curse et al., 2006; Onori et al., 2006; Domínguez-Romero et al., 2007).

One of the main disadvantages of the RUSLE *R* factor is the need for high frequency – typically lower than 15 min or pluviograph data –, long data series. Information of this nature is rarely available with adequate spatial and temporal coverage, and daily records are usually the best available data at most locations. Attempts to estimate rainfall erosivity from daily rainfall records or storm event data have been based largely on exponential relationships (Richardson et al., 1983; Bagarello and D'Asaro, 1994; Petkovsek and Mikos, 2004; Angulo-Martínez and Beguería, 2009). Due to the high temporal and spatial variability of rainfall erosivity (Angulo-Martínez et al., 2009), accurate records based on long data series are required in order to evaluate temporal or spatial trends that may be linked with atmospheric circulation patterns.

Previous work has shown that the spatial and interannual variability of rainfall intensity can be explained by atmospheric circulation patterns (Hurrell, 1995; Jones et al., 1997; Hurrell et al., 2003). The main atmospheric circulation pattern that affects the climate of Europe is the North Atlantic Oscillation (NAO). The NAO is characterized by a north–south sea-level pressure dipolar pattern, with one of the centres located over Iceland and the other one approximately over the Azores Islands. This dipolar pattern reflects the strong contrast in meridional pressure over the North Atlantic region. The positive phase of the NAO reflects below normal heights and pressure across the high latitudes of the North Atlantic, and above-normal heights and pressure over the central North Atlantic. The NAO negative mode produces high-pressure blocking in the northeast Atlantic, and a more meridional circulation than the opposite mode (Jacobeit, 1987; Moses et al., 1987). Upper-air troughs and incursions of polar air over the Mediterranean are more frequent during negative NAO, and the Atlantic storm tracks are displaced south. All these factors are responsible of wetter conditions in the western Mediterranean (Jacobeit, 1987; Moses et al., 1987; Maheras, 1988; Kutiel et al., 1996). In contrast, the NAO positive mode is characterized by moisture transport across the North

Atlantic that has a more southwest to northeast orientation and extends further into northern Europe and Scandinavia. Hence, moisture transport and rainfall are reduced over southern Europe and the Mediterranean (Hurrell, 1995; Hurrell and van Loon, 1997; Moulin et al., 1997; Marshall et al., 2001). The NAO shows a clear signal throughout the whole year, but with stronger intensity and extension during winter due to the stronger meridional gradients.

The influence of the NAO on precipitation over the Iberian Peninsula has been recognized in previous studies (Rodó et al., 1997; Esteban-Parra et al., 1998; Rodríguez-Puebla et al., 1998; Serrano et al., 1999), with a clear difference between the western (Atlantic) and eastern (Mediterranean) areas of the region. It is possible to identify a large part of central and southern Spain and southern Portugal for which the influence of the NAO is dominant. In contrast, the southeastern Mediterranean coast of Spain is dominated by more easterly Mediterranean influences.

There are no previous studies relating rainfall erosivity to atmospheric teleconnection patterns in the Iberian Peninsula. Some works have addressed the influence of the El Niño-Southern Oscillation (ENSO) on rainfall erosivity for some areas of United States (D'Odorico et al., 2001) and Peru (Romero et al., 2007). Since the North Atlantic Oscillation controls the interannual variability of rainfall over the region, there is a need to know in which way rainfall erosivity is related to NAO. Since rainfall erosivity is more related to single extreme precipitation events than to average events, the relationship between the NAO and rainfall erosivity is not necessarily explained by the negative correlation between the NAO and total precipitation amounts. In addition, this knowledge may help agricultural managers to take soil conservation measures to protect the soil.

The aim of this study is to assess the influence of the NAO on daily rainfall erosivity in NE Spain, during the period 1955–2006. A second objective is to identify the changes in the probability of occurrences of extreme daily erosivity events in relation with daily NAO phases.

2 Study Area and Methods

2.1 Study Area

The study area covers the northeast of Spain, encompassing an area of about 147,000 km^2 that corresponds to the administrative territorial province demarks which conform the Ebro Basin. The study area is geographically complex. It encloses several mountain ranges and a main central valley, the Ebro basin. The area is limited to the north by the Cantabric Sea, the Cantabrian Range and the Pyrenees, with maximum elevations above 3000 m a.s.l. At the south and southwest the Iberian Range closes the Ebro valley, with maximum elevations in the range of 2000–2300 m a.s.l. To the east valley is closed by the Catalan Prelittoral Range, with maximum elevations of 1000–1900 m a.s.l., and then continues to the Mediterranean sea.

The climate is influenced by both the Atlantic and the Mediterranean, and the effect of the relief on precipitation and temperature. The bordering mountain ranges isolate the central valley, blocking the maritime influence and resulting in a continental climate with arid conditions (Lana and Burgueño, 1998; Vicente-Serrano, 2005). A climatic gradient in the NW–SE direction is notable, determined by strong Atlantic influences in the north and northwest of the area during much of the year, and the influence of the Mediterranean to the east. The mountain ranges add complexity to the climate of the region, with the Pyrenees extending the Atlantic Ocean influence to the east by increasing precipitation. Precipitation in inland areas is characterized by alternating wet and dry periods as a consequence of the seasonal displacement of the polar front and its associated pressure systems. Inter-annual variability in precipitation can be very high, and prolonged dry periods can be followed by torrential rainfall events that last for several days. The most extreme precipitation events have been recorded along the Mediterranean seaside (Romero et al., 1998; Llasat, 2001; Peñarrocha et al., 2002).

Rainfall erosivity reproduces the characteristics of the precipitation in the study area, as revealed by station-data (Fig. 1). Three different areas can be recognized, following a NW–SE gradient. The NW zone, influenced by the Atlantic, shows the highest monthly rainfall values and minimum rainfall erosivity; the highest erosivity is attained at the beginning of the summer, coinciding with late spring storms. The central zone includes the majority of the observatories, with precipitation amounts that are lower than in the NW zone (although still abundant), though erosivity is

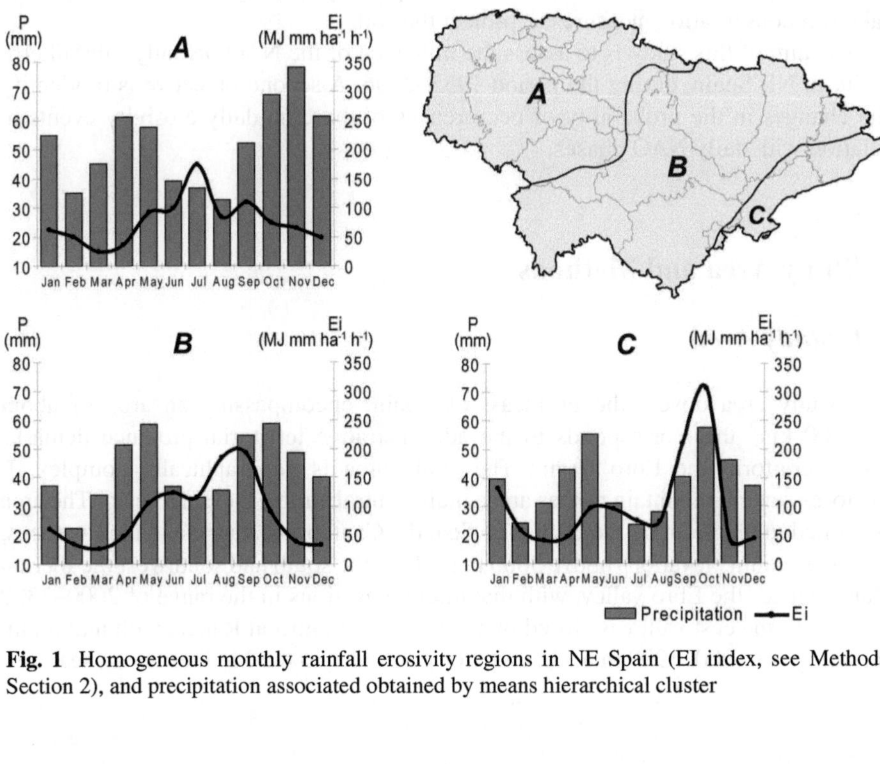

Fig. 1 Homogeneous monthly rainfall erosivity regions in NE Spain (EI index, see Methods, Section 2), and precipitation associated obtained by means hierarchical cluster

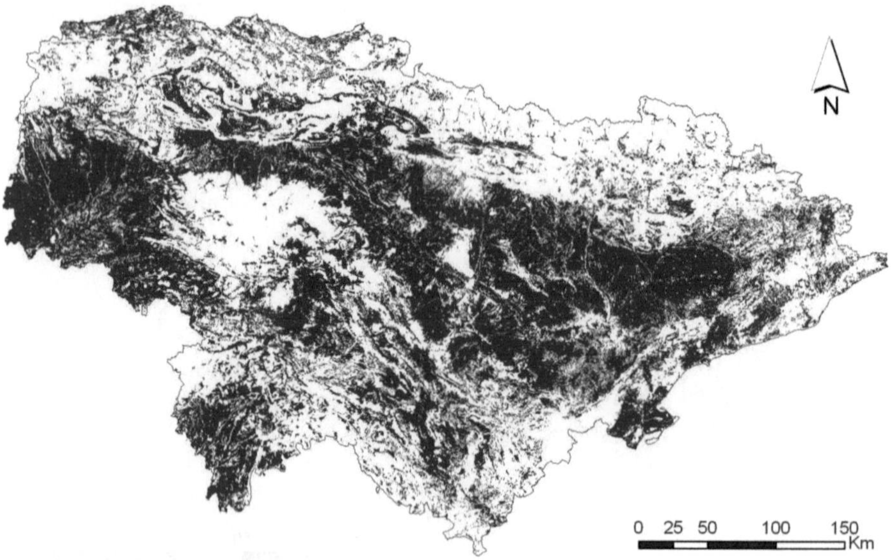

Fig. 2 Spatial surface of the cultivated soil in the NE Spain (68,000 km^2); data: Corine Land Cover version 8/2005 (EEA, 2000)

greater and shows two annual peaks, one in late spring (May–June) and a second (highest) at the end of summer (August–September). The NE zone has a typical Mediterranean rainfall behavior, with maxima in spring and autumn and highest erosivity in autumn. It is important to note that the spring rainfall peaks are not as erosive as those of the autumn, due to differences between these seasons in rainfall generation mechanisms. Spring rainfall events are usually frequent, but tend to have low intensity. In contrast, precipitation in autumn is usually attributable to a few very intense events.

Due to its complex climatology (as a consequence of being a meteorological border region) and the contrasted relief, the Ebro Basin has a long history of social, economic and environmental damage caused by extreme rainfall events (García-Ruiz et al., 2000; Lasanta, 2003; Llasat et al., 2005). The main activity in this area is agriculture, representing approximately 46% of the land use. It is extended around the Ebro valley occupying a broader area close to the Mediterranean coast (Fig. 2). During winter agricultural soils remain uncovered. Consequently, rainfall erosivity is the principal cause of soil erosion with no limiting factor at that time. Mountain areas represent approximately 20% of the study region.

2.2 Daily Rainfall Erosivity Database

The database consisted of 156 daily rainfall series from the Spanish Meteorological Agency (AEMET) for the period 1955–2006 (Fig. 3). Data on cumulative precipitation were taken every day at 0600 LT. The series were obtained via a

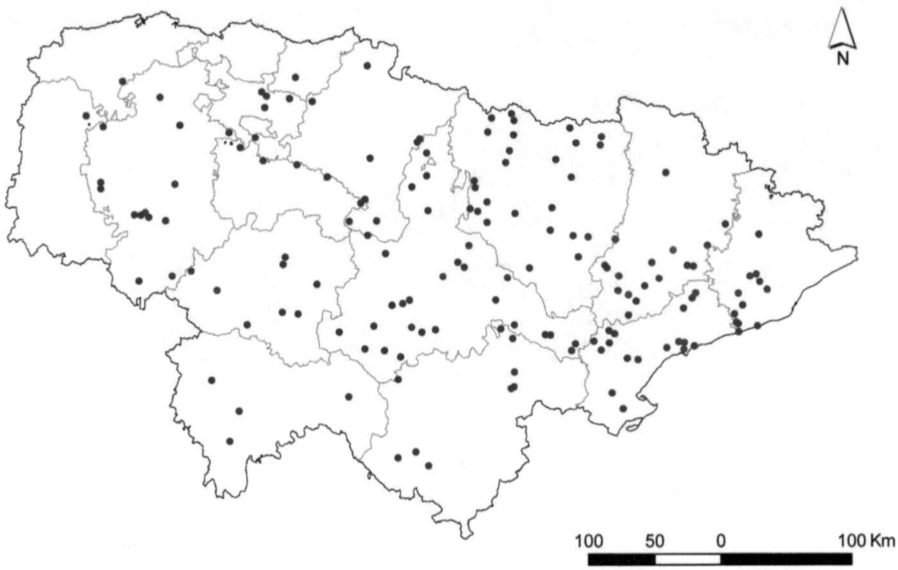

Fig. 3 Spatial distribution of the 156 daily rainfall observatories in the study area for the period 1955–2006

process that included reconstruction, gap filling, quality control and homogeneity testing (Vicente-Serrano et al., 2010).

Daily rainfall series were transformed into daily rainfall erosivity records based on an exponential relationship between the rainfall volume, P (mm/d), and rainfall erosivity index, EI (MJ mm/ha/h/d; Richardson et al., 1983):

$$EI = a\,P^b + \varepsilon, \tag{1}$$

where a and b are empirical parameters and ε is a random, normally distributed error. Parameters a and b are adjusted month-by-month to take account of intra-annual variations in rainfall characteristics. This leads to the more general expression:

$$EI_m = a_m\,P^{b_m} + \varepsilon, \tag{2}$$

where $m = \{1, ..., 12\}$ represents the month of the year. Parameter estimation in the model is achieved by weighted least squares (WLS) regression after a logarithmic transformation of the terms in Equation (2). Weights were assigned to the observations in order to reduce the excessive influence of small erosive events during parameterization (Angulo-Martínez and Beguería, 2009).

For obtaining the values of the parameters a and b it is necessary to count with data on daily EI and P. This data was provided by the automated system for hydrologic information of the Ebro basin water authority (SAIH-Ebro). This dataset

consisted on 110 series of rainfall data with 15 min resolution, covering the period 1997–2007. Daily *EI* values were calculated from these series following the RUSLE methodology, and were used for estimating the parameters in Equation (2). Given the good spatial coverage of the dataset, spatial fields of the parameters a and b were obtained by spatial interpolation, and their values at the location of the AEMET observatories were recorded. These values where then used to obtain daily series of *EI* for the period 1955–2006, using the AEMET daily rainfall data.

2.3 Atmospheric Circulation Events: North Atlantic Oscillation Index

Following Jones et al. (1997) a North Atlantic Oscillation index (NAOI) was calculated as the normalized difference between time series of sea level pressure recorded at two points in the southwest Iberian Peninsula (Gibraltar, 35°N 5°W) and southwest Iceland (Reykjavik, 65°N, 20°W). Daily sea-level pressure grids from the ds010.0 Daily Northern Hemisphere Sea Level Pressure Grids data set (University Corporation for Atmospheric Research, 1979) were used for calculating the NAOI. We selected those grid points located closest to the mentioned points.

Although the NAO influence is more intense during winter time, in the present study we considered the whole year. Positive and negative events of the NAO index were identified as those days with values of the NAO index higher than 0.5 and lower than –0.5, respectively.

2.4 Differences in Daily Rainfall Erosivity Between Positive and Negative NAO Phases

To evaluate the relationship between the phase of the NAO_i and monthly rainfall erosivity we generated different daily rainfall erosivity (*EI*) series regarding the value of NAO_i – positive (negative) daily EI_{NAOi} series were created with the daily *EI* values which were greater (inferior) of 0.5 (−0.5) NAO_i values, during the period 1955–2006. Daily rainfall erosivity was summed to obtain monthly positive – EI_{NAOi+} – and negative – EI_{NAOi-} – rainfall erosivity for NAO_i. We computed the relative difference EI_{dif} between median annual and monthly and in every observatory by:

$$EI_{dif} = (EI_{NAO-} - EI_{NAO+}) / EI_{NAO+}, \tag{3}$$

where EI_{NAOi-} and EI_{NAOi+} are the median annual and monthly rainfall erosivity during negative and positive days of the NAO_i. Month maps of EI_{dif} were produce to help visualize spatial differences in the effect of NAO_i on rainfall erosivity.

The statistical significance of the relative differences was evaluated using the Wilcoxon-Mann-Whitney (WMW) rank test (Siegel and Castelan, 1988). The non-parametric WMW test, though less powerful than the *t* test, was preferred due to

its robustness against non-normality of the variables (Helsel and Hirsch, 1992). The monthly rainfall erosivity data series during daily negative NAO were compared with the monthly rainfall erosivity data series with the opposite NAO sign, month by month. The significance level was established at $\alpha = 0.05$.

2.5 Extreme Value Analysis of Rainfall Erosivity During Positive and Negative NAO Phases

Changes in the probability distributions of extreme rainfall erosivity depending on the NAO phase were analyzed using the extreme value theory (Hershfield, 1973). Peaks-over-threshold (POT) series of daily erosivity (*EI*) were obtained for positive and negative NAO phases, selecting only the observations for which *EI* exceeded a value corresponding to the 95th percentile of the respective series. The resulting series were fitted to a Generalized Pareto (GP) distribution by the maximum likelihood approach. GP is the limit distribution of a POT variable, provided that the POT occurrences fit a Poisson process, i.e. that they are time independent. The appropriateness of the GP distribution to model the daily erosivity POT series was checked visually by means of the L-moment ratios diagram and by a Kolmogorov-Smirnov test (Beguería, 2005; Beguería et al., 2009).

The GP distribution is described by a shape parameter k and a scale parameter α, with probability density function:

$$f(x) = \frac{1}{\alpha}\left[1 - \frac{k}{\alpha}(x - \varepsilon)\right] \tag{4}$$

and distribution function:

$$P(X \leq x) = 1 - \left(1 - \kappa\frac{(x - \varepsilon)}{\alpha}\right)^{\frac{1}{\kappa}}, \tag{5}$$

where x is the daily rainfall erosivity exceeding the threshold value ε, which acts as a location parameter. Estimates of α, κ can be obtained by maximum likelihood (Rao and Hamed, 2000). The highest expected precipitation X_T over a period of T years is obtained as:

$$X_T = \varepsilon + \frac{\alpha}{\kappa}\left[1 - \left(\frac{1}{\lambda T}\right)^{\kappa}\right], \tag{6}$$

where λ is the average number of events per year.

Maps of the extreme events corresponding to return periods of 10 and 50 years were produced for positive and negative NAO phased based on the GP models fitted to the POT series. These maps were compared in order to evaluate differences

in expected extreme events depending on the NAO. Though NAO mostly influences during winter season, some influence can be recognised during autumn and spring. Therefore the POT analysis was applied to the period between October and February.

3 Results

3.1 Differences in Average Daily Rainfall Erosivity Between Negative and Positive NAO Phases

The results of the WMW test comparing the distributions of daily erosivity during positive and negative NAO phases yielded 92 (59%) of significant series globally. Monthly results are shown in Table 1. These differences were present almost all the year, however being especially evident from autumn to early spring (October–March). Most observatories reflected this behaviour. During summer (June–August) the influence of NAO on rainfall erosivity was almost non-existent.

The spatial distribution of the significant relative differences between the daily rainfall erosivity recorded during negative and positive NAO phases for every month are shown in Fig. 4. The NAO influence can be noted from November to April, with different spatial patterns. From June to October we found no influence of NAO in rainfall erosivity, in addition, the relative differences were not significant for most of the area during June to September.

Higher differences were found closer to the Mediterranean. The highest differences between positive and negative NAO phases occurred from November to April, though March did not show any difference, most noticeable in the Mediterranean area where rainfall erosivity under negative NAO phases was between 6 and 30 times higher to the values registered under positive NAO. November showed the highest differences, affecting almost one half of the area extending from the Mediterranean coast to the west towards the centre of the Ebro valley.

Table 1 Difference in monthly erosivity between positive and negative NAO phases (Wilcoxon-Mann-Witney test): number and proportion of series with significant differences

Month	Significant series
Jan	144 (92%)
Feb	136 (87%)
Mar	68 (44%)
Apr	147 (94%)
May	109 (70%)
Jun	27 (17%)
Jul	40 (26%)
Aug	36 (23%)
Sep	11 (7%)
Oct	94 (60%)
Nov	152 (97%)
Dec	145 (93%)

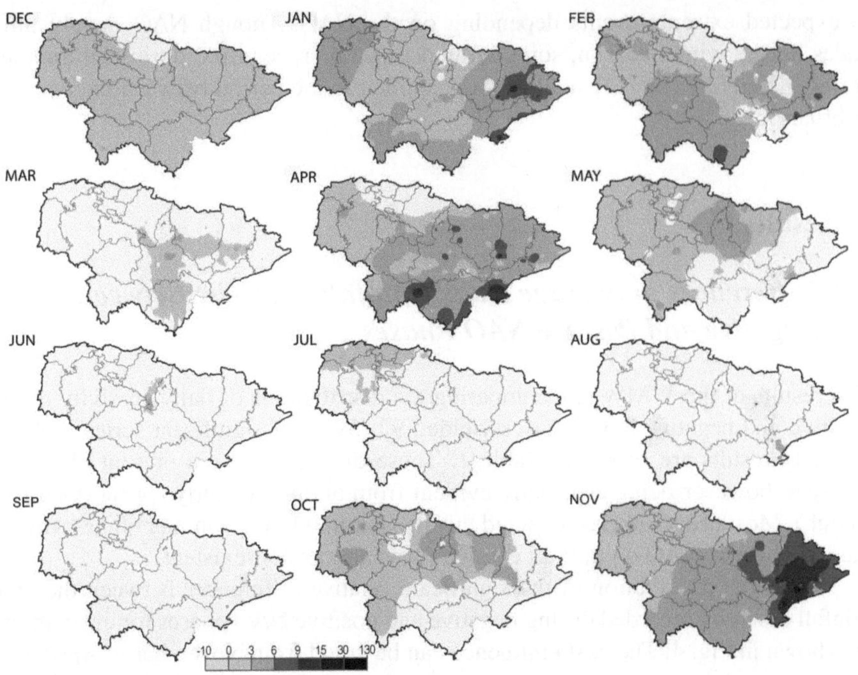

Fig. 4 Relative differences in median monthly rainfall erosivity during negative and positive NAO phases. *White masked areas* indicate no significant difference

3.2 Differences in Extreme Rainfall Erosivity Between Negative and Positive NAO Phases

The GP model provided a very good fit to the POT series of rainfall erosivity, as shown by the L-moment diagrams (Fig. 5). Independently of the sign of the NAO, the empirical L-moments of the rainfall erosivity series plotted closest to the theoretical curve of the GP distribution. The results of the Kolmogorov-Smirnov test allowed accepting the GP distribution for the data, since only in very few cases the null hypothesis that the data came from a GP distribution was rejected. This result agrees with recent studies that demonstrated the high performance of the GP distribution in fitting extreme hydrological variables using partial duration series (Hosking and Wallis, 1987; Madsen and Rosbjerg, 1997; Beguería, 2005). Here we also found that the GP distribution has a good performance in fitting daily rainfall erosivity data.

By adjusting the daily rainfall erosivity series for the selected months to the GP probability distribution we obtained the probability of occurrence of daily rainfall erosivity records associated with negative and positive NAO conditions. Results show greatest rainfall erosivity values expected under negative NAO conditions. Differences could be found between observatories. Those located inland to the south of the region did not show significant differences, however, as expected by results

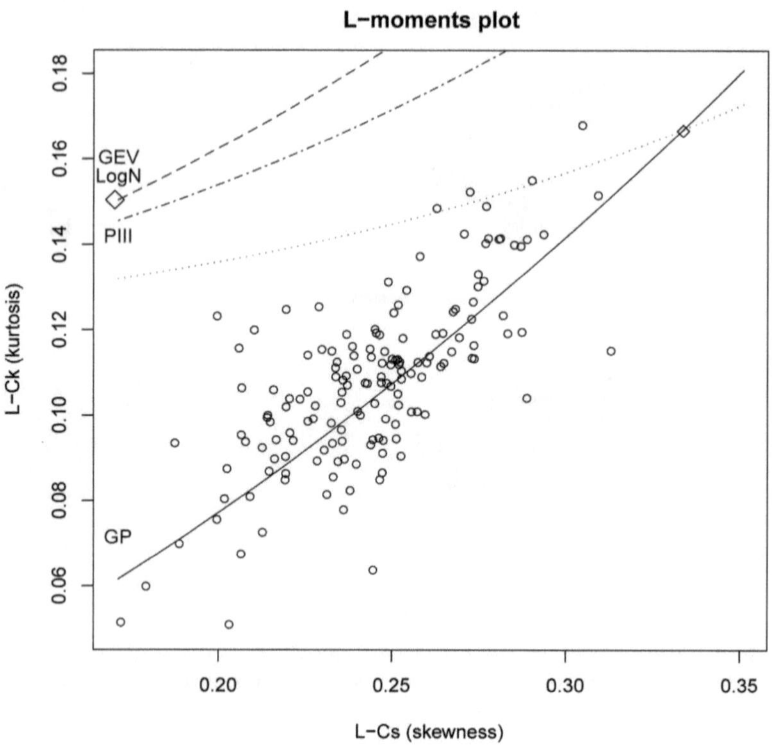

Fig. 5 L-moments ratio diagram, comparing empirical L-moment ratios (L-skewness and L-kurtosis) with the ones of several widely used extreme value distributions: Generalized Extreme Value (GEV), Log-Normal (LN), Pearson III (PIII) and Pareto (GP)

obtained in the precedent section. The mountainous areas and the Mediterranean coast showed a high contrast between the expected rainfall erosivity values under negative or positive conditions.

The rainfall erosivity values expected for a return period of 10 years related to negative and positive NAO conditions during winter and the respective standard error are shown in Fig. 6. Under negative NAO conditions daily rainfall erosivity values of 50 up to 385 (±8 to 22) MJ mm ha/h/d may be recorded in the northern and eastern half of the region. The highest values are concentrated at the north-east perimeter, close to the Mediterranean coast from the Ebro Delta upwards to the north and in the Central Pyrenees. Under positive NAO most of the region shows low daily rainfall erosivity values ranging between 5 and 50 (±0.5 to 8) MJ mm/ha/h/d. High erosivity values are only expected in localized areas such as the Ebro Delta and in the Central Pyrenees.

The results are very similar for a return period of 50 years (Fig. 7), although the spatial pattern of the very high erosivity values excludes the Pyrenees region and the highest erosivity values are thus confined to the Ebro Delta.

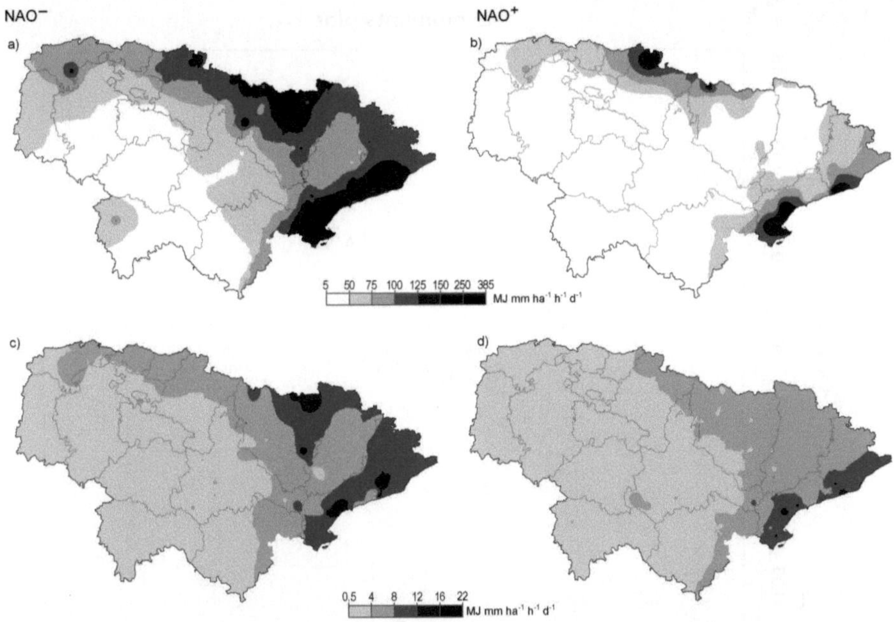

Fig. 6 Spatial distribution of expected daily rainfall erosivity during winter (October–March) corresponding to a return period of 10 years under negative (**a**) and positive (**b**) NAO phases, and their respective standard errors (**c** and **d**, respectively)

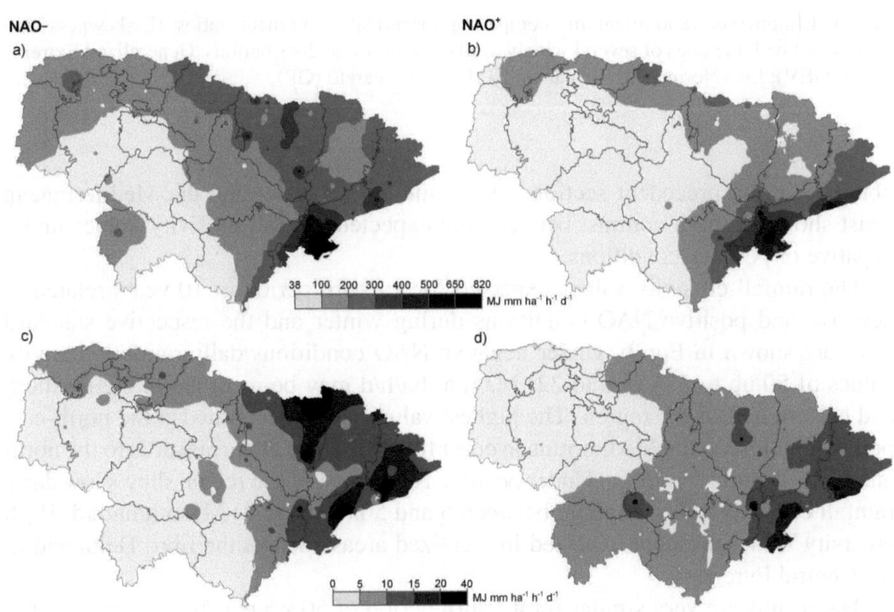

Fig. 7 Spatial distribution of expected daily rainfall erosivity during winter (October–March) corresponding to a return period of 50 years under negative (**a**) and positive (**b**) NAO phases, and their respective standard errors (**c** and **d**, respectively)

4 Discussion and Conclusions

Rainfall erosivity bridges the gap between rainfall dynamics and its direct impact over the surface of the soil, taking into account the energy of the raindrops hitting the soil, and the runoff which will remove the disaggregated soil particles. There is an increasing interest regarding the effects of the climate dynamics, especially in a context of climate change, in the environment. However, very few studies have related the atmospheric circulation patterns to rainfall erosivity due mostly to the lack of accurate rainfall erosivity series long enough to make this analysis possible.

The NAO is responsible for most of the climatic variability in the North Atlantic, modifying direction and intensity of the westerly winds, the track of the polar depressions and the location of the anticyclones. During the negative phase cyclones move southward increasing precipitation on the western Mediterranean. On the contrary, during the positive phases, the Azores subtropical high is reinforced, leading to a drier weather in the Mediterranean region.

This study has shown that atmospheric dynamics, in this case the NAO, have influenced interannual variability of rainfall erosivity at NE of Spain for the period 1955–2006. Higher rainfall erosivity values were related to negative NAO conditions whereas the high pressures associated with positive NAO conditions resulted in lower rainfall erosivity values. This difference was highest towards the east of the study area (Mediterranean region), and the centre of the Ebro valley, and during the winter months (November–April). Agriculture is a major land use in this area, and during this time of the year the agricultural soils are mostly uncovered and exposed to the erosive power of the rain, intensifying the impact of the NAO on soil erosion.

Knowledge of the spatial and temporal patterns of rainfall erosivity will help deepening our understanding of the processes of soil erosion, and might prove helpful for designing soil protection measures.

Acknowledgements We thank the Spanish Meteorological Agency (Agencia Española de Meteorología, AEMET) and the Ebro Basin Water Authority (Confederación Hidrográfica del Ebro, CHE) for providing the data used in this study. The research of M.A. was supported by a JAE-Predoc Research Grant from the Spanish National Research Council (Consejo Superior de Investigaciones Científicas; CSIC).

References

Angulo-Martínez M, Beguería S (2009) Estimating rainfall erosivity from daily precipitation records: a comparison among methods using data from the Ebro Basin (NE Spain). J Hydrol 379:111–121

Angulo-Martínez M, López-Vicente M, Vicente-Serrano SM, Beguería S (2009) Mapping rainfall erosivity at a regional scale: a comparison of interpolation methods in the Ebro Basin (NE Spain). Hydrol Earth Syst Sci 13:1907–1920

Bagarello V, D'Asaro F (1994) Estimating single storm erosion index. Trans Am Soc Agric Eng 37:785–791

Beguería S (2005) Uncertainties in partial duration series modelling of extremes related to the choice of the threshold value. J Hydrol 303:215–230

Beguería S, Vicente-Serrano SM, López-Moreno JI, García-Ruiz JM (2009) Annual and seasonal mapping of peak intensity, magnitude and duration of extreme precipitation events across a climatic gradient, North-east Iberian Peninsula. Int J Climatol 29:1759–1779

Brown LC, Foster GR (1987) Storm erosivity using idealized intensity distributions. Trans Am Soc Agric Eng 30:379–386
Coutinho MA, Tomás PP (1995) Characterization of raindrop size distributions at the Vale Formoso Experimental Erosion Centre. Catena 25:187–197
Curse R, Flanagan J, Frankenberger B, Gelder D, Herzmann D, James D, Krajenski W, Kraszewski M, Laflen J, Opsomer J, Todey D (2006) Daily estimates of rainfall, water runoff and soil erosion in Iowa. J Soil Water Conserv 61:191–199
Diodato N (2004) Estimating RUSLE's rainfall factor in the part of Italy with a Mediterranean rainfall regime. Hydrol Earth Syst Sci 8:103–107
D'Odorico P, Yoo J, Over TM (2001) An assessment of ENSO-Induced patterns of rainfall erosivity in the Southwestern United States. J Clim 14:4230–4242
Domínguez-Romero L, Ayuso Muñoz JL, García Marín AP (2007) Annual distribution of rainfall erosivity in western Andalusia, southern Spain. J Soil Water Conserv 62:390–401
EEA (2000) CORINE land cover 2000. European Environment Agency. http://image2000.jrc.it
Esteban-Parra MJ, Rodrigo FS, Castro-Diez Y (1998) Spatial and temporal patterns of precipitation in Spain for the period 1880–1992. Int J Climatol 18:1557–1574
García-Ruiz JM, Arnáez J, White SM, Lorente A, Beguería S (2000) Uncertainty assessment in the prediction of extreme rainfall events: an example from the Central Spanish Pyrenees. Hydrol Process 14:887–898
Helsel DR, Hirsch RM (1992) Statistical methods in water resources. Elesevier, New York, 522 pp
Hershfield DM (1973) On the probability of extreme rainfall events. Bull Am Meteorol Soc 54:1013–1018
Hosking JRM, Wallis JR (1987) Parameter and quantile estimation for the Generalized Pareto distribution. Technometrics 29:339–349
Hoyos N, Waylen PR, Jaramillo A (2005) Seasonal and spatial patterns of erosivity in a tropical watershed of the Colombian Andes. J Hydrol 314:177–191
Hurrell J (1995) Decadal trends in North Atlantic Oscillation and relationship to regional temperature and precipitation. Science 269:676–679
Hurrell J, Kushnir Y, Ottersen G, Visbeck M (eds) (2003) The North Atlantic Oscillation: climate significance and environmental impacts. Geophys Monogr Ser, vol 134. AGU, Washington, DC
Hurrell JW, van Loon H (1997) Decadal variations in climate associated with the North Atlantic oscillation. Clim Change 36:301–326
Jacobeit J (1987) Variations of trough positions and precipitation patterns in the Mediterranean area. J Climatol 7:453–476
Jones PD, Jónsson T, Wheeler D (1997) Extension to the North Atlantic Oscillation using early instrumental pressure observations from Gibraltar and South-West Iceland. Int J Climatol 17:1433–1450
Kutiel H, Maheras P, Guika S (1996) Circulation indices over the Mediterranean and Europe and their relationship with rainfall conditions across the Mediterranean. Theor Appl Climatol 54:125–138
Lana X, Burgueño A (1998) Spatial and temporal characterization of annual extreme droughts in Catalonia (Northeast Spain). Int J Climatol 18:93–110
Lasanta T (2003) Gestión agrícola y erosión del suelo en la cuenca del Ebro: el estado de la cuestión. Zubía 21:76–96
Llasat MC (2001) An objective classification of rainfall events on the basis of their convective features. Application to rainfall intensity in the North-East of Spain. Int J Climatol 21: 1385–1400
Llasat MC, Barriendos M, Barrera A, Rigo T (2005) Floods in Catalonia (NE Spain) since the 14th century. In: Benito G, Ouarda TBMJ, Bárdossy A (eds) Palaeofloods, historical data & climate variability: applications in flood risk assessment. J Hydrol 313:32–47
Madsen H, Rosbjerg D (1997) The partial duration series method in regional index-flood modeling. Water Resour Res 33:737–746
Maheras P (1988) Changes in precipitation conditions in the western Mediterranean over the last century. J Climatol 8:179–189

Marshall J, Kushnir Y, Battisti D, Chang P, Czaja A, Dickson R, Hurrell J, McCartney M, Saravanan R, Visbeck M (2001) North Atlantic climate variability: phenomena, impacts and mechanisms. Int J Climatol 21:1863–1898

Moses T, Kiladis GN, Diaz HF, Barry RG (1987) Characteristics and frequency of reversals in mean sea level pressure in the North Atlantic sector and their relationship to long-term temperature trends. J Climatol 7:13–30

Moulin C, Lambert CE, Dulac F, Dayan U (1997) Control of atmospheric export of dust from North Africa by the North Atlantic Oscillation. Nature 387:691–694

Onori F, De Bonis P, Grauso S (2006) Soil erosion prediction at the basin scale using the revised universal soil loss equation (RUSLE) in a catchment of Sicily (southern Italy). Environ Geol 50:1129–1140

Petkovsek G, Mikos M (2004) Estimating the R factor from daily rainfall data in the sub-Mediterranean climate of southwest Slovenia. Hydrol Sci J 49:869–877

Peñarrocha D, Estrela MJ, Millán M (2002) Classification of daily rainfall patterns in a Mediterranean area with extreme intensity levels: the Valencia region. Int J Climatol 22: 677–695

Rao AR, Hamed KH (2000) Flood frequency analysis. CRC Press, Boca Raton, FL

Renard KG, Foster GR, Weesies GA, McCool DK, Yoder DC (1997) Predicting soil erosion by water: a guide to conservation planning with the Revised Universal Soil Loss Equation (RUSLE). Handbook #703. US Department of Agriculture, Washington, DC

Richardson CW, Foster GR, Wright DA (1983) Estimation of erosion index from daily rainfall amount. Trans Am Soc Agric Eng 26:153–160

Rodó X, Baert E, Comin FA (1997) Variations in seasonal rainfall in southern Europe during the present century: relationships with the North Atlantic oscillation and the El Niño–Southern oscillation. Clim Dyn 19:275–284

Rodríguez-Puebla C, Encinas AH, Nieto S, Garmendia J (1998) Spatial and temporal patterns of annual precipitation variability over the Iberian Peninsula. Int J Climatol 18: 299–316

Romero CC, Baigorría GA, Stroosnijder L (2007) Changes of erosive rainfall for El Niño and La Niña years in Northern Andean highlands of Peru. Clim Change 85:343–356

Romero R, Guijarro JA, Ramis C, Alonso S (1998) A 30-year (1964–1993) daily rainfall data base for the Spanish Mediterranean regions: first exploratory study. Int J Climatol 18: 541–560

Serrano A, Garcia AJ, Mateos VL, Cancillo ML, Garrido J (1999) Monthly modes of variation of precipitation over the Iberian Peninsula. J Clim 12:2894–2919

Shi ZH, Cai CF, Ding SW, Wang TW, Chow TL (2004) Soil conservation planning at the small watershed level using RUSLE with GIS. Catena 55:33–48

Siegel S, Castelan NJ (1988) Nonparametric statistics for the behavioral sciences. McGraw-Hill, New York

Van der Knijff JM, Jones RJA, Montanarella L (2000) Soil erosion risk assessment in Italy. European Commission—European Soil Bureau, 52 pp

Van Dijk AIJM, Bruijnzeel LA, Rosewell CJ (2002) Rainfall intensity – kinetic energy relationships: a critical literature appraisal. J Hydrol 261:1–23

Vicente-Serrano SM (2005) Las sequías climáticas en el valle medio del Ebro: Factores atmosféricos, evolución temporal y variabilidad espacial. Consejo de Protección de la Naturaleza de Aragón, 277 pp

Vicente-Serrano SM, Beguería S, López-Moreno JI, García-Vera MA, Stepanek P (2010) A complete daily precipitation database for north-east Spain: reconstruction, quality control and homogeneity. Int J Climatol 30:1146–1163

Wischmeier WH (1959) A rainfall erosion index for a universal soil-loss equation. Soil Sci Soc Am Proc 23:246–249

Wischmeier WH, Smith DD (1978) Predicting rainfall erosion losses: a guide to conservation planning. USDA Handbook 537. USDA, Washington, DC

Marshall J, Kushnir Y, Battisti D, Chang P, Czaja A, Dickson R, Hurrell J, McCartney M, Saravanan R, Visbeck M (2001) North Atlantic climate variability: phenomena, impacts and mechanisms. Int J Climatol 21:1863–1898

Moses T, Kiladis GN, Diaz HF, Barry RG (1987) Characteristics and frequency of reversals in mean sea level pressure in the North Atlantic sector and their relationship to long-term temperature trends. J Climatol 7:13–30

Moulin C, Lambert CE, Dulac F, Dayan U (1997) Control of atmospheric export of dust from North Africa by the North Atlantic Oscillation. Nature 387:691–694

Onori F, De Bonis P, Grauso S (2006) Soil erosion prediction at the basin scale using the revised universal soil loss equation (RUSLE) in a catchment of Sicily (Southern Italy). Environ Geol 50:1129–1140

Petkovšek G, Mikoš M (2004) Estimating the R factor from daily rainfall data in the sub-Mediterranean climate of southwest Slovenia. Hydrol Sci J 49:869–877

Peñarrocha D, Estrela MJ, Millán M (2002) Classification of daily rainfall patterns in a Mediterranean area with extreme intensity levels: the Valencia region. Int J Climatol 22:677–695

Rao AR, Hamed KH (2000) Flood frequency analysis. CRC Press, Boca Raton, FL

Renard KG, Foster GR, Weesies GA, McCool DK, Yoder DC (1997) Predicting soil erosion by water: a guide to conservation planning with the Revised Universal Soil Loss Equation (RUSLE). Handbook #703. US Department of Agriculture, Washington, DC

Richardson CW, Foster GR, Wright DA (1983) Estimation of erosion index from daily rainfall amount. Trans Am Soc Agric Eng 26:153–160

Rodó X, Baert E, Comín FA (1997) Variations in seasonal rainfall in southern Europe during the present century: relationships with the North Atlantic oscillation and the El Niño-Southern oscillation. Clim Dyn 13:275–284

Rodríguez-Puebla C, Encinas AH, Nieto S, Garmendia J (1998) Spatial and temporal patterns of annual precipitation variability over the Iberian Peninsula. Int J Climatol 18:299–316

Romero CC, Baigorria GA, Stroosnijder L (2007) Changes of erosive rainfall for El Niño and La Niña years in the northern Andean highlands of Peru. Clim Change 85:343–356

Romero R, Guijarro JA, Ramis C, Alonso S (1998) A 30-year (1964–1993) daily rainfall data base for the Spanish Mediterranean regions: first exploratory study. Int J Climatol 18:541–560

Serrano A, García JA, Mateos VL, Cancillo ML, Garrido J (1999) Monthly modes of variation of precipitation over the Iberian Peninsula. J Clim 12:2894–2919

Shi ZH, Cai CF, Ding SW, Wang TW, Chow TL (2004) Soil conservation planning at the small watershed level using RUSLE with GIS. Catena 55:33–48

Siegel S, Castellan NJ (1988) Nonparametric statistics for the behavioral sciences. McGraw-Hill, New York

Van der Knijff JM, Jones RJA, Montanarella L (2000) Soil erosion risk assessment in Italy. European Commission – European Soil Bureau, 52 pp

Van Dijk AIJM, Bruijnzeel LA, Rosewell CJ (2002) Rainfall intensity – kinetic energy relationships: a critical literature appraisal. J Hydrol 261:1–23

Vicente-Serrano SM (2005) Las sequías climáticas en el valle medio del Ebro. Factores atmosféricos, evolución temporal y variabilidad espacial. Consejo de Protección de la Naturaleza de Aragón, 277 pp

Vicente-Serrano SM, Beguería S, López-Moreno JI, García-Vera MA, Stepanek P (2010) A complete daily precipitation database for north-east Spain: reconstruction, quality control and homogeneity. Int J Climatol 30:1146–1163

Wischmeier WH (1959) A rainfall erosion index for a universal soil-loss equation. Soil Sci Soc Am Proc 23:246–249

Wischmeier WH, Smith DD (1978) Predicting rainfall erosion losses: a guide to conservation planning. USDA Handbook 537. USDA, Washington, DC

Impacts of the North Atlantic Oscillation on Landslides

José Luís Zêzere and Ricardo M. Trigo

Abstract Western Iberia landslides are mostly triggered by rainfall, as, in fact, are most landslides worldwide. Results obtained using empirical relationships between rainfall amount and duration, and slope instability show that critical rainfall conditions for failure are not the same for different types of landslides. While rapid debris flows are usually triggered by very intense showers concentrated in just a few hours, shallow translational soil slips are most commonly triggered by intense precipitation falls within the 1–15 days long range. On the contrary, activity of the more deeply-seated landslides of rotational, translational and complex types is related to successive weeks of nearly constant rainfall, over periods of 30–90 days. Large-scale patterns such as the El Niño and the North Atlantic Oscillation (NAO) change slowly and have been shown to have an impact in both the precipitation regime and the temporal occurrence of different landslide types in different areas of the world. In this work a particular attention is devoted to the impact of NAO on the landslide events that have occurred in the region located just north of Lisbon between 1956 and 2010. Results show that the large inter-annual variability of winter precipitation observed in Portugal is largely modulated by the NAO mode. The application of a 3-month moving average to both NAO index and precipitation time series allowed the identification of many months with landslide activity as being characterized by negative average values of the NAO index and high values of average precipitation (above 95 mm/month). Landslide activity in the study area is related to both intense, short duration precipitation events (1–15 days) and long-lasting rainfall episodes (1–3 months). The former events trigger shallow translational slides while the later episodes are usually associated with deeper and larger slope movements. The association between the NAO and landslide activity is shown to be more evident for the group of deep seated landslide events.

Keywords Landslides · Rainfall triggering · Thresholds · NAO

J.L. Zêzere (✉)
Centre for Geographical Studies, Institute of Geography and Spatial Planning, University of Lisbon, Lisbon, Portugal
e-mail: zezere@campus.ul.pt

S.M. Vicente-Serrano, R.M. Trigo (eds.), *Hydrological, Socioeconomic and Ecological Impacts of the North Atlantic Oscillation in the Mediterranean Region*, Advances in Global Change Research 46, DOI 10.1007/978-94-007-1372-7_14,

1 Introduction

According to Cruden (1991) a landslide can be defined as the movement of a mass of rock, earth or debris down a slope. Furthermore, the centre of gravity of the affected material in a landslide always moves in a downward and outward direction (Terzagui, 1950). The landslide classification proposed by Cruden and Varnes (1996) is the most widely adopted and it considers five landslide types, distinguished according to the type of rupture and displacement mechanisms: fall, topple, slide, lateral spread and flow.

Landslide factors are multiple and usually occur at the same time making it difficult to define in any particular case "what is the cause of the landslide". Glade and Crozier (2005) based on previous work by Crozier (1986) and Popescu (1994), proposed the classification of landslide causes in the following classes: predisposing factors, preparatory factors and triggering factors. This classification crosses with three different stability states on slopes defined by the same authors: stability, marginal stability and active instability (Glade and Crozier, 2005).

Landslide predisposing factors are those inherent terrain characteristics (e.g., slope geometry, soil and rock characteristics) that can be assumed as static at the short and mid-term and that control the likelihood of landslide occurrence in space. Landslide preparatory and triggering factors are dynamic and can be subdivided in geomorphologic processes, physical processes and anthropogenic processes (Popescu, 1994). When acting as preparatory factors, these processes promote the decrease of the stability margin of the slope without starting the slope movement. When acting as triggering factors, processes are the immediate cause of slope instability, and it is possible to establish a cause-effect relationship between triggering factor occurrence and landslide occurrence. Therefore, it is clear that geomorphologic, physical and anthropogenic factors may act both as landslide preparatory factors and landslide triggering factors depending on process intensity and temporal extent, as well as on the previous stability state of the slope (Glade and Crozier, 2005).

Rainfall is the most important physical process responsible for the triggering of landslides in the Mediterranean region, as in most regions worldwide (Wieczorek, 1996; Corominas, 2001; Guzzetti et al., 2007). Although the decisive role played by rainfall, there is no general "universal rule" regarding rainfall thresholds related to slope instability, despite the enormous effort done with this goal (e.g. Caine, 1980; Fukuota, 1980; Crozier, 1986; Van Asch et al., 1999; Corominas, 2001; Polemio and Petrucci, 2000; Guzzetti et al., 2007).

The control of rainfall on landslides differs substantially depending upon landslide depth and kinematics and the affected material. Shallow soil slips and rapid debris flows are typically activated by short period of very intense rain while deep-seated rotational and translational slides are usually associated with less intense rainfall occurring in period lasting from several weeks to several months (Van Asch et al., 1999). Therefore, even within a single region, different types of slope movements may be associated with different rainfall characteristics that produce different

hydrological triggering conditions (Van Asch et al., 1999; Polemio and Petrucci, 2000; Corominas, 2001; Zêzere and Rodrigues, 2002; Trigo et al., 2005).

Low frequency atmospheric patterns are associated with anomalous precipitation at the seasonal scale, and this is the main rationale to study the relationships between the atmospheric circulation patterns and landslide activity triggered by rainfall. Previous studies have related the Southern Oscillation and landslide activity in different parts of the world (Coe et al., 1998; Godt, 1999; Ngecu and Mathu, 1999). More recently, the authors have found a significant relationship between the North Atlantic Oscillation and the recent landslide-triggered events occurred in the Lisbon Region (Trigo et al., 2005; Zêzere et al., 2005). Apparently, this relationship is associated with the control that the NAO exerts, at the monthly and seasonal scales, on the storms coming from the North Atlantic Ocean that directly affect the Lisbon region (Zêzere et al., 2005, 2008). In this regard we are particularly interested to verify if this link holds for the recent unusual winter of 2009–2010 that has been characterised by high intense precipitation (Vicente-Serrano et al., 2011) and extreme negative NAO values (Cattiaux et al., 2010).

2 Hydrological Triggering Conditions of Landslides and Relation with Rainfall Patterns

Debris flows, shallow soil slips and deep-seated rotational and translational slides are among the most common of landslide types triggered by rainfall.

The generation of surface run-off and high peak discharges in first-order mountain catchments is a critical triggering mechanism for debris flows (Van Asch et al., 1999). This run-off occurs both on saturated and unsaturated soil conditions and provides water to debris material that have accumulated in channels, thus increasing the pore pressure within the debris mass which initiates the debris flow (Van Asch et al., 1999). In such circumstances, the maximum fluid pressure that can be generated in the debris mass is controlled by the infiltration capacity of the soil and the geometry of the slope (gradient, shape and roughness). The activation of debris flow is typically associated with short-duration (minutes to hours) peak rainfall events.

Shallow translational soil slips are failures occurring on steep slopes at a critical depth in general not more than 1–2 m. These landslides are typically triggered by the rapid infiltration and the unsaturated percolation of water in the thin soil material (colluvium deposits) which overlies impermeable rocks. The critical reduction of the soil shear resistance and the consequent slope failure is due to the rapid and temporary rise of the pore water pressure and the loss of soil apparent cohesion resulting from soil saturation (Gostelow, 1991; Iverson, 2000). Activity of rainfall-triggered shallow translational soil slips are usually associated with intense rainfall accumulating within the 1–15 days period.

Deep-seated rotational and translational slides are slope failures occurring along a slip surface located typically at a depth ranging from 5 to 20 m. Materials mobilized by these landslides are soils and rocks that yield small effective hydraulic

diffusivities (Iverson, 2000). Deep-seated landslides are triggered by the reduction of shear strength of involved rocks and soils, resulting from the steady rise on the groundwater level (Gostelow, 1991; van Asch et al., 1999; Iverson, 2000). Therefore, when compared with shallow soil slips deep landslides need larger absolute amounts of water for triggering conditions, and are associated with larger windows of antecedent rainfall (Van Asch et al., 1999). Activity of this type of landslides is usually related to successive weeks of nearly constant rainfall, over periods lasting from 30 to 90 days.

3 Temporal Occurrence of Rainfall-Triggered Landslides in Lisbon Region and Rainfall Triggering Thresholds

3.1 Rainfall Regime and Landslide Events

The rainfall regime in the Lisbon Region is characterized by high variability at the inter-seasonal and the inter-annual scales. The mean annual precipitation (MAP) at the reference rain gauge of S. Julião do Tojal is 730 mm where a daily rainfall data series is available since 1956 (SJT, Fig. 1). Summer months are noticeably dry and rainfall concentrates in the period lasting from October to March (78% of the total amount; 72% of the total rainy days).

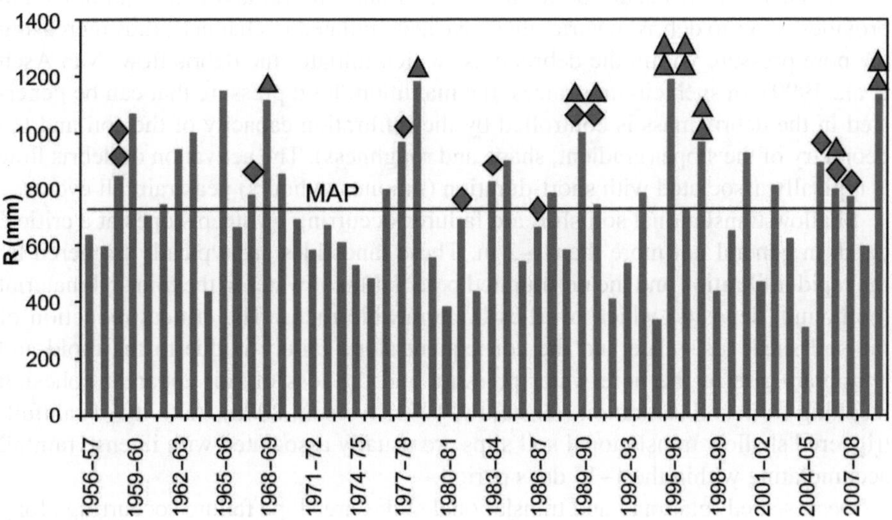

Fig. 1 Annual precipitation (climatological year) at the reference rain gauge of S. Julião do Tojal (period: 1956/1957–2009/2010). *Diamonds*: Shallow landslides events; *triangles*: Deep landslides events

Although the term "landslide event" may be used to characterize the occurrence of a single landslide, this term has a wider connotation as it is used to describe a number of individual landslides triggered in a wide area by a single triggering event, such as an earthquake or an intense rainstorm (Crozier and Glade, 1999). Within this concept, a landslide event in the Lisbon region is a period (a date) for which at least 5 individual landslides are known to have occurred on natural slopes. This number is in accordance with the typical landslide activity observed in the study area (Zêzere et al., 1999).

The dates of landslide activity were reconstructed from field work, archive investigation and interviews with the population living in the study area. A particular attention was given to the particular type(s) of landslide(s) occurred in each event, and a distinction was made between shallow and deep-seated landslides. From 1956–1957 to 2009–2010, 25 landslide events were recognized corresponding to a frequency of one event each 2.16 years, although some years were marked by the occurrence of multiple events (e.g. 1989–1990, 1995–1996, 2009–2010). The total area affected by the above mentioned landslide types within each landslide event

Table 1 Landslide events triggered by rainfall in the Lisbon region from 1956 to 2010

Landslide event #	Date (yyyy/mm/dd)	Critical rainfall amount/duration mm (dd)	Rainfall intensity/ duration mm/day (dd)	Return period (years)	Dominant landslide type
1	1958/12/29	149 (10)	14.9 (10)	2.7	Shallow
2	1959/03/09	175 (10)	17.5 (10)	4.4	Shallow
3	1967/11/15	137 (1)	137.0 (1)	62	Shallow
4	1968/11/15	350 (30)	11.7 (30)	7.6	Deep
5	1978/03/04	204 (15)	13.6 (15)	3.9	Shallow
6	1979/02/10	694 (75)	9.2 (75)	22	Deep
7	1981/12/30	174 (5)	34.7 (5)	14	Shallow
8	1983/11/18	164 (1)	164.0 (1)	194	Shallow
9	1987/02/25	52 (1)	52.0 (1)	2.1	Shallow
10	1989/11/22	164 (15)	11.0 (15)	2.2	Shallow
11	1989/11/25	217 (15)	14.4 (15)	4.7	Shallow
12	1989/12/05	333 (30)	11.1 (30)	6.3	Deep
13	1989/12/21	495 (40)	12.4 (40)	18	Deep
14	1996/01/09	544 (60)	9.1 (60)	11	Deep
15	1996/01/23	686 (75)	9.1 (75)	21	Deep
16	1996/01/28	495 (40)	12.4 (40)	18	Deep
17	1996/02/01	793 (90)	8.8 (90)	27	Deep
18	2001/01/06	447 (60)	7.4 (60)	5.1	Deep
19	2001/01/09	467 (60)	7.8 (60)	6	Deep
20	2006/03/20	156 (4)	39.0 (4)	13	Shallow
21	2006/10/27	219 (10)	21.9 (10)	11	Shallow
22	2006/11/28	427 (40)	10.7 (40)	9	Deep
23	2008/02/18	141 (1)	140.9 (1)	73	Shallow
24	2010/01/14	332 (30)	11.1 (30)	6	Deep
25	2010/03/09	632 (90)	7.0 (90)	8	Deep

was the criterion used to distinguish between shallow and deep landslide events reported in Fig. 1 and Table 1. Shallow (deep) landslide events are characterized by more than 50% of landslide area associated to slip surfaces depth < 1.5 m (depth > 1.5 m). According to this criterion, 13 deep and 12 shallow landslide events were recorded. Interestingly, the nature of the events is non-stationary, with significant changes in frequency of deep and shallow events between the first and second halves of the period considered. Thus, shallow landslide events were more frequent from 1956 until 1989 (9 in 11 cases), while deep landslide events were dominant from 1989 to 2010 (11 in 14 cases).

3.2 Absolute Antecedent Rainfall Threshold

Following the approach used previously by the authors (Trigo et al., 2005; Zêzere et al., 2005), the relationship between rainfall and landslide activity was assessed by computing the cumulative absolute antecedent rainfall for the 1–5, 10, 15, 30, 40, 60, 75, and 90 consecutive days anteceding each landslide event. The Gumbel distribution (Gumbel, 1958) was used to derive the return period for each rainfall amount/duration combination. The critical rainfall combination (quantity-duration) responsible for each landslide event is shown in Table 1, and it was obtained, assuming as critical pair the combination with the higher return period. We acknowledge this assumption lacks a physical basis, nevertheless it provides the best discrimination between landslide events and other rainfall periods not related with slope instability (Zêzere et al., 2005).

The combination between rainfall intensity and critical rainfall duration for the 25 landslide events registered from 1956 to 2010 is shown in Fig. 2. The regression analysis shows that rainfall intensity increases exponentially as duration decreases, following the equation $\mathrm{Ri} = 92.03\mathrm{D}^{-0.594}$ ($r^2 = 0.902$), where Ri is the rainfall intensity in mm/day and D is the duration of rainfall in days. Figure 2 also shows the maximum yearly values for durations of 1, 5, 10, 15, 30, 40, 60, 75 and 90 days that were computed for those years without landslide activity. The vast majority of these values lie below the fitted curve, thus confirming that regression curve may be used as a reliable rainfall intensity – duration threshold for the study area.

The above mentioned regression model was used to derive the minimum daily rainfall needed to reach the precipitation triggering threshold at the reference rain gauge (S. Julião do Tojal), for any of the following consecutive days: 1–5, 10, 15, 30, 40, 60, 75 and 90 (Zêzere et al., 2008). This calculation accounts for the continuous cumulative absolute antecedent rainfall, and it is shown in Fig. 3 for the time span lasting from 2000 to 2010. During this period, our model correctly predicted 7 landslide events (Landslide events # 18, 19, 20, 21, 22, 23 and 24), and the landslide event # 25 is the solely "false negative" (i.e., unpredicted landslide event). In addition, there is no evidence of "false positives" (i.e., predicted landslide events that did not occur).

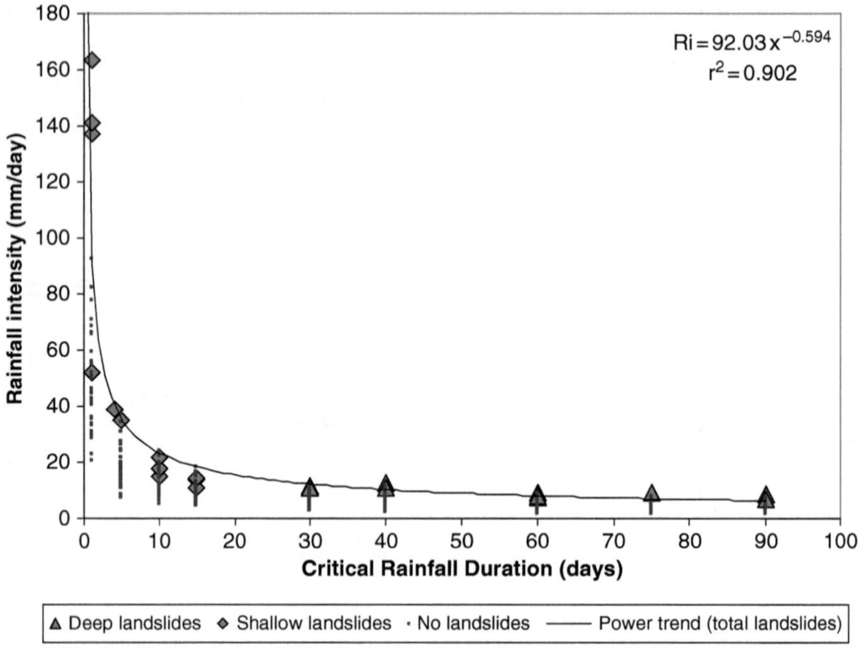

Fig. 2 Regression line between rainfall intensity and rainfall duration for landslide events in the Lisbon region (period 1956–2010)

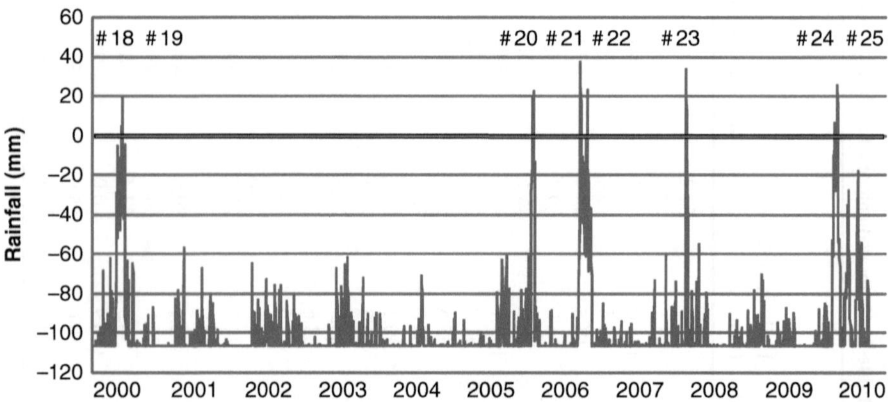

Fig. 3 Daily rainfall needed to reach the rainfall threshold in S. Julião do Tojal for landslide initiation in the Lisbon region (period: September 2000–August 2010)

Table 2 Calibrated antecedent rainfall (CAR) for landslide events in the Lisbon region from 1956 to 2010

Landslide event #	Date (yyyy/mm/dd)	Daily rainfall (mm)	3 days (mm)	5 days (mm)	10 days (mm)	15 days (mm)	30 days (mm)	No. of previous consecutive days with 30-days CAR >50 mm	Dominant landslide type
1	1958/12/29	22.0	54.2	73.6	91.7	99.2	108.5	6	Shallow
2	1959/03/09	45.5	27.5	55.4	82.2	82.2	83.6	4	Shallow
3	1967/11/15	137.0	4.1	4.1	10.3	16.3	25.2	0	Shallow
4	1968/11/15	45.5	43.4	47.3	74.3	92.3	107.7	18	Deep
5	1978/03/04	14.5	43.9	67.1	93.8	105.0	109.8	8	Shallow
6	1979/02/10	26.0	89.3	93.6	110.1	119.7	133.6	17	Deep
7	1981/12/30	78.0	75.5	75.5	102.6	109.5	111.3	3	Shallow
8	1983/11/18	163.7	42.1	59.8	88.2	96.7	103.8	10	Shallow
9	1987/02/25	51.8	22.2	35.4	36.0	46.5	52.1	1	Shallow
10	1989/11/22	33.5	19.9	50.7	62.9	70.9	72.1	4	Shallow
11	1989/11/25	38.0	36.9	42.7	75.7	88.5	89.1	7	Shallow
12	1989/12/05	30.1	58.1	58.1	78.8	93.7	109.7	17	Deep
13	1989/12/21	26.7	29.1	40.5	65.3	78.2	100.9	33	Deep
14	1996/01/09	66.5	17.8	29.5	54.2	79.5	88.4	17	Deep
15	1996/01/23	23.6	36.5	38.7	53.2	78.9	96.8	31	Deep
16	1996/01/28	25.5	9.2	30.9	53.7	62.3	85.6	36	Deep
17	1996/02/01	24.5	34.5	48.6	69.1	79.9	97.9	40	Deep
18	2001/01/06	24.5	22.7	39.1	53.7	79.1	87.4	17	Deep
19	2001/01/09	16.0	22.1	38.6	51.9	67.1	85.3	20	Deep
20	2006/03/20	42.1	89.5	89.6	89.7	95.9	107.4	2	Shallow
21	2006/10/27	26.0	64.4	99.1	126.8	131.4	132.1	6	Shallow
22	2006/11/28	36.5	33.4	58.0	62.8	74.5	79.6	3	Deep
23	2008/02/18	140.9	0.0	0.0	0.3	6.2	6.6	0	Shallow
24	2010/01/14	16.6	55.6	56.1	70.4	78.3	99.7	19	Deep
25	2010/03/09	29.7	36.1	44.8	62.4	72.9	83.4	15	Deep

3.3 Calibrated Antecedent Rainfall Thresholds

Due to superficial and sub-superficial drainage, the impact of a particular rainy event on slope hydrology decreases in time (Canuti et al., 1985; Crozier, 1986). In order to account this effect in the analysis of rainfall-triggered landslide events, we reconstructed the calibrated antecedent rainfall (CAR) for 3, 5, 10, 15 and 30 days using the algorithm proposed by Crozier (1986):

$$\text{CARx} = \text{KP1} + \text{K}^2\text{P2} + \ldots \text{K}^n\text{Pn} \tag{1}$$

where CARx is the calibrated antecedent rainfall for day x; P1 is the daily rainfall observed for the day before day x; Pn is the daily rainfall registered for the n-th day before day x; and K (assumed to be 0.9) is the empirical parameter that accounts for the decrease of the impact of a particular rainy event in time due to drainage processes.

Table 2 summarizes the calibrated antecedent rainfall computed with Equation (1) for the 25 landslide events registered in the Lisbon Region. The distinction between shallow and deep landslide events concerning the number of days relevant to the antecedent rainfall was confirmed by combining the daily rainfall and the CAR (Fig. 4). The best result for shallow landslide events was obtained with the combination between the daily rainfall and the 5 days CAR through the exponential rule $\text{Dr} = 151.57e^{-0.0336\text{CAR}}$ (where Dr is the daily rainfall), whereas deep landslide events are better discriminated by a combined threshold of daily rainfall = 16 mm and 30 days CAR = 80 mm.

Differences on the relevance of antecedent rainfall for shallow and deep landslide events are also remarkable regarding the number of previous consecutive days with 30 days CAR>50 mm (Table 2). This number ranges from 0 to 10 days for dates of shallow movements occurrence, and increases up to a minimum of 15 days for dates characterised by deep landslide activity, with the single exception of landslide event #22.

4 Impact of NAO on Precipitation and on Landslide Occurrence in Lisbon Region

Previous works have shown that the phase of the NAO pattern is the most important to model the temporal precipitation distribution over western Iberia (Hurrell, 1995; Trigo et al., 2004; Zêzere et al., 2005). In fact the authors have shown that the NAO is (statistically) associated with the occurrence of landslide events, particularly those triggered by prolonged periods (up to 3 months) of precipitation (Trigo et al., 2005; Zêzere et al., 2005, 2008).

The NAO index used here was developed at the Climatic Research Unit (University of East Anglia, UK) and is defined, on a monthly basis, as the difference between the normalized surface pressure at Gibraltar (southern tip of Iberian Peninsula) and Stykkisholmur in Iceland (Jones et al., 1997). The NAO index for

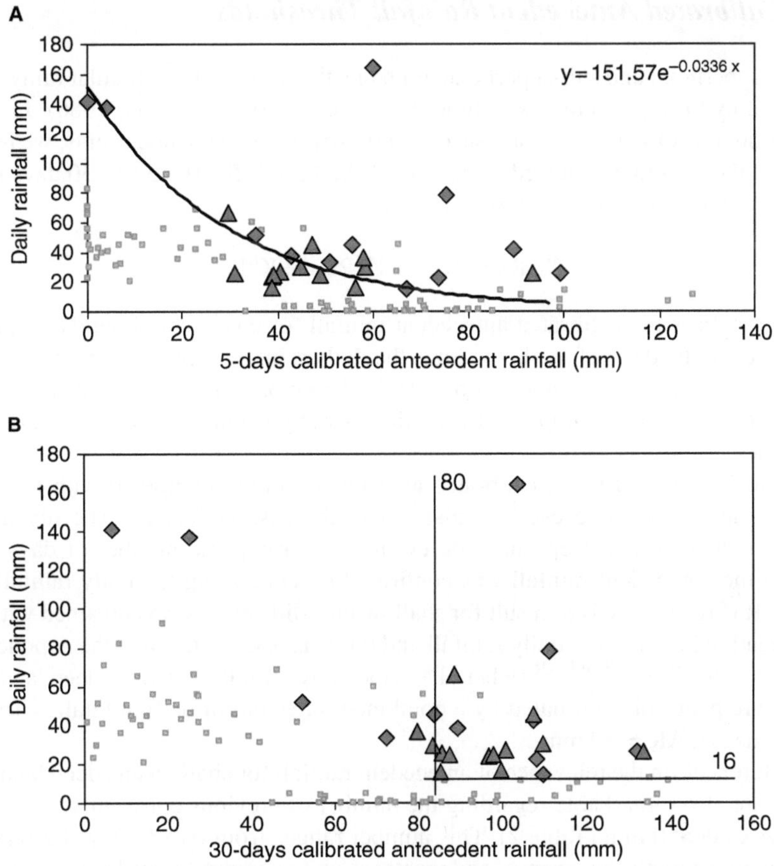

Fig. 4 Relationship between daily rainfall and 5-days (**a**) and 30-days (**b**) calibrated antecedent rainfall (CAR) for landslide events in the Lisbon region (period 1956–2010). *Diamonds*: values associated with shallow landslides events; *triangles*: values associated with deep landslides events; *small dots*: values obtained from the yearly maximum daily rainfall and from the yearly maximum CAR, computed for years without landslide events

winter months presents a positive trend for the last 3 decades of the twentieth century, thus the corresponding distribution is dominated by positive values. In order to remove this positive bias we have normalized all values of NAO index, at the monthly scale, between 1938 and 2010 (the period with precipitation data for the nearby SJT series). We apply the same approach developed previously (Trigo et al., 2005; Zêzere et al., 2008) but updating it to the year 2010. Thus a 3-month moving average was applied to filter the original NAO index and SJT precipitation time series, restricting the analysis to monthly data from the wet season (NDJFM), with the single exception of the landslide event #21 for which October and September values were considered. October values are used to compute November and December averages. Thus, values for February of *year n*, correspond

to the rainfall and NAO index averages computed between December of *year n*–1 and February of *year n*, while values for November of any year are restricted to the October and November values for that same year. For landslides episodes that have occurred in the first 5 days of the month we only considered the NAO index and precipitation values from the previous 2 months (events 5, 12 and 17). The scatter plot obtained with both filtered time series is shown in Fig. 5, being limited to the winter (NDJFM) months. Small open diamonds represent months where landslide events did not occur (or were not reported). Black circles correspond to months with only one recognized landslide event (e.g. February 1996, event 17), while black triangles represent those months with more than one landslide event (e.g. December 1989, events 12 and 13). A number of relevant remarks can be raised in relation to Fig. 5:

1. The linear correlation between both time series is –0.58 (statistically significant at the 1% level) increasing slightly to –0.61 if we restrict the analysis to those months affected by landslide activity represented by the numbered black circles and triangles (regression line is also represented).
2. All months characterized with landslide events are located above the 3-month average precipitation (horizontal line indicating the threshold of 95 mm/month). Furthermore, with the exception of event #23 all cases are characterized by negative values of the averaged NAO index. Interestingly, this event occurred on the 18 February 2008 corresponding to the new all time record of daily precipitation observed in Lisbon (since 1865), Capital of Portugal located just 14 km south of SJT (Fragoso et al., 2010).

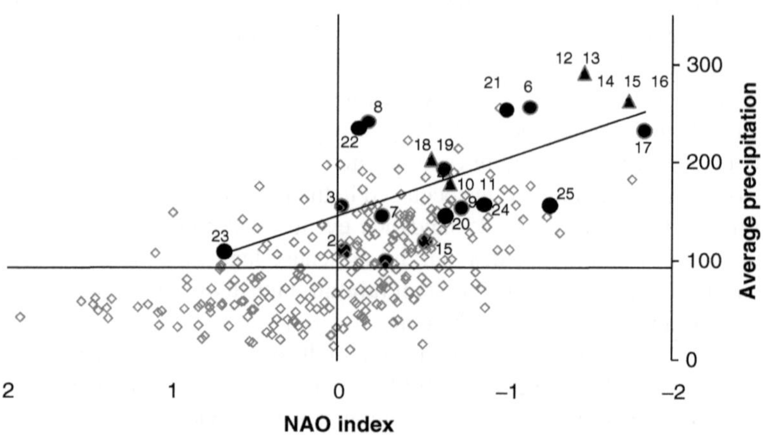

Fig. 5 Scatter plot between 3-month moving average NAO index and SJT precipitation time series (November–March), for the period ranging between 1956 and 2010 (270 months). *Small open diamonds*: months with no landslide events reported. *Black circles*: months with one recognized landslide event. *Black triangles*: months with more than one recognized landslide event. *Numbers* next to *gray symbols* represent the landslide episode number as given in Tables 1 and 2

3. The majority of the multiple landslide events episodes (#12 to #16) are characterized by extremely negative average values of the NAO index (lower than –1) and extremely high values of average precipitation (clearly above 200 mm/month). This prolonged anomalous precipitation regime favors the occurrence of multiple deep landslide events (Table 1).

5 Conclusion

Landslide events occurred since 1956 in the Lisbon region were triggered by rainfall. Events dominated by shallow soil slips have been related to short periods (1–15 days) of intense rainfall, whereas events characterized by the occurrence of deep seated slope movements have been associated to longer periods (30–90 days) of less intense rain. The distinction between landslide types concerning the number of days relevant to the antecedent rainfall was confirmed combining the daily rainfall and the calibrated antecedent rainfall (CAR) from 3 to 30 days for each landslide event. An exponential rule was identified ($Dr = 151.57e^{-0.0336CAR}$) for the 5 days CAR, which is a threshold for shallow landslide events, while deep landslide events were better discriminated by a combined threshold of daily rainfall = 16 mm and 30 days CAR = 80 mm.

These different rainfall triggering conditions are consistent with the different hydrologic processes associated with different types of slope movements. Intense rainfall allows the rapid growth of pore water pressure and the drop of capillarity forces that sustain the apparent cohesion of thin soils. Therefore, shallow soil failures occur within the soil material or at the contact with the underlying less permeable bedrock. Long lasting rainfall episodes enable the steady rise of the groundwater table and the development of positive pore water pressures into the soil. Consequently, deep seated failures occur in relation to the reduction of shear strength of affected materials.

Despite the different relationships existing between the activity of shallow and deep seated landslides and the rainfall triggering conditions, it was possible to derive a single exponential law relating rainfall intensity and the rainfall critical duration for the total set of landslide events occurred in the Lisbon region from 1956 to 2010. The fitting of the curve to the 25 landslide events is impressive ($r^2 = 0.9$) and the vast majority of rainfall episodes that did not cause landslides lie below the curve, thus confirming the curve as a consistent rainfall intensity/duration threshold for the triggering of landslides in the study area.

The linear correlation between the 3 month average NAO index and precipitation is statistically significant (–0.61) for those months characterized by the occurrence of landslide events. The complete set of landslide events lies above the 3-month average precipitation, and 24 in 25 events present negative values of the 3-month average NAO index. Taking into account the different rainfall periods associated with shallow and deep seated landslides events, the association between the NAO and landslide activity is more evident for the group of deep seated landslide events. In addition, the majority of months characterized by the occurrence of

multiple landslide events exhibit extremely high values of average precipitation and extremely negative average values of the NAO index.

However, it must be stressed that it is not so clear-cut to establish the sequence of physical phenomena that link the NAO configuration with local landslide activity. Without pretending to describe the entire chain of physical processes involved these must include; (1) the NAO control of the preferred path for north Atlantic storm tracks, (2) how these low pressure systems influence the precipitation regime of western Iberia, (3) to what extent the precipitation anomalies impact local soil moisture at different temporal scales, and (4) how these changes can be critical for surpassing water pressure thresholds that can trigger landslides.

Acknowledgements This work was supported by Projects Maprisk – Methodologies for assessing landslide hazard and risk applied to municipal planning (PTDC/GEO/68227/2006) and DISASTER – GIS database on hydro-geomorphologic disasters in Portugal: a tool for environmental management and emergency planning (PTDC/CS-GEO/103231/2008) funded by the Portuguese Foundation for Science and Technology (FCT).

References

Caine N (1980) The rainfall intensity–duration control of shallow landslides and debris flows. Geografiska Annaler 62:23–27

Canuti P, Focardi P, Garzonio CA (1985) Correlation between rainfall and landslides. Bull Int Assoc Eng Geol 32:49–54

Cattiaux J, Vautard R, Cassou C, Yiou P, Masson-Delmotte V, Codron F (2010) Winter 2010 in Europe: a cold extreme in a warming climate. Geophys Res Lett 37:L20704. doi:10.1029/2010GL044613

Coe JA, Godt JW, Wilson RC (1998) Distribution of debris flows in Alameda County, California triggered by 1998 El Niño rainstorms: a repeat of January 1982? EOS 79:266

Corominas J (2001) Landslides and climate. Keynote lectures from the 8th international symposium on landslides 4:1–33

Crozier M (1986) Landslides: causes, consequences and environment. Croom Helm, London

Crozier MJ, Glade T (1999) Frequency and magnitude of landsliding: fundamental research issues. Z Geomorph NF Suppl-Bd 115:141–155

Cruden DM (1991) A simple definition of a landslide. Bull Int Assoc Eng Geol 43:27–29

Cruden DM, Varnes DJ (1996) Landslide types and processes. In: Turner AK, Schuster RL (eds) Landslides: investigation and mitigation. National Research Council, Transportation Research Board, Washington, DC, pp 36–75

Fragoso M, Trigo RM, Zêzere JL, Valente MA (2010) The exceptional rainfall episode registered in Lisbon in 18 February 2008. Weather 65(2):31–35

Fukuota M (1980) Landslides associated with rainfall. Geotech Eng 11:1–29

Glade T, Crozier MJ (2005) The nature of landslide hazard and impact. In: Glade T, Anderson MG, Crozier MJ (eds) Landslide hazard and risk. Wiley, Chichester, pp 43–74

Godt JW (1999) Maps showing locations of damaging landslides caused by El Niño rainstorms, winter season 1997–98, San Francisco Bay region, CA. USGS http://pubs.usgs.gov/mf/1999/mf-2325/

Gostelow P (1991) Rainfall and landslides. In: Almeida-Teixeira M et al (eds) Prevention and control of landslides and other mass movements. CEC, Bruxels, pp 139–161

Gumbel EJ (1958) Statistics of extremes. Columbia University Press, New York

Guzzetti F, Peruccacci S, Rossi M, Stark CP (2007) Rainfall thresholds for the initiation of landslides in central and southern Europe. Meteorol Atmos Phys 98:239–267

Hurrell JW (1995) Decadal trends in the North Atlantic Oscillation: regional temperatures and precipitation. Science 269:676–679

Iverson RM (2000) Landslide triggering by rain infiltration. Water Resour Res 36:1897–1910

Jones PD, Jónsson T, Wheeler D (1997) Extension to the North Atlantic Oscillation using early instrumental pressure observations from Gibraltar and south-west Iceland. Int J Climatol 17:1433–1450

Ngecu WM, Mathu EM (1999) The El-Niño triggered landslides and their socioeconomic impact in Kenya. Environ Geol 38:277–284

Polemio M, Petrucci O (2000) Rainfall as a landslide triggering factor: an overview of recent international research. In: Bromhead E, Dixon N, Ibsen M (eds) Landslides in research, theory and practice. Thomas Telford, London, pp 1219–1226

Popescu M (1994) A suggested method for reporting landslide causes. Bull Int Assoc Eng Geol 50:71–74

Terzagui K (1950) Mechanisms of landslides. Geological Society of America (Berkey Volume), pp 83–123

Trigo RM, Pozo-Vázquez D, Osborn TJ, Castro-Díez Y, Gámiz-Fortis S, Esteban-Parra MJ (2004) North Atlantic Oscillation influence on precipitation, river flow and water resources in the Iberian Peninsula. Int J Climatol 24:925–944

Trigo RM, Zêzere JL, Rodrigues ML, Trigo IF (2005) The influence of the North Atlantic Oscillation on rainfall triggering of landslides near Lisbon. Nat Hazards 36:331–354

Van Asch T, Buma J, Van Beek L (1999) A view on some hydrological triggering systems in landslides. Geomorphology 30:25–32

Vicente-Serrano SM, Trigo RM, Liberato ML, López-Moreno JI, Lorenzo-Lacruz J, Beguería S, Morán-Tejeda H, El Kenawy A (2011) The 2010 extreme winter North Hemisphere Atmospheric variability in Iberian precipitation: anomalies, driving mechanisms and future projections. Clim Res 46:51–65

Wieczorek GF (1996) Landslides triggering mechanisms. In: Turner AK, Schuster RL (eds) Landslides: investigation and mitigation. National Research Council, Transportation Research Board, Washington, DC, pp 76–90

Zêzere JL, Ferreira AB, Rodrigues ML (1999) Landslides in the north of Lisbon region (Portugal): conditioning and triggering factors. Phys Chem Earth, Part A 24:925–934

Zêzere JL, Rodrigues ML (2002) Rainfall thresholds for landsliding in Lisbon area (Portugal). In: Rybar J, Stemberk J, Wagner P (eds) Landslides. A.A. Balkema, Lisse, pp 333–338

Zêzere JL, Trigo RM, Fragoso M, Oliveira SC, Garcia RAC (2008) Rainfall-triggered landslides in the Lisbon region over 2006 and relationships with the North Atlantic Oscillation. Nat Hazards Earth Syst Sci 8:483–499

Zêzere JL, Trigo RM, Trigo IF (2005) Shallow and deep landslides induced by rainfall in the Lisbon region (Portugal): assessment of relationships with the North Atlantic Oscillation. Nat Hazards Earth Syst Sci 5:331–344

The Impact of the NAO on the Solar and Wind Energy Resources in the Mediterranean Area

David Pozo-Vazquez, Francisco Javier Santos-Alamillos, Vicente Lara-Fanego, Jose Antonio Ruiz-Arias, and Joaquín Tovar-Pescador

Abstract The influence of the NAO on the solar and wind energy resources in the Mediterranean area, in particular, and over the whole North Atlantic area, in general, has been explored based on the analysis of 20 years of reanalysis and satellite data. The analysis was carried out for the winter and annual periods and proved the existence of a marked influence of the NAO pattern on the spatial and temporal variability of the solar and wind energy resources in the study area. Particularly, the NAO signature on the solar/wind energy resources was found to be a north-south dipolar pattern, with the positive/negative centre over the Mediterranean area. Although the spatial patterns obtained are similar, the nature of the relationship obtained for wind energy was found to be less linear and more complex than that attained for solar energy. A composite analysis revealed that interannual variability of the solar and wind energy resources in the Mediterranean area can reach values above 20% in winter and 10% in the annual case associated with changes in the NAO phase. Finally, the analysis of the spatial patterns of the NAO influence on the solar and wind energy resources revealed the existence of spatial and local balancing between these two energy resources in the study area. Results are of interest regarding the estimation of the expected interannual variability of the wind farms and solar plants production in the study region.

Keywords NAO · Balancing renewable energy · Solar energy · Wind energy

1 Introduction

According to the latest estimates, it is expected that the worldwide primary energy demand will increase by 45% and the demand for electricity will increase by 80% between 2006 and 2030 (IEA, 2008). As a consequence, and without a decisive action, energy-related greenhouse gases (GHG) emissions will more than double by 2050 and the increased oil demand will intensify the concerns over the security of supplies. There are different pathways to stabilize GHG concentrations, but a key

D. Pozo-Vazquez (✉)
Department of Physics, University of Jaén, Jaén, Spain
e-mail: dpozo@ujaen.es

S.M. Vicente-Serrano, R.M. Trigo (eds.), *Hydrological, Socioeconomic and Ecological Impacts of the North Atlantic Oscillation in the Mediterranean Region*, Advances in Global Change Research 46, DOI 10.1007/978-94-007-1372-7_15,

issue in all of them is the replacement of fossil fuels by renewable energy sources. European Union (EU) regulations, for example, set a binding target for 20% of the EU's total energy supply to come from renewable by 2020 (corresponding to a sharp increase from the modest 6.5% observed in 2007). Furthermore, they set a firm target of cutting 20% of the EU's greenhouse gases emissions by 2020 relative to 1990. In this scenario, solar and wind energies -the renewables with the most mature- will necessarily play a key role.

There are three main solar technologies. The first one is the direct conversion of sunlight into electricity (Photovoltaic, PV effect), the concentrating solar power (CSP) and the solar thermal collectors for heating and cooling (SHC). PV and CSP technologies are meant to produce electricity, while SHC can only be used for heating and cooling purposes. PV systems enable direct conversion of sunlight into electricity through semi-conductor devices, through the global horizontal irradiance (GHI) they receive. Today, PV provides 0.2% of total global electricity generation, being Germany the leading country in the world, with 5 GW installed capacity, followed by Spain with above 4 GW (about 4% of the installed capacity). While its use is small today, PV power has a particularly promising future. Actually, global PV capacity has been increased at an average annual growth rate of more than 40% since 2000 and it has significant potential for long-term growth over the next decades. Particularly, PV is projected to provide around 5% of global electricity consumption by 2030, rising to 11% by 2050 (IEA, 2010a). In this context, the Mediterranean area (France, Italy, Greece) are expected to play a major role (IEA, 2010a). The basic concept of CSP is relatively simple: CSP devices concentrate energy from the sun's rays to heat a receiver to high temperatures. This heat is then transformed first into mechanical energy (by turbines or other engines) and then into electricity. The solar energy that CSP plants use is measured as direct normal irradiance (DNI), which is the energy received on a surface tracked perpendicular to the sun's rays. CSP developers typically set a bottom threshold for DNI of about 2000 kWh/m^2/year. Below that, other solar electric technologies that take advantage of both direct and diffuse irradiance, such as PV, are assumed to have a competitive advantage. Good DNI is usually found in arid and semi-arid areas with reliably clear skies, which typically lies at latitudes from 15 to 40° North or South. As a consequence, one of the most favorable areas for the CSP technology is the Mediterranean basin. In 2010, the global stock of CSP plants neared 1 GW capacity, mainly in Spain (about 0.5 GW). Projects now in development or under construction in more than a dozen countries (including China, India, Morocco, Spain and the United States) are expected to totalize 15 GW, and CSP is expected to contribute about 5% (1500 GW) of the annual global electricity production in 2050 (IEA, 2010b). Particularly, potential in the Middle East and North Africa would cover about 100 times the current consumption of the Middle East, North Africa and the European Union combined (IEA, 2010b). Finally, wind energy is probably the most cost-effective of the renewable energy technologies. It must be stressed that wind energy was the renewable source with the fastest growing rate: since 2000, cumulative installed capacity has grown at an average rate around 30% per year (Global Wind Energy Council, 2007). In 2008, wind energy provided nearly 20% of electricity consumption in Denmark,

more than 11% in Portugal and Spain, over 4% of all EU electricity and nearly 2% in the United States (IEA, 2009). By 2030, wind electricity is estimated to be produced annually from over 1000 GW of wind installed capacity, corresponding to 9% of global electricity production with an expected rising to 12% in 2050 (IEA, 2009).

The Mediterranean area presents some special features regarding the transition from a fossil era to a greener one. In fact the region presents considerable wind and outstanding solar energy resources that would make this transition easier. According to different studies (DLR, 2007; DESERTEC, 2010; Price Water House Cooper, 2010), Europe and North Africa together could produce by 2050 about 80% of their electricity needs from renewables, if their respective grids were sufficiently interconnected.

1.1 Energy and Meteorology

The expected huge development of the renewable energies in the coming decades will demand a corresponding development of the so-called *energy meteorology* science. Energy meteorology includes all the aspects of the meteorology and climatology relevant for the renewable energies. In fact, the meteorological and climate information is a key issue at the various stages of a renewable energy project, as outlined in Fig. 1. Particularly, in the first stages (project development), the climate information is required in terms of the resource evaluation. This evaluation includes data obtained based on direct measurement, satellites and models

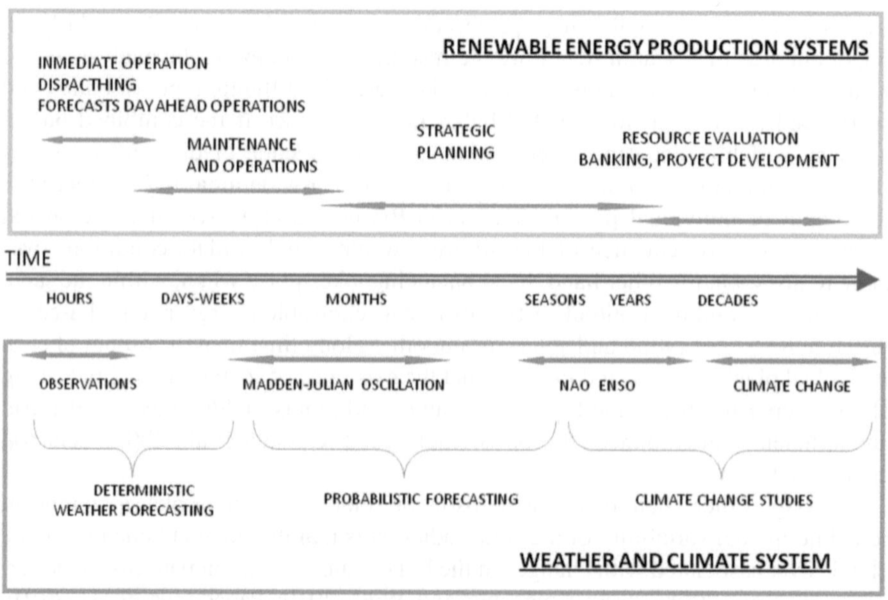

Fig. 1 The relationship between the weather and climate and different activities related to the renewable energy

estimations. Once operating, weather forecasts are needed for the yield integration in the power grid. This requires information from both satellite (nowcasting) and forecasts from Numerical Weather Prediction models (forecasting) (Lara-Fanego et al., 2011). Accurate information on expected solar irradiance or wind speed is used for the management of the electricity grids, the scheduling of conventional power plants and also for decision making on the energy market. Monthly and seasonal forecasts are also of interest for maintaining operation and strategic planning, as they can potentially allow the electricity system operators to estimate the energy production availability and the electric network operator to conveniently adapt demand and resources (Weisser and Foxon, 2003).

Accurate resource evaluation is a key issue in the first stages of a renewable energy project. Particularly, as the life span of wind farms or solar plants (both PV and STPP) is about 20 years, the economic feasibility study of the project should be considered an adequate estimation of the interannual variability of the resources. Usually, the resource evaluation is mainly based on field measurements collected at meteorological stations. Since these measures are expensive and time consuming, typically only a few years of data are available for the investment decision making (Barbour and Walker, 2008). However, a few years of measurements do not generally account for the natural climate variability of the resources. Consequently, the projected production could significantly vary from 1 year to another and through decades. Therefore, some complementary studies, usually based on satellite estimates or model simulations, are necessary for a reliable renewable energy resources evaluation. Aside from these very important economic aspects, the evaluation of interannual variability is crucial regarding the reliability of a power system with a high penetration of renewable energy. Particularly, an important area of research regarding the resources evaluation is the analysis of the balancing between different renewables. There are two different types of balancing: spatial and local balancing. Spatial balancing takes place if the combined output of many variable renewable energy plants, based on same or different resources (mainly wind and solar), at different locations over a wide area, is smoother than the output of individual power plants (von Bremen, 2008). Note that the spatial balancing is a consequence of the different weather and climate conditions over large regions. On the other hand, local balancing takes place when, within the same location, the combined output of two different renewable energy plants, based on different resources (wind and solar), is smoother along time than the output of each individual plant. Several studies show that the existence of spatial or local balancing allows increasing the reliability of solar and wind plants yields, provided the grid are sufficiently interconnected (von Bremen, 2008; Cassola et al., 2008; Kempton et al., 2010).

Although other phenomena, as aerosols or solar cycles, may play a significant role, interannual variability of the solar radiation is mainly related to changes in the cloud cover associated with changes in the large scale circulation patterns (Sanchez-Lorenzo et al., 2008, 2009; Chiacchio and Wild, 2010; Papadimas et al., 2010). Concerning the wind resources, surface winds are also mainly driven by large scale circulation. Nevertheless, local features such as surface roughness, local thermal

contrast and orography can modify their spatial and temporal features, especially in areas of complex topography as the Mediterranean region (Burlando, 2009; Li et al., 2010). On the other side, the North Atlantic Oscillation (NAO) is recognised as a major actor regarding the interannual variability of the large scale circulation in the Mediterranean area, especially during winter (Hurrell et al., 2003). Particularly, the impact of the NAO on the cloud cover variability and on solar radiation at the surface at the whole North Atlantic area has been previously described (Trigo et al., 2002; Pozo-Vázquez et al., 2004; Lohmann et al., 2006; Sanchez-Lorenzo et al., 2008, 2009; Chiacchio and Wild, 2010; Papadimas et al., 2010). However, the NAO impact on the wind and, therefore, wind energy resources is almost unexplored. Therefore, the NAO can plausibly have significant effects on the wind and solar energy resources in the Mediterranean basin which are worth to be analyzed.

In this work we present an analysis of the impact of the NAO on the solar and wind energy resources in the Mediterranean area. Particularly, the study aims to evaluate the role of the NAO in the interannual variability of the wind and solar energy. Monthly means of daily integrated wind and solar energy data, covering the whole North Atlantic and the Mediterranean area, including the north of Africa, are analyzed. The dataset was derived from satellite retrievals (solar energy) and meteorological reanalysis (wind energy and cloud cover) and extends over a period of about 20 years. We first carried out a correlation analysis to explore the linear relationship between the NAO and the solar and wind energy. An analysis of the cloud cover-NAO correlation is also undertaken. In a second part, the composite analysis is used to study the linear and non-linear component of the solar and wind energy anomalies response to the NAO phases. Finally, the eventual existence of balancing between the wind and solar energy resources associated with the NAO cycle is explored. The work is organized as follows: Section 2 describes the dataset and the processes used to derive the wind and solar energy data. Section 3 deals with the results. Finally, in Section 4 some concluding remarks are outlined.

2 Data Description

Two datasets have been used in this work. Firstly, the ERA-Interim reanalysis dataset was used to derive cloud cover data and the wind energy resources. The ERA-interim global reanalysis are the third generation of reanalysis of the European Center for Medium Range Weather Forecasting (ECMWF). Original data have a spatial resolution of about 0.7° and a temporal step of 3 h (Simmons et al., 2007). In this work we have analyzed the period 1989–2008. Wind speed at 10 m above the surface at 3 h interval was used to estimate daily integrated values of the wind energy based on:

$$E_{\text{daily-wind}} = \sum_{1}^{8} \left(\frac{1}{2} \rho v^3 \right) 3, \tag{1}$$

where v is the wind speed every 3 h and ρ is the density of the air at sea level. Note the energy units are (W·h). Based on these daily integrated wind energy estimates, monthly means values were computed. Finally, winter season and annual mean values were computed based on the monthly values. December through February were used for winter, while annual values were computed from December of the preceding year to November of the considered year. Regarding the cloud cover, daily mean values were computed based on the 3 h interval values. Then, monthly mean values and winter and annual mean values were computed based on the same procedure that for the wind data.

Solar energy was derived from the Irradiance at the Surface derived from ISCCP cloud (ISIS) dataset (Lohmann et al., 2006). It consists of 3-hourly values of direct normal and global irradiances at the surface that have been derived by means of one-dimensional radiative transfer computations. The ISIS dataset covers the time period 1984–2004, with a spatial resolution of 280 by 280 km. Note that 3-h temporal resolution of ISIS enables the study of daily cycles. The daily integrated solar energy estimates were obtained along the period 1985–2004 based on the global irradiance data:

$$E_{daily-solar} = \sum_{1}^{8} GHI \times 3, \tag{2}$$

where GHI is the global irradiance every 3 h. As for the wind energy, and based on these daily integrated solar energy estimates, monthly means values were computed and, then, winter season and annual mean values were obtained.

Both the ERA-interim and the ISIS dataset have a global coverage. In this work, although the analysis has been focused in the Mediterranean region, the study area has been extended to the whole North Atlantic region, including the north of Africa, from latitude 20º northward.

Finally, the winter and annual NAO index as defined by Hurrell et al. (2003) was used to monitor the NAO.

3 Results

3.1 NAO-Solar Energy Correlation Analysis

Figures 2a and 2b show the sample correlation coefficients between the NAO index and the solar energy for the winter and annual cases, respectively. During the winter, a dipolar pattern can be observed: while almost the whole Mediterranean region presents positive correlations, most part of central and northern Europe presents negative values. The highest positive correlation values (0.8) are located over the southwest of the Iberian Peninsula. Also over most part of the central and eastern Mediterranean area high positive values (0.6) are observed. On the other hand, the highest negative correlations are located over the north west of Russia (–0.6) and British Islands (–0.4). Note also the relatively weak, but significant, negative

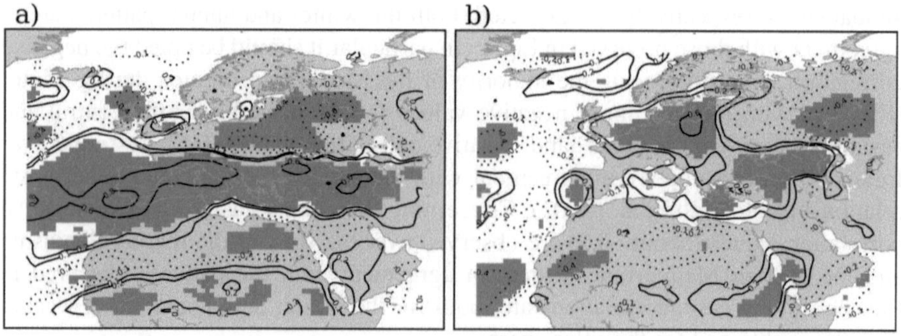

Fig. 2 (**a**) Correlation between the NAO index and the solar energy for the winter season. The period of analysis covers 2005–2004. *Dots* indicate negative correlation while *continuous line* indicates positive correlation. *Shading* indicates local statistical significance of the difference at 90% confidence level based on a t-test. (**b**) as in (**a**) but for the annual data

correlation obtained over Egypt. Results are in agreement with those recently presented by Papadimas et al. (2010), which showed the existence of a Mediterranean solar radiation spatial variability mode associated with the NAO during the winter. Regarding the annual correlation map, it resembles winter case but displaced north-eastward and with weaker values. Particularly, the highest positive correlations (0.6) are found over Poland and the Baltic area. Regarding the Mediterranean area, weak but statistically significant positive correlations (0.2) are found over the southwest of the Iberian Peninsula and the eastern Mediterranean area and weak negative correlations over the north of Egypt.

Since interannual variability of the solar radiation is mainly associated with cloud cover, the former NAO-solar energy correlation patterns should resemble the NAO-cloud cover spatial correlation over the study area. Figures 3a and 3b shows the correlation between the NAO index and the total cloud cover for the winter and

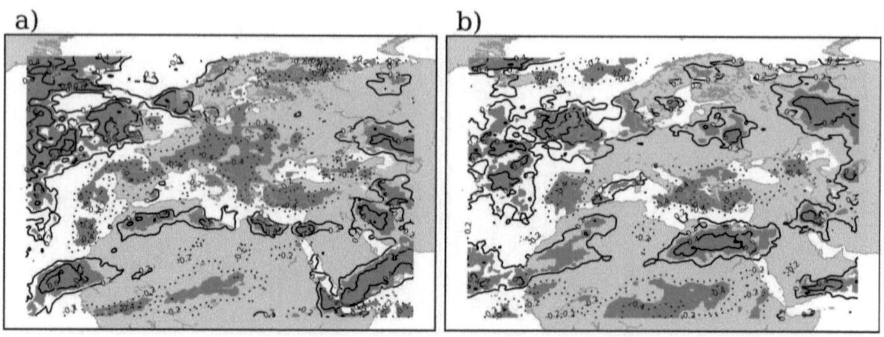

Fig. 3 (**a**) Correlation between the NAO index and the cloud cover for the winter season. The period of analysis covers 1989–2008. *Dots* indicates negative correlation while *continuous line* indicates positive correlation. *Shading* indicates local statistical significance of the difference at 90% confidence level based on a t-test. (**b**) as in (**a**) but for the annual data

annual cases, respectively. As expected, both the winter and annual patterns are in agreement with those observed in Fig. 2. In particular it should be noted the negative correlation pattern covering the majority of the Mediterranean area during winter (Fig. 3a), with the maximum negative values (–0.3) located over the central and eastern sectors. On the other hand, positive correlation values (0.6) are found over Ireland. These results are in agreement with the analysis of the cloud-cover NAO relationship of Trigo et al. (2002). Concerning the annual case (Fig. 3b), negative correlation values (about –0.3) are observed over the west of the Iberian Peninsula and the eastern Mediterranean area, in agreement with the patterns observed in Fig. 2b. In any case, correlations are, both for the winter and annual case, weaker than those obtained for the solar energy.

3.2 NAO-Wind Energy Correlation Analysis

Figure 4a, b show the correlation between the NAO index and the wind energy for the winter and annual cases, respectively. In both cases, a dipolar pattern can be observed, with negative correlation values over the whole Mediterranean region and positive correlations concentrated over central and northern Europe. Noteworthy, the correlations are weaker than those obtained for the solar energy, indicating a weaker linear relationship between the NAO and the wind energy. Note that annual dipolar pattern seems to be shifted south-eastward compared to the winter case. Particularly, for winter, the maximum positive correlations (0.6) are found over the north of the British Islands and the Scandinavian Peninsula. For the annual case, these high positive values are also observed in central Europe. On the other hand, the maximum negative values are found over Greece in winter and over the north of Algeria in the annual case. Note the striking feature over the Gibraltar Strait area during winter. Over this area, high positive correlation values (0.6) are found while

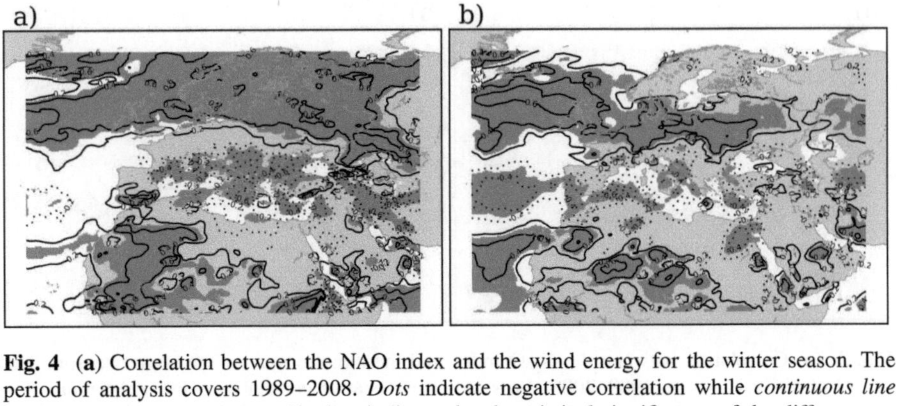

Fig. 4 (**a**) Correlation between the NAO index and the wind energy for the winter season. The period of analysis covers 1989–2008. *Dots* indicate negative correlation while *continuous line* indicates positive correlation. *Shading* indicates local statistical significance of the difference at 90% confidence level based on a t-test. (**b**) as in (**a**) but for the annual data

negative values are found in the rest of the Mediterranean area. This feature, which deserves more research in the future, is probably related to the special topographic characteristic of this area.

3.3 Solar Energy Composite Analysis

In this section, composite patterns of winter solar energy anomalies are obtained as a function of the state of the NAO. Data are represented as anomalies (in percentage) from the mean of complete study period (1985–2004). Figure 5a shows the composite of solar energy anomalies for winters in which the value of NAO index was positive (NAO>1) and Fig. 5b for winters in which the NAO index was negative (NAO<–1). The differences between positive NAO and normal NAO (i.e., –1<NAO<1) winters (Fig. 5c) and negative NAO and normal NAO winters (Fig. 5d) are also shown. Finally, the differences between the anomalies during positive and negative NAO winters (i.e., the differences between Fig. 5c, d) are represented in Fig. 5e. Note that Fig. 5c, d, f represent the spatial patterns of the maximum interannual changes (from one winter to other) that can be expected in the solar energy resources. During positive NAO winters (Fig. 5a), the solar energy presents positive anomalies over the whole Mediterranean area and negative anomalies over central Europe and the north of the British Islands. The spatial pattern of anomalies observed during negative NAO winters are similar but of reverse sign (Fig. 5b). Both patterns are in agreement with the correlation patterns presented in Fig. 2a. When comparing the anomalies during positive NAO and normal NAO winters (Fig. 5c), the whole Mediterranean area presents statistically significant positive anomalies. The maximum values (10–15%) are observed over the Iberian Peninsula, Greece and Turkey. On the other hand, the north of the British Islands presents the highest negative anomalies (about –10%). The comparison between negative and normal NAO winters (Fig. 5d) shows the existence of statistically significant negative anomalies (–4 to –8%) over the Iberian Peninsula, the south of France and over certain areas of Greece and Turkey. In addition, statistically significant positive anomalies (+4 to +8%) are found over eastern and northern Europe. By comparing the anomalies during positive and negative NAO winters (Fig. 5d), noticeable interannual changes in the availability of solar energy during the winter season can be expected in certain areas of the study region. Particularly, over the Iberian Peninsula, the northern part of Greece and Turkey, central Europe and the British Islands, interannual changes in the solar resources of values up to 15% can be expected associated with interannual changes in the NAO phase.

Similarly to Figs. 5 and 6 presents the solar energy anomalies associated with the NAO phases but on annual basis. In general, the annual spatial patterns of anomalies resemble their winter counterparts but with weaker anomalies. As for the winter case, statistically significant positive anomalies are found over the whole Mediterranean area when comparing positive NAO and normal NAO years (Fig. 6c), but anomalies are considerable weaker (about 4% compared to 10% of the winter case). When comparing negative and normal NAO years (Fig. 6d) statistically

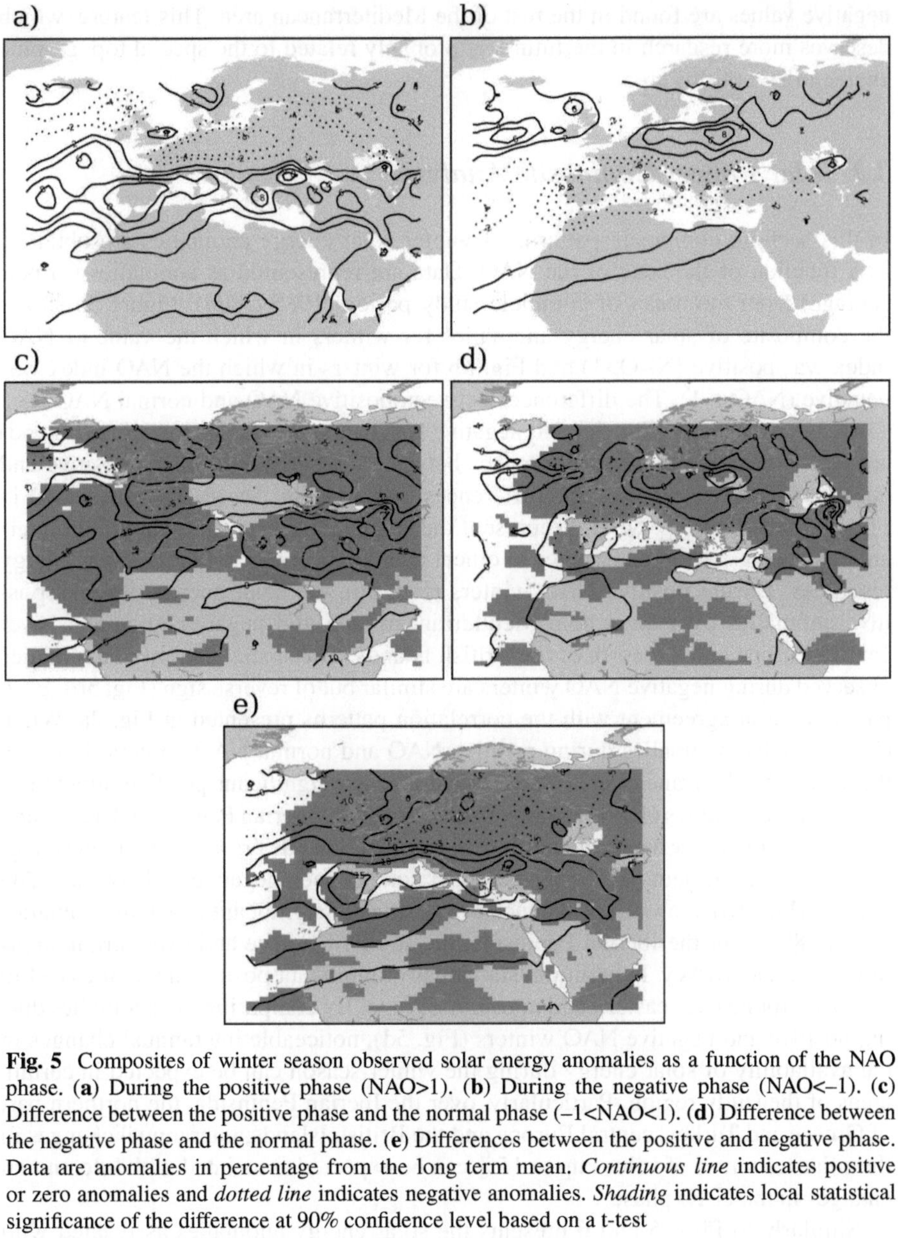

Fig. 5 Composites of winter season observed solar energy anomalies as a function of the NAO phase. (**a**) During the positive phase (NAO>1). (**b**) During the negative phase (NAO<–1). (**c**) Difference between the positive phase and the normal phase (–1<NAO<1). (**d**) Difference between the negative phase and the normal phase. (**e**) Differences between the positive and negative phase. Data are anomalies in percentage from the long term mean. *Continuous line* indicates positive or zero anomalies and *dotted line* indicates negative anomalies. *Shading* indicates local statistical significance of the difference at 90% confidence level based on a t-test

significant negative anomalies (about –3%) are found over the South of France and Italy and positive anomalies (about 2%) over the southeast of the Iberian Peninsula, Tunis and the east of Turkey. The differences between positive and negative NAO years (Fig. 6e) show that important interannual changes in the availability of solar energy can be expected in certain areas of the study region, but of a lower magnitude

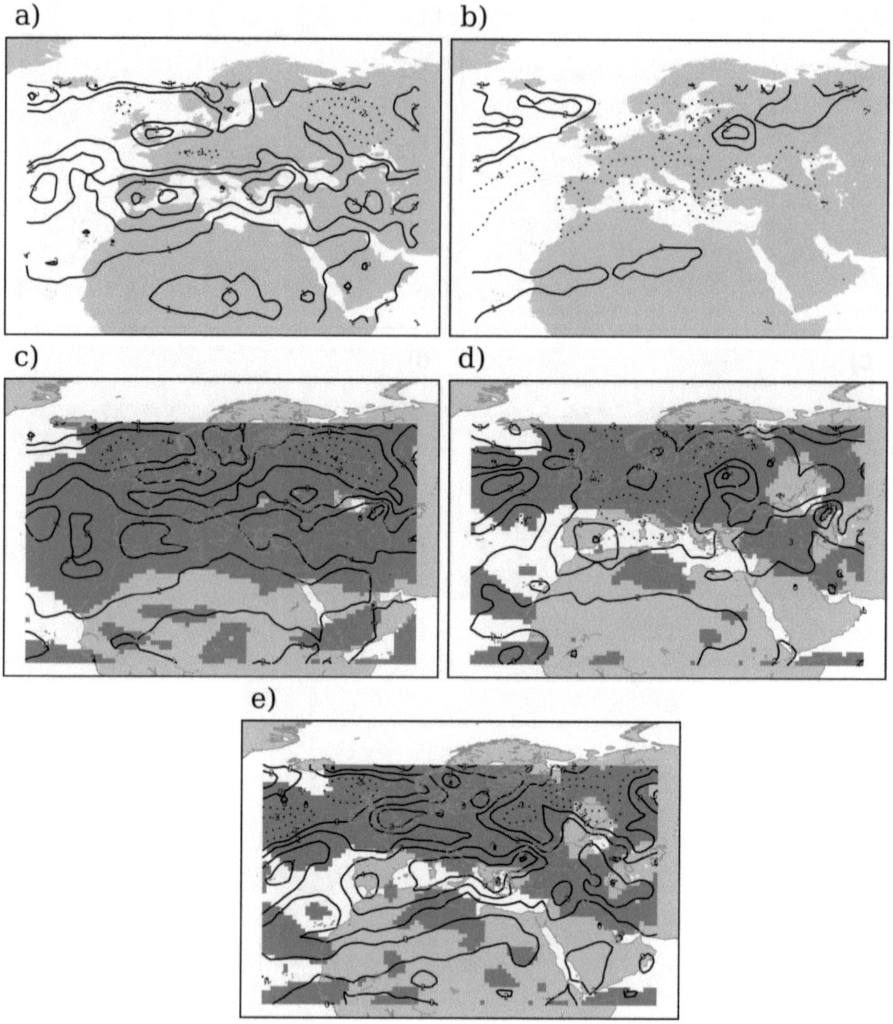

Fig. 6 As in Fig. 5 but for the annual data

than those observed for the winter case. Particularly, in the Mediterranean region, interannual changes of values, between 4 and 8%, can be expected associated with interannual changes in the NAO. Results are in agreement to Lohmann et al. (2006) analysis of the interannual variability of the GHI in the Mediterranean area.

3.4 Wind Energy Composite Analysis

As in Fig. 5 for the solar energy, Fig. 7 presents the composite patterns of wind energy anomalies during the winter associated with the different phases of the NAO.

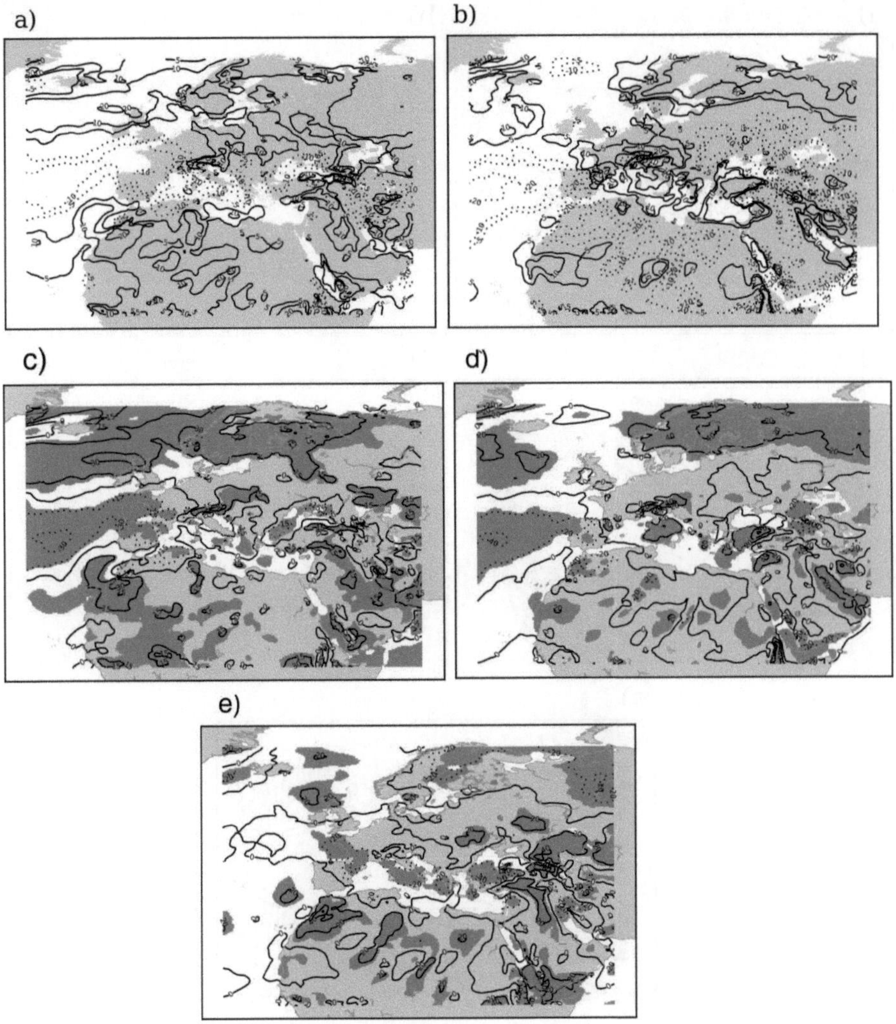

Fig. 7 As in Fig. 5 but for the wind energy anomalies

During positive NAO winters (Fig. 7a), positive anomalies are observed over the British Isles and the Scandinavian Peninsula, reaching up to 20%. Also positive anomalies but weaker (+10%) are observed over central Europe, the northwest of Morocco/Gibraltar Strait area, the north of Egypt and the eastern of Turkey. On the other hand, over the Iberian Peninsula, the central Mediterranean and the Black Sea areas, negative anomalies of values between –5 and –10% are found. When comparing these anomalies with those observed during normal NAO winters (Fig. 7c), some interesting features are obtained. Particularly, statistically significant and positive values of up to 30% are observed over the British Islands and the Scandinavian

area. Positive differences of values between 15 and 30% are observed also over the northwest of Morocco, central Europe and certain areas of the south and north of Turkey. On the other hand, statistically significant negative differences (–15 to –30%) are found over the Iberian Peninsula, north of Algeria and the northwest of Turkey. The pattern observed during negative NAO winters (Fig. 7b) is not the opposite of that observed during the positive NAO winters. This result is indicative of an important non-linear relationship between the wind energy and the NAO in the study area during winter. Particularly, during negative NAO winters, positive anomalies are observed over the whole Mediterranean basin (10–20%) and the Scandinavian area (about 20%), while negative anomalies are observed over the Iberian Peninsula (–10 to –20%). When comparing these anomalies with those found during normal winters (Fig. 7d), statistically significant positive differences are observed over Italy, Greece and Turkey (+20%) and over the Scandinavian area (+40%). On the other hand, negative differences are observed over the Iberian Peninsula and the north of Morocco (–20 to –40%). Finally, the patterns in Fig. 7e indicate that interannual changes in the winter NAO from positive to negative phase (or vice-versa) may lead to important changes in the wind resources availability. Particularly, over the central Mediterranean expected changes from one winter to other can reach about 20%, while over Turkey and the northwest of Morocco these changes can reach up 40%. Remarkably, and as a consequence of the non-linear influence of the NAO on the wind resource, anomalies observed in Fig. 7e in the Mediterranean area are of the same order of magnitude of those of Fig. 7c, d. Therefore, it can be concluded that in many areas of this region, as Turkey, Greece or the northwest of Morocco, interannual variability of the wind resources from one winter to other can be really high. This reflects the marked interannual variability of the NAO: over these areas, wind resources experiment noticeable changes from one winter to another. This happens not only when the NAO changes from the positive to the negative phase, but also when the NAO changes from a positive or negative phase to a normal phase or vice-versa (Fig. 7c–e).

The composite anomaly patterns for the annual case (Fig. 8) are similar to their winter counterparts but displaced south-eastward and with weaker anomalies. Particularly, as for the winter case, and as may be expected from Fig. 4b, the annual case also presents an important non-linear relationship between the NAO and the wind energy resources. Notably, Fig. 8a, b do not present opposite patterns. Statistically significant negative anomalies (about –8%) are found over the central Mediterranean area when comparing positive NAO and normal NAO years (Fig. 8c). Positive anomalies are observed over central and northern Europe and over the Iberian Peninsula, with values in the order of 8%. When comparing negative and normal NAO years (Fig. 8d) statistically significant and positive differences (with values about 10%) are found over the central and eastern Mediterranean areas, the north of the Iberian Peninsula and some areas in central and northern Europe. On the other hand, statistically significant negative differences are observed over the south of the Iberian Peninsula and the north of Morocco and Algeria. Finally, Fig. 8f indicates the existence of an important interannual variability of the wind resources in the study region associated with interannual changes in the NAO from

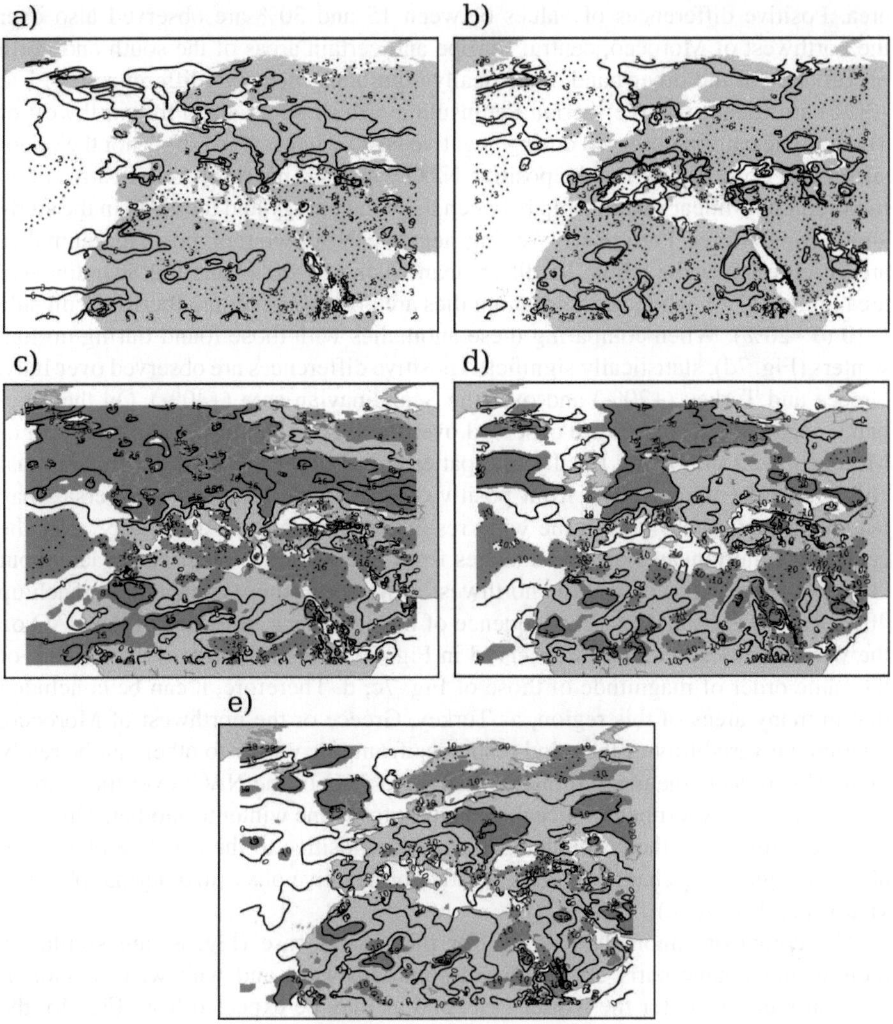

Fig. 8 As in Fig. 7 but for the annual data

positive to negative phase (or vice-versa). Particularly, in the central and eastern part of Mediterranean region, over the Gibraltar Strait area and over the northwest of Morocco, these changes can reach up to 10%. Nevertheless, and as a consequence of the important non-linear relationship between the wind energy resources and the NAO, these changes are of a lower magnitude than the changes observed when the NAO changes from a positive or negative phase to the normal phase (or vice-versa) (Fig. 8c, d). Similarly to the winter analysis, it can be concluded that in certain areas of the Mediterranean (namely the northwest of Morocco, the Gibraltar Strait and Greece), interannual variability of the wind resources can be really high. This situation reflects the fact that wind resources over these areas experiment marked

changes from 1 year to other not only when the NAO changes from the positive to the negative phase, but also when the NAO changes from a positive or negative phase to a normal phase or vice-versa (Fig. 8c–e).

3.5 The NAO and the Balancing of the Solar and Wind Energy Resources

The previous sections revealed that the variability of the solar and wind energy resources in the study region associated with the NAO presents considerable spatial differences. Notably, the spatial patterns of the NAO influence on the solar and wind resources reveal in some cases the existence of spatial balancing. For instance, the dipolar pattern of the spatial correlation in Fig. 2a indicates the existence of spatial balancing of the solar resources between the central and northern Europe, on one side, and the Mediterranean basin, on the other. Even more, there exists also a weaker but more interesting (because of the availability of the solar resources) balancing between the north of Egypt and the rest of the Mediterranean area. Unfortunately this spatial balancing is not present in the annual analysis (Fig. 2b). Regarding the wind, the dipolar pattern in Fig. 4a also reveals the existence of spatial balancing between northern Europe and the Mediterranean area as earlier highlighted by Pryor et al. (2006). Compared to the winter solar energy case, the strength of this balancing seems to be weaker. In addition, during winter, there exists also spatial balancing between the western Mediterranean (Gibraltar strait) and the rest of the Mediterranean region. Regarding the annual case, and unlike the solar resources, the spatial balancing is also present but displaced south-west-ward (Fig. 4b). Finally, the comparison of Figs. 2a and 4a reveals that the NAO impact on the solar and wind resources over the study area presents a similar north-south dipolar pattern but the sign of the dipoles are just opposite. Therefore, it can be concluded the existence of a notably local balancing between the solar and wind resources almost over the whole study region during winter. Regarding the Mediterranean region, the strength of this balancing seems to be especially high in the central part, over Greece, where correlation between the NAO and the solar energy is higher than 0.6 (Fig. 2a) and between the NAO and the wind energy with *circa* –0.6 (Fig. 4a). For the annual case (Figs. 2b and 4b) the balancing is also present but weaker. Concerning the Mediterranean area, the southwest of the Iberian Peninsula, Greece and Turkey present a remarkable balancing. For instance, over the southwest of Iberian Peninsula the NAO-solar energy correlation reaches about 0.2 while NAO-wind energy correlation is about –0.4.

4 Summary and Conclusions

The influence of the NAO on the solar and wind energy resources in the Mediterranean area has been explored based on the analysis of 20 years of reanalysis and satellite data. Particularly, the role of the NAO cycle in the interannual

variability of the solar and wind energy resources was analyzed based on correlation and composite analysis. Although the study is focused on the Mediterranean area, for the sake of completeness, the analysed data covered the whole North Atlantic and Mediterranean area, including the north of Africa. The analysis was carried out twice for the winter and annual periods and aimed to contribute to a better understanding of renewable energy resources in the study region.

Regarding the winter solar energy, a north south dipolar pattern of correlation was found, with the positive centre of the dipole over the Mediterranean and the negative one over central and northern Europe. Correlation coefficient over the whole Mediterranean area reached values above 0.6, with the highest value (0.8) over the Iberian Peninsula. The annual case presented a correlation map resembling the winter case but displaced north-eastward and with weaker values. A similar analysis for the cloud cover instead of for the solar energy was able to explain the former results. In any case, the NAO-cloud cover correlation was found, both for the winter and annual case, to be weaker than those obtained for the solar energy. The composite analysis revealed some additional features of the NAO-solar energy relationship. Particularly, positive anomalies (10–15%) were found over the Iberian Peninsula, Greece and Turkey associated with the positive phase NAO winters. On the other hand, during negative NAO winters, negative anomalies (–4 to –8%) were found over the Iberian Peninsula, the south of France, Greece and Turkey. As a consequence, over these areas, interannual changes in the winter solar energy resources of values of up to 15% can be expected associated with interannual changes in the NAO phase. Regarding the annual analysis, the composite analysis showed similar results but anomalies were considerable weaker. For instance, the positive NAO years were found associated with positive anomalies (4%) over the whole Mediterranean area. Negative NAO years were found associated with negative anomalies over the South of France and Italy (–3%) and positive anomalies (2%) over the southeast of the Iberian Peninsula, Tunis and the east of Turkey. As a consequence, for the entire Mediterranean region, interannual changes of values between 4 and 8% were found associated with interannual changes in the NAO phase.

The NAO-wind energy relationship was found to be considerably more complex than the NAO-solar energy one. Particularly, the winter correlation analysis proved the existence of north-south dipolar pattern, with positive values over central and northern Europe and negative values over the Mediterranean area. For the latter area, negative correlation values (–0.4 to –0.6) were observed, except over the Gibraltar Strait area, where positive correlation values were found. The annual correlation analysis showed a similar pattern, but shifted south-eastward. Thus over the Mediterranean area, the maximum negative values were observed over Greece (–0.6) in the winter case and over Algeria (–0.6) in the annual case. Noteworthy, in general the correlations were weaker than for the solar energy case, indicating a weaker linear relationship between the NAO and the wind energy. The composite analysis revealed additional features regarding the complex relationship between the NAO and the wind energy in the study area. For instance, the winter analysis proved the existence of positive anomalies (15–30%) over the northwest of Morocco and the south and north of Turkey, and negative anomalies (–15 to –30%)

over the Iberian Peninsula, north of Algeria and the northwest of Turkey associated with the positive phase of the NAO. The composite pattern observed during negative NAO winters was not the opposite of the observed during the positive NAO winters. Particularly, the main features during this negative phase were the presence of positive anomalies (+20%) over Italy, Greece and Turkey and negative anomalies (–20 to –40%) over the Iberian Peninsula and the north of Morocco. As a consequence, notable interannual changes in the winter wind energy resources associated with interannual changes in the NAO phase were found over the central Mediterranean (20%), Turkey and the northwest of Morocco (+40%). The annual composite patterns showed to be similar to their winter counterparts but the patterns were displaced south-eastward and the anomalies were weaker. Particularly, over the central Mediterranean, negative anomalies (–8%) were found associated with the positive phase of the NAO. During the negative phase, positive anomalies (10%) were found over the central and eastern Mediterranean area and the north of the Iberian Peninsula, while negative anomalies (–10%) were found over the south of the Iberian Peninsula and the north of Morocco and Algeria. Consequently, interannual changes in the annual wind resources of values over 10% were found in the central and eastern part of Mediterranean region, over the Gibraltar Strait area and over the northwest of Morocco associated with interannual changes in the NAO phase.

Finally, the analysis of the spatial patterns of the NAO influence on the solar and wind energy resources revealed the existence of spatial and local balancing. Among others, the dipolar pattern of the spatial correlation obtained for the solar energy during winter indicated the existence of spatial balancing of the solar resources between the central and northern Europe and the Mediterranean. However, it should be noted that this spatial balancing was not present in the annual analysis. The dipolar pattern of the winter NAO-wind energy relationship also revealed the existence of spatial balancing between northern Europe and the Mediterranean area for the wind energy. Nevertheless, compared to the winter solar energy case, the strength of this balancing was found to be weaker. In addition, during winter, spatial balancing between the western Mediterranean (Gibraltar Strait) and the rest of the Mediterranean region was also found. Regarding the annual case, and unlike the solar resources, the spatial balancing was also present but displaced south-westward. In addition to the spatial balancing found for the wind and the solar energy, a significant local balancing between the solar and wind resources almost over the whole study region was found associated with the NAO cycle. Regarding the Mediterranean region and during winter, the strength of this balancing was found to be especially high in the central section. In the annual case the most important local balancing was found over the southwest of the Iberian Peninsula, Greece and Turkey, although the magnitude was weaker than in the winter case.

The authors believe that the results above summarized are of interest regarding the solar and wind energy resources evaluation in the Mediterranean area. In this regard we would like to emphasise that, in the light of results obtained here, a more in depth assessments of solar and wind resources should be required and that measuring solar and wind data covering only a few years should be avoided. This will

provide a better estimation of the expected interannual variability of the renewable resources and, therefore, of the plant production interannual variability.

Acknowledgements The Spanish Ministry of Science and Technology (grant ENE2007-67849-C02-01) and the Andalusian Ministry of Science and Technology (grant P07-RNM-02872) and FEDER funds financed this study. Support is also obtained from the Plan Andaluz de Investigación (Grupo TEP-220). The DLR-ISIS data images were obtained from the Institute of Atmospheric Physics, German Aerospace Center (DLR) (Lohmann, 2006).

References

Barbour P, Walker S (2008) Wind resource evaluation: Eola Hills. Technical report, Energy Resources Research Laboratory, Department of Mechanical Engineering, Oregon State University, Corvallis, OR

Burlando M (2009) The synoptic-scale surface wind climate regimes of the Mediterranean Sea according to the cluster analysis of ERA-40 wind fields. Theor Appl Climatol 96:69–83

Cassola F, Burlando M, Antonelli M, Corrado FR (2008) Optimization of the regional spatial distribution of wind power plants to minimize the variability of wind energy input into power supply systems. J Appl Meteorol Climatol 47:3099–3116

Chiacchio M, Wild M (2010) Influence of NAO and clouds on long-term seasonal variations of surface solar radiation in Europe. J Geophys Res 115:D00D22. doi:10.1029/2009JD012182

DESERTEC (2010) DESERTEC Red paper. An overview of the DESERTEC concept. DESERTEC organization. http://www.desertec.org/

DLR (2007) TRANS-CSP. DLR-ITT technical report, Köln, Germany

Global Wind Energy Council (2007) Global wind energy council report. http://www.gwec.net/uploads/media/Global_Wind_2007_Report_final.pdf

Hurrell JW, Kushnir Y, Visbeck M, Ottersen G (2003) An overview of the North Atlantic Oscillation. In: Hurrell JW et al (eds) The North Atlantic Oscillation: climate significance and environmental impact. Geophys Monogr Ser, vol 134. AGU, Washington, DC, pp 1–35

IEA (2008) World energy outlook. International Energy Agency, OECD Publication

IEA (2009) Technology roadmap. Wind energy. International Energy, OECD Publication

IEA (2010a) Technology roadmap. Solar photovoltaic energy. International Energy, OECD Publication

IEA (2010b) Technology roadmap. Concentrating solar power. International Energy, OECD Publication

Kempton W, Pimenta FM, Veron DE, Colle BA (2010) Electric power from offshore wind via synoptic-scale interconnection. Proc Natl Acad Sci 107:7240–7245

Lara-Fanego V, Ruiz-Arias JA, Pozo-Vázquez D, Santos-Alamillos F, Tovar-Pescador J (2011) Evaluation of the WRF model solar irradiance forecasts in Andalusia (southern Spain). Sol Energy, doi:10.1016/j.solener.2011.02.014

Li X, Zhong S, Blan X, Hellman WE (2010) Climate and climate variability of the wind power resources in the Great Lakes region of the United States. J Geophys Res 115:D18107. doi:10.1029/2009JD013415

Lohmann S, Schillings C, Mayer B, Meyer R (2006) Long-term variability of solar direct and global radiation derived from ISCCP data and comparison with reanalysis data. Solar Energy 80:1390–1401

Papadimas CD, Fotiadi AK, Hatzianastassiou N, Vardavas I, Bartzokas A (2010) Regional co-variability and teleconnection patterns in surface solar radiation on a planetary scale. Int J Climatol 30:2314–2329

Pozo-Vázquez D, Tovar-Pescador J, Gámiz-Fortis SR, Esteban-Parra MJ, Castro-Díez Y (2004) NAO and solar radiation variability in the European North Atlantic region. Geophys Res Lett 31. doi:10.1029/2003GL018502

Price Water House Coopers (2010) 100% *Renewable electricity: a roadmap to 2050 for Europe and North Africa*. Price Water House Coopers, London

Pryor S, Barthelmie R, Shoof JT (2006) Inter-annual variability of wind indices across Europe. Wind Energy 9:27–38

Sanchez-Lorenzo A, Calbó J, Brunetti M, Deser C (2009) Dimming/brightening over the Iberian Peninsula: trends in sunshine duration and cloud cover and their relations with atmospheric circulation. J Geophys Res 114:D00D09. doi:10.1029/2008JD011394

Sanchez-Lorenzo A, Calbó J, Martin-Vide J (2008) Spatial and temporal trends in sunshine duration over Western Europe (1938–2004). J Clim 21:6089–6098

Simmons A, Uppala S, Dee D, Kobayashi S (2007) ERA-I: new ECMWF reanalysis products from 1989 onwards. ECMWF Newsletter 110:25–35

Trigo RM, Osborn TJ, Corte-Real JM (2002) The North Atlantic Oscillation influence on Europe: climate impacts and associated physical mechanisms. Clim Res 20:9–17

von Bremen L (2008) Large-scale variability of weather dependent renewable energy sources. In: Troccoli A (ed) Management of weather and climate risk in the energy industry. Springer, The Netherlands, pp 189–205

Weisser D, Foxon T (2003) Implications of seasonal and diurnal variations of wind velocity for power output estimation of a turbine: a case study of Grenada. Int J Energy Res 27:1165–1179

Price Water House Coopers (2010) 100% renewable electricity: a roadmap to 2050 for Europe and North Africa. Price Water House Coopers, London
Pryor S, Barthelmie R, Schoof JT (2006) Inter-annual variability of wind indices across Europe. Wind Energy 9:27–38
Sanchez-Lorenzo A, Calbó J, Brunetti M, Deser C (2009) Dimming/brightening over the Iberian Peninsula: trends in sunshine duration and cloud cover and their relations with atmospheric circulation. J Geophys Res 114:D00D09. doi:10.1029/2008JD011394
Sanchez-Lorenzo A, Calbó J, Martin-Vide J (2008) Spatial and temporal trends in sunshine duration over Western Europe (1938–2004). J Clim 21:6089–6098
Simmons A, Uppala S, Dee D, Kobayashi S (2007) ERA-Interim: new ECMWF reanalysis products from 1989 onwards. ECMWF Newsl 110:25–35
Trigo RM, Osborn TJ, Corte-Real JM (2002) The North Atlantic Oscillation influence on Europe: climate impacts and associated physical mechanisms. Clim Res 20:9–17
von Bremen L (2010) Large-scale variability of weather dependent renewable energy sources. In: Troccoli A (ed) Management of weather and climate risk in the energy industry. Springer, The Netherlands, pp 189–206
Weisser D, Foxon T (2003) Implications of seasonal and diurnal variations of wind velocity for power output estimation of a turbine: a case study of Grenada. Int J Energy Res 27:1165–1179

Index

S.M. Vicente-Serrano, R.M. Trigo (eds.), *Hydrological, Socioeconomic and Ecological Impacts of the North Atlantic Oscillation in the Mediterranean Region*, Advances in Global Change Research 46, DOI 10.1007/978-94-007-1372-7,

MIX
Papier aus verantwortungsvollen Quellen
Paper from responsible sources
FSC® C105338

If you have any concerns about our products,
you can contact us on
ProductSafety@springernature.com

In case Publisher is established outside the EU,
the EU authorized representative is:
Springer Nature Customer Service Center GmbH
Europaplatz 3, 69115 Heidelberg, Germany

Printed by Libri Plureos GmbH
in Hamburg, Germany